Holography MarketPlace

5th Edition

*The reference text
and sourcebook for
holography worldwide.*

The original version of this book contained holograms from the vendors and many of them are no longer in business. Therefore this version of the book contains everything that was in the original version but it has no holograms.

Edited by Brian Kluepfel, Alan Rhody and Franz Ross

Copyright h 1996 Ross Books
Llbrary Catalogue Information
Holography MarketPlace - Sixth Edition
Bibliography:
1. Includes Index.
2. Holography
3. Directories - Holography Industry
4. Photography
5. Physics
ISBN 9780894960987
Printed in Canada

Foreward

Almost 14 years ago, in late 1981, our publishing business was getting ready to publish a book called the Holography Handbook. As luck would have it, a man named Steve McGrew, who lived close by, was just getting ready to test run a new technology that had never been commercially used before called the embossed hologram. One of our authors, Fred Unterseher, knew Steve and, after some discussion, we decided to test the technology with the Holography Handbook, thus creating the first book published with an embossed hologram in it.

We knew it was one of those minor milestones in the development of holography when we published the book. We notified publishing trade magazines, sold book club rights, and distributed the book widely in book shops. As you might expect, within a year or so there were one or two more publications displaying embossed holograms and within three to four years it seemed that everybody was using the new technology (many of whom claimed to be the first to present the product in published form).

We have that same feeling regarding the hologram you see on the cover of the 5th Edition Holography MarketPlace. Here are some of the reasons why:

Until now, holographic stereograms provided the viewer with a very limited angle of view. Basically, you could tilt the image from side to side to see the image move or see around the object (called left to right parallax). What the customer would ideally like to see, however, is full parallax; the ability to move the plate in any direction and have a view of the image from that vantage point, as well as seeing motion. In other words, true-to-life parallax. This is kind of like being able to fly around the image like a bird

and see the image from any perspective while motion is going on - just as it would happen in real life. This ability to have a full parallax view has been discussed theoretically in the literature but, until now, no one has been able to create and market in published form an example of this technique.

Another item being demonstrated by the cover hologram is the ability to have sophisticated computer animation programs integrated into a hologram. Without full parallax, the technology available in today's computers cannot really be utilized. The holographic image on the cover contains twelve separate animated sequences! Each sequence is a video clip that was sampled from ordinary televised footage. We tested slow-moving scenes and fast ones.

Thus we have the ability to record, play back, and publish, "mini movies" while "moving around" the images (being played back. This is similar to the great animation sequences that you sometimes see before important televised sports programs where television sets, running foot- age of sports contests, etc. are turned about in the air. The difference now is that you can print this kind of image in a book or on a cereal box. The commercial possibilities are astounding.

The hologram you have in your hands is the first example of this process and, as with all prototypes, there is room for improvement. A larger image area, and a number of other things can and will be done to improve the play- back of the image by holographers interested in getting in on this financially lucrative new medium.

To make the cover hologram demonstrate full parallax

ability, various views of the image were used. As an example, the view of the image "straight on" is like this:

Whereas the image used to create the view from approximately 70 degrees up and 30 degrees right is like this:

And the view from approximately 70 degrees up and 30 degrees to the left is like this:

Theoretically, there is no reason why you can not have an image rotate a full 360 degrees in any direction. There are limitations, of course, on the number of images used in a sequence and discussion of this subject is presented in Chapter 2, where this process is discussed in more detail.

The bottom line is that this new process opens the door to a whole new range of creative applications and makes for a very strong marriage between the holographer and the well established (and well financed) world of electronic imaging. In essence, holography can offer the computer community the ability to take their real-life full parallax images and print them on a piece of paper - motion and all. This basically offers full parallax printed cinema using a new output device - the holographic "printer".

Now a computer graphics person can generate a hard copy of three dimensional scenes and/or animated images that were previously trapped on a monitor and hang them on a wall or attach them to a printed page.

Conversely, the digital origination of artwork offers holographers a new input device. Imagine the range of visual effects that are possible on a computer screen. This collaboration drastically expands the palate of images that we can generate - thereby expanding artistic and commercial opportunities, as well.

This unique hologram was a collaborative effort between ourselves (Alan Rhody instigated and coordinated the project), a computer graphics artist (Brian Kane), the developer of the system that transferred the digital artwork into a full parallax holographic stereogram (Chuck Hassen), a holographer (Bob Hess), a replicator (Hughes Power Products), and a film supplier (DuPont). We extend our thanks to everyone that contributed to the success of this project.

This edition of the Holography Marketplace also includes some very exciting holograms submitted by our advertisers. We feel that the selection of holograms displayed in the book truly represents the industry as it exists today.

Lastly, we have updated our business database and have improved our indexing system. We have attempted to contact each business listed in order to update our records. Please inform us of any errors or omissions. We welcome any suggestions you may have for future editions.

Franz Ross
Berkeley, California
October: 1995
Voice Phone. 510 841 2474
FAX 510 841 2695
e-mail: ROSSBOOKS@aol.com

The Cover Hologram

Viewing Instructions

The hologram on the cover of this edition of the Holography MarketPlace is best viewed when it is illuminated from above by a single, bright light source. Any light source that casts a sharp, well defined shadow is optimal-such as a halogen spotlight, an unfrosted light bulb or direct sunlight. Fluorescent lights and other diffuse sources typically found in offices and stores will not playback this image clearly. Avoid multiple light sources, which will create multiple (blurry) images. Experiment with different lights to achieve the clearest, highest resolution image.

Hold the hologram in front of you and slowly tilt it from side to side. You can see around each side of the image! Now, slowly tilt the hologram up and down. Watch each of the "televised" images change! Look closely at the hologram as you continue to move it around to see additional visual effects.

Production Methods

In brief, the steps involved in producing this hologram were:

1) Digitizing the artwork and creating an ordered sequence of graphic cells;

2) Displaying the digitized artwork using LCD technology and specialized optical components, including a diffusion screen;

3) Recording the sequential graphic cells on a holographic master using a custom designed apparatus and conventional recording and processing methods. (The 30 x 40 cm transmission master was exposed using an Argon ion laser. A 17.5 " collimating minor directed the reference beam. The object beam traveled through the LCD and the projection screen before exposing the plate.);

4) Creating a production tool for the replication facility (In this instance, 12 image planed reflection transfers were ganged on a single plate using a Helium-Neon laser. Registration marks were added.);

5) Replicating the image on photopolymer;

6) Finishing the holograms (lamination and die cutting). Additional details can be found in Chapter Two.

For further technical information, please contact:
Bob Hess
Point Source Productions
Boulder Creek. California
October, 1995
Telephone 408 338 1304
FAX 408 338 3438
Fax 408 338 3438

Index to Advertisers

Table of Contents

1
Sales & Distribution

This chapter discusses the commercial development of artistic holography, especially the sale and distribution of holograms and related products by the giftware industry.

The Commercialization of Artistic Holography

In its early stages of development holography often remained unseen by the general public. Only scientists and researchers had access to the lasers and other specialized equipment that were needed to create and view a hologram. When methods were developed in the late 1960s that enabled a hologram to be created and viewed in more practical ways, holography slowly left the laboratory and began a journey that has resulted in the creation of a multi-million-dollar, worldwide industry. A great portion of this industry deals with "artistic" holography, (i.e. three-dimensional images of things) which is also commonly referred to as "pictorial" or "display" holography.

During the 1970s and early ' 80s holograms were mostly made by individual holographic "artists" on a one-by-one basis. The production process was labor-intensive and time-consuming. Production techniques were developed through trial and error. Raw materials such as film emulsions were scarce, equipment was often homemade, and production quality often inconsistent. Unfortunately, the individuals and small companies that were capable of making high-quality holograms generally did not have the money or marketing expertise needed to get their works into wide-spread distribution. Therefore, holograms were still out of view of the public-at-large.

The handful of galleries and stores that did show holograms proved that the public was fascinated by this emerging medium. Although most holograms were treated as futuristic artworks

or novelty items - and were relatively expensive - the public constantly asked for more affordable ones. Enterprising gallery owners, retailers and holographers recognized this demand for holographic merchandise, and a small industry began to slowly evolve. Holographers and their entrepreneurial partners began to create products rather than artworks; well connected retailers began to distribute holograms to other retailers; and hologram aficionados became customers. Everyone recognized the potential of this new industry, yet manufacturing and display/lighting problems still needed to be overcome before holograms and related products could enter the mainstream marketplace. Limited production runs kept prices high.

Over the next decade technological advances enabled holograms to be mass-produced in a variety of ways. This ability to reproduce a holographic image made it feasible for artists, technicians and businessmen to join together to create facilities dedicated solely to producing large runs of affordable, high-quality holograms. These holograms were intended for a variety of commercial applications including security, packaging and advertising - as well as products for the giftware industry.

Art holographers copied their most popular images onto film (which was less expensive and easier to handle than holograms produced on sheets of glass) and began to use assembly-line production methods in their labs. Whole catalogues of images soon became available, intended for sale as wall decor. Retail price points dropped considerably. Other holographers perfected methods of mass producing very bright dichromate holograms for use as jewelry. Still others concentrated on developing high-speed automated replication technologies capable of embossing holograms on very inexpensive foils and plastics - perfect for use on toys, optical novelties and paper products.

Once reliable supply lines were established, it became feasible for other companies to package and market these holograms in a variety of ways and integrate them into the normal chain of giftware distribution. Businesses in the United States and England quickly grew into major distributors. Film holograms were matted and/or framed and marketed as high-tech 3D. Holographic fashion accessories (including watchfaces, pendants and earrings) were developed. Executive gifts and desktop accessories were cre- ated. Rolls and rolls of kids' stickers were produced. New toys were invented. Holograms started to appear at national gift shows, in giftware catalogs, and in the media.

Savvy retailers soon realized that holograms and related products were very popular with the buying public, and if displayed correctly, could prove quite profitable to sell. A good display of holograms drew a crowd, generated cus- tomer excitement and more importantly, generated dollars! (Most giftware items sold in stores are priced at twice their cost.) Holographic merchandise spread from science museum gift shops to more mainstream outlets, and even included a number of specialty stores set up to sell only hologram products.

Increased visibility created greater public awareness of the product and demand for new and better holograms. Artists added color and motion to their images. Manufacturers automated further and invented materials especially suited for holographic applications. Holograms became

brighter and easier to see under typical viewing conditions. Distributors created new product lines by integrating holograms into existing merchandise. Packaging was brought up to commercial standards. Wholesalers adopted more sophisticated marketing techniques, while retailers offered a wider selection of goods.

Today, a wide variety of holograms are manufactured around the world for the giftware industry, with the highest concentration of factories located in North America and Europe. English holographers have traditionally dominated the silver halide film replication business. American holographers are actively developing photopolymer replication factories. The production of dichromate holograms seems to have slowed, while facilities capable of producing embossed holograms have multiplied significantly, especially throughout Asia. Surprisingly, very few holograms are exported from Japan.

The number of distributors and wholesalers dealing exclusively in holograms has dwindled as the market has diversified. To stay profitable several major distributors have developed their own custom images in order to target specific consumer groups The sale of "licensed" holographic images featuring popular sports figures, cartoon characters and movie scenes has grown steadily, while the sale of more mundane images has stagnated.

It seems that there are fewer holography specialty shops in business now than a few years ago. Those that continue to do well have increased the variety of goods that they carry and often include related optical novelties in their product mix. Sales of holographic artwork are practically nonexistent. However, more stores than ever before are carrying some sort of hologram-related product. It is not uncommon to find an inexpensive hologram item at the corner store.

The Chain of Distribution

Let's examine the chain of distribution as it typically exists for a holographic product.

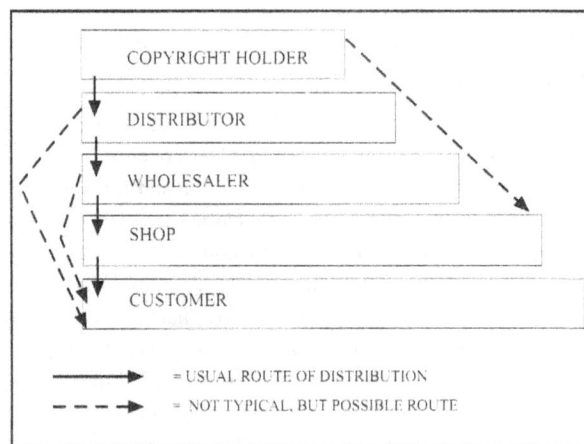

Chain of Holography Distribution

The Copyright Holder

The distribution process starts with the copyright holder. Any unique work of art, including paintings, photographs, or computer-generated graphics, can be protected from unauthorized duplication (in most countries) by registering the image in the appropriate manner. In the case of holography, the original work of art is either a model, a graphic, or a computer program designed to generate a holographic image. Whoever creates the unique work of art that later becomes a hologram is considered the copyright holder of the image. It is also possible to copyright the finished hologram as a unique work of art itself, provided that none of the components that appear in the holographic image be- long to another party. There are, however, statutory limits stating that after a number of years, a piece of art can become public domain and may be used freely.

Every hologram, if properly copyrighted, has only one owner, the copyright holder. The copyright holder therefore controls all subsequent distribution and is positioned at the top of the distribution chain. The copyright holder can be an individual or a group of people such as a business. Most commonly, a business commissions an artist to make a model or to design graphics and the artist turns over all copyright privileges to the business as part of the arrangement. This is legally known as "work for hire" and each party's responsibilities and rights must be documented to avoid problems concerning ownership. Holograms that are not copyrighted can be copied by whoever owns the "master" hologram.

The Manufacturer

Second in line is the manufacturer. Different companies specialize in manufacturing specific types of holograms. The manufacturing process generally involves three processes - mastering (creating the original hologram), reproduction (producing some quantity of copies), and finishing (lamination, cutting, sorting, etc.). Some companies do everything - others subcontract out some part of the job. Often a company that manufacturers holograms also owns copy- rights in order to have a selection of stock images to offer their customers. A few manufacturers bypass the normal chain of distribution and sell directly to retailers.

The copyright holder needs to know the exact cost of each unit produced, since the manufacturer's charge will obviously influence the final price billed to the end user. To figure the unit cost, one would take the total bill from the manufacturer (including any additional shipping and handling charges) and divide it by the number of usable copies actually delivered. As in most manufacturing businesses , prices decrease as quantity increases . In order to figure the suggested retail price of their product, it is very common for a copyright holder to multiply the manufacturer's unit cost five times. For instance, a product that costs the copyright holder $4.00 will be resold to a distributor for $6.00, and will end up selling in a store for $20.00.

If you are having holograms made to your specifications, choose a company that produces holograms appropriate to your final application. Be aware that manufacturers have

Typical costs, margins, and profits for hologram with retail price of $20				
Business	Buys For	Resells For	Discount Off Retail	Markup (% of Cost)
Copyright Holder	$4	$6	80%	50%
Distributor	$6	$8	70%	33%
Wholesaler	$8	$10	60%	25%
Retailer	$10	$20	50%	100%
Customer	$20	N/A	N/A	N/A

not yet standardized their pricing - some itemize production processes, others quote a finished price. Some quote by the square inch, others according to a sliding scale based on quantities ordered.

The Distributor

The distributor is a business that specializes in buying large quantities of product from a copyright holder and distributing it to other businesses that cannot afford to, or are not interested in, stocking a quantity of inventory. Distributors are often contractually obligated to order large amounts of merchandise, carry an entire line of their supplier's products, and not sell competing products. This alleviates many problems for the copyright holder and the manufacturer who are not usually set up to market their own products to numerous customers.

In return, the distributor commonly receives the sole rights to sell the product in a particular geographic region or to a particular group of customers and pays less than any other customer down the line. Distributors commonly pay 70% below the suggested retail price, resell these products for 60% below suggested retail, and depend upon a large volume of sales to make their profit. For example, if they pay $6.00 for an item from a copyright holder, they would resell it to a wholesaler for $8.00.

Many distributors also repackage goods under their own names, deal with import/export procedures and constantly work to expand the marketplace. A popular product can make a distributor a lot of money, due to the fact that potential customers have no alternative supplier. The distributor sells mostly to wholesalers.

The Wholesaler

The wholesaler is the link that connects distributors to retailers. The essential function of a wholesaler is to get the product into shops. Most wholesalers use a combination of in-house salespersons or independent sales reps working on commission to persuade buyers to try a product. Many rely on telemarketing departments, catalogue mailings and trade shows to establish new accounts and service existing ones. A good wholesaler will teach a shop owner how to best merchandise a product, provide point-of-purchase materials, restock displays, update product selection, and generally keep the customer happy.

A wholesaler normally carries many different product lines which is a convenience to retailers that want to consolidate the number of their suppliers. Also, wholesalers stock far less merchandise than a distributor, which allows them to react quickly to changes in the marketplace.

Wholesalers do not generally have exclusive rights to a product. They often extend payment terms to their customers (net 30 is common) after a probationary period or credit check. For their efforts, wholesalers receive special pricing that allows them to make a profit when they resell the goods to retailers. Wholesalers typically work off 25% profit margins, i.e. if they buy an item for $8.00, they will resell it for $10.00. Some distributors offer additional dis- counts to wholesalers for larger orders.

The Retailer

Retailers are the point of contact with the public, the place where merchandise is displayed and purchased by the customer. Holograms and related products have been sold in a variety of settings, ranging from temporary table-top setups (flea markets, trade show booths) to art galleries and department stores. Many entrepreneurial businessmen start with a small cart or kiosk in a busy mall during holiday season and graduate to a bigger store that runs year round. Several single store operations have expanded to multi-store chains.

Wholesalers have placed merchandise in obvious locales like museum gift shops, technology stores, and poster shops; less obvious locations include airport shops, nature stores and stationery stores. Other dealers have targeted specific interest groups such as hobbyists and collectors, and sell licensed products to comic book shops, trading card stores, and the like. Larger suppliers have cut deals with amusement parks, national chains, and event promotion agencies. Holograms have even been sold by mail

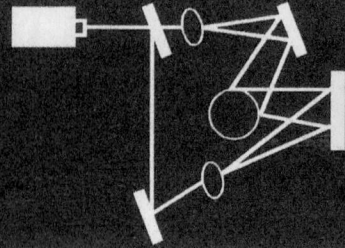

order using catalogs and classified ads, even though a written description or photograph does not adequately capture the wonder of a three dimensional image.

All successful hologram retail businesses have several things in common - they are located in high-traffic (pedestrian) areas in places where people go to buy interesting items (often tourist destinations), they display the merchandise correctly and with flair, and they offer a high level of friendly customer service. Although the retail price suggested by the copyright holder is only a guideline, most retailers double their costs to establish the final price a customer sees in the store. Therefore, an item that costs them $10.00 will end up on the shelf for $20.00.

Categories of Merchandise

Most successful hologram stores sell a selection of holographic merchandise, including pictures (stock wall decor images and limited-edition fine art), jewelry (and related fashion accessories), executive gifts (desktop accessories), toys and optical novelties. There are several ways to categorize this merchandise - by price, by size, by manufacturer, and so on. For now, we'll discuss selection by "format" (which refers to the type of material the hologram is produced on).

Silver Halide Glass Plate

Traditionally, most holographers have produced their holograms on sheets of glass (glass plates) coated with a high resolution light sensitive emulsion called silver halide. This type of emulsion is similar, but not identical to, the emulsions used in conventional photographic films.

These glass plates are rather costly, but can be used to make the highest quality holograms - mostly because the glass plates are quite rigid and will not move during the exposure period (movement will ruin the hologram). Work- ing with these glass plates is quite time-consuming, as each plate has to be individually handled with great care during each production step.

Due to the time and the cost involved in making holo- grams on silver halide glass plates, they are mainly used for limited-edition holographic wall art, archival images or custom work. A finished 8" x 10" glass plate hologram typically retails at $300.00 or more for an open edition work; more for a signed and numbered artwork.

It is possible to produce striking "deep image" holograms on silver halide glass plates -that is, images which dis- play considerable projection in front of the hologram's surface, and/or appear to be a considerable distance behind the hologram's surface. To achieve such dramatic effects, glass plate holograms require the best possible illumination, which usually takes some foresight.

Glass plate holograms are extremely fragile and should always be securely wrapped and handled with care. When the hologram is framed, the actual emulsion is facing toward the back and is protected. The surface which we can see (and touch) is ordinary glass and can be cleaned by gentle polishing with a soft cloth.

Silver Halide Film

Since glass plates are too expensive and impractical to mass produce, holographers developed methods to produce holograms on silver halide holographic film, which is cheaper and easier to handle. It is thin and flexible, similar to the film used in an ordinary camera. After exposure and processing, each piece of film is usually sandwiched in an cardboard matte, ready to package and/or frame as wall decor.

An 8" x 10" silver halide film hologram typically retails for $150.00 (less than half the cost of a comparable hologram produced on glass), making this format much more suitable for the giftware market. The bestselling size has traditionally been a 4" x 5" film mounted in an 8" x 10" matte due to the retail price (which has hovered around $35.00). Other standard sizes include a 2" x 2.5" film mounted in a 5" x 7" matte (which typically sell from $15.00 - $20.00), and a 5" x 8" film mounted in an 8" x 10" matte (which typically sells for $70.00).

Silver halide film holograms do not display quite the resolution, projection and depth of their glass plate counterparts, however, under proper illumination they are quite striking.

The emulsion of a film hologram is usually covered by a cardboard matte, however, the front surface can scratch easily. Fingerprints and dust can be cleaned off by gently polishing with a soft cloth. Framing these pieces behind clear glass is recommended. Like all photographic films, they should be kept away from excessive heat and moisture.

Photopolymer Film

Several companies (most notably Polaroid and DuPont) have developed a plastic film material that is especially suited for reproducing holograms. These photopolymers are extremely bright and durable, which makes them appropriate for a wide variety of commercial applications. The automated production process used to reproduce these holograms cuts costs considerably. An 8" x 10" matted photopolymer film hologram can retail for well under $100 .00.

Image depth is a bit shallower and projection distance is a bit less than in silver halide films. However, photopoly- mer holograms are a magnitude brighter and generally have a much wider viewing angle. Although this allows for more latitude when using a less-than-ideal light source, a single overhead light is still recommended for illuminating wall decor. The material works exceedingly well for items that will be illuminated by normal outdoor lighting, such as jewelry, bookmarks and postcards. The plastic surface of the photopolymer films can be cleaned with a soft cloth.

Dichromate Glass

Another photosensitive material used to make holograms is called "dichromate." Unlike silver halide glass plates or rolls of film, this material is actually a chemical/gelatin mix that is coated onto a piece of glass. After exposure, a cover glass is securely attached to the base plate, creating a permanent seal - since moisture or excessive heat can dissolve the dichromate gel and the hologram will eventually disappear. Sometimes plastic is used instead of glass, lowering production costs.

Dichromate holograms are very bright and can be viewed under less-than-ideal lighting conditions, such as exist outdoors. Therefore, they are most frequently used as jewelry items and fashion accessories. The glass can be cut into any shape, but it is often cut into small discs. These holo- gram discs have traditionally been used in watchfaces, broaches, key-rings and belt buckles. For years, the $20.00 dichromate glass pendant has been a staple for retailers. An 8" x 10" framed dichromate plate usually retails for over $100.00, but they are seldom made anymore.

Embossed Foils

These holograms are reproduced by a process that is not optical and does not use costly photosensitive materials. In this case, the holographic image is transferred onto a mechanical stamping die, which is then used to emboss the image on rolls of very thin plastic films or metal foils. These holograms can be reproduced for cents per square inch, making it the least expensive way to copy an image. These foils are commonly hot-stamped on paper or plastic, or backed with an adhesive layer. The material is very durable.

The embossing process creates very shallow images that can be viewed under poor illumination. They are usually silver-backed and reflect a rainbow of colors. Some of the best-selling holographic products are animated 3D embossed images which display fluid motion and realistic colors. These commonly retail for $15.00 - $20.00, matted and packaged, ready to frame. Most of the embossed holograms on the market are called 2D/3D. They display multi- level graphics and are usually used as inexpensive stickers, or part of keychains, pins, and magnets.

Often a repeating prismatic pattern is embossed onto the plastic or foil, which creates a ever-changing rainbow effect. This colorful material is often used in toys and optical novelties.

Opening a Shop - Basic Guidelines

1) Choose the best location you can afford. Tourist destinations in "festival" type shopping districts located near dining and drinking establishments traditionally do very well. High-traffic locations ensure a steady stream of curious shoppers. Malls do very well around the holidays, but can be slow during the off-season. Do extensive market research.

2) Shop for a reliable supplier that can stock you in a timely manner. A good supplier will be able to replace the merchandise you have sold quickly, thereby reducing the number of units you need to backstock. A good supplier will also assist you with displays, inventory selection, and point-of-purchase materials.

3) Stock a wide selection of merchandise in various price ranges. Many hologram stores cover their overhead with the sales of inexpensive items in the $1.00 - $10.00 range. Most retail sales average $5.00 - $25.00. Wall art typically

sells well at the $35.00 price point, as do $65.00 watches. Sales of more expensive pieces can be icing on the cake.

4) Plan your displays carefully. Due to lighting considerations and limited viewing angles, holograms need to be merchandised more carefully than many other products. Installing adjustable track lighting is more practical than using stationary fixtures. Halogen spot lights (50 - 100 watts) work very well. Avoid fluorescent and recessed in- candescent lights.

5) Create an entertaining atmosphere. No one needs to buy a hologram. A friendly, informative staff will boost sales dramatically.

Start-up Costs

The amount of capital you need depends on how large an operation you want to have. On the low end, you can stock a "cart" operation, as opposed to an actual shop, for as little as $5,000. Buying starting stock for a small shop (several hundred square feet) will cost several times that. A larger space of 800 - 1,000 square feet can easily hold $50,000.00 worth of goods. Plan on spending $ 10,000 - $15,000.00 for high-quality lighting and displays for this size store.

Based on a "double markup for inventory," $200,000 worth of annual gross sales should support an owner/manager making $30,000.00, a small staff (dividing another $30,000.00), rent/ overhead in a high-traffic location ($25,000.00), and first year build-out ($15,000.00).

2

Holographic Stereograms

This chapter explores one of the most exciting types of holograms being created today - the holographic stereogram. We interview several innovators in this field to discuss their latest work.

Introduction

This chapter discusses a very popular type of three dimensional image - the Holographic Stereogram. This kind of hologram is produced in a variety of formats. The holograms on the front and back cover of this book are two examples of stereograms which are discussed in more detail in this chapter.

The holographic stereogram is one of the most exciting compositions in holography today. It allows artists to incorporate a wide range of visual effects in their images - especially cinema, animation and dimension. Today's computer technology is making the production process more accessible and the finished products more refined.

In this chapter , we will provide an brief overview of the evolution of holographic stereograms, with an emphasis on current methodologies. We will discuss the techniques used to produce the holograms that appear on the front and back covers of this book in detail with the people that created them. We have also included related information from a cinematographer and a model maker who have extensive experience in the production process.

Trade Jargon

We should point out, before we get started, that several techniques are used to make holographic stereograms. Sort- ing out the jargon can become confusing. Some of the names you will hear that refer to holographic stereograms are:

Holographic Stereograms - Probably the most common,

safest, and most inclusive name. It is used to name any hologram that belongs to the group of holograms that are designed to achieve their effect by utilizing a human's capacity for stereo vision.

Integral Holograms - In general, this is a term that refers to a finished image which is constructed from many discrete units.

Cross Hologram - Describes a holographic stereogram process developed by Lloyd Cross that utilizes stereographic and integral techniques. This is probably the first commercial holographic stereogram process developed and it involves filming a subject on a rotating stage.

Multiplex Hologram - Describes a hologram produced using a system developed by Lloyd Cross and refined by the Multiplex Company.

"Printed" Movies - Refers to a flat hologram that features series of pictures that appear animated when the viewer moves in relation to the hologram. These holograms are typically produced in a way that they can be affixed to a page or other flat surface.

Embossed Stereogram - Embossing is simply a method of mass producing the stereogram. It is typically the most affordable way to replicate a moving holographic image that is relatively shallow.

Because of the variety of terms, we allow the innovators we interview to define the nomenclature they use. Steve Smith, for example, even provides a short glossary of terms used in his work. To better understand the aforementioned terms, let's look at how the holographic stereogram evolved.

Early Development

Most historians credit Lloyd Cross and his cohorts in San Francisco with the development of a process that resulted in the first reliable method for producing a holographic stereogram - it resulted in a three-dimensional cinema that appeared to "float in space". Their method, developed in the early 1970's, allowed live subjects, life-size models and special visual effects to be incorporated into their holograms in a practical and affordable way. Here is a simplified description of the process:

Stage: Make a rotating stage.

Scene: Place an object or a scene with live actors on the stage.

Cinema Camera: Set up a stationary movie camera in front of the stage.

Shoot: Film the subject as the stage rotates 360 degrees, making sure to shoot at least three frames for each degree of rotation. In addition to the stage moving, the subject is allowed to move slightly in a manner that will result in a smooth animated sequence. Rapid or uneven motion, however, will create undesired "blurring" effects.

Develop: Develop the movie footage in a normal manner.

HOP Transfer: We now want to make a hologram of each frame of the movie footage . These holograms will be sequentially exposed onto a sheet of film using a hologra-

phy setup whose elements are collectively referred to as the Holographic Optical Printer (HOP - See the diagram below).

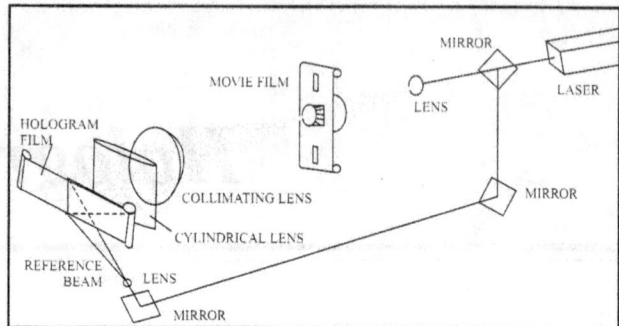

The HOP setup illuminates each individual frame of movie footage with laser light from the object beam in this setup. The reference beam, of course, meets the object beam at the emulsion by another path to create the hologram . Each frame is optically "condensed" into a narrow strip on the film using lenses and a mechanical slit aperture which restricts the image to one, narrow, vertical slit. The film is advanced and the process is repeated. A series of vertical slit holograms, running the length of the film, results.

Process and View: After the process is complete, you will have a length of film with hundreds of thin vertical holograms on it. Once processed, take the film and wrap it into a cylinder shape. When the film is illuminated from inside the cylinder (behind the film) with an appropriate light source, the viewer will see an apparently solid image floating in space inside the cylinder! As the cylinder rotates, or the viewer walks around it, the image looks fully dimensional and appears to move! (See diagram below.)

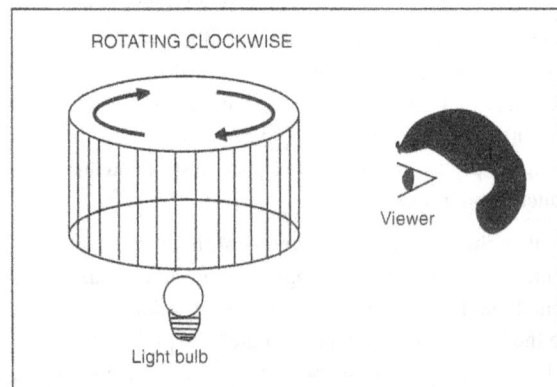

These dramatic effects result from the fact that each of the viewer's eyes sees a slightly different image at the same time. Our brain then combines these images to give us a "stereogram" effect. One limitation to Cross' approach is that this technique creates images that display horizontal parallax only (i.e. you can't see above and below the image). This is very adequate in most situations because in life we generally inspect images by looking side to side and not over and under the image.

In order to commercialize the endeavor, Cross and his colleagues manufactured a motorized display unit for their free-standing 360 degree version. They also developed a stationary wall-mounted unit that displayed 120 degrees of viewing angle as the person moved around. The idea of creating a self-contained holographic display device was quite revolutionary and very admirable. The complete units, which incorporated an inexpensive light source (an unfrosted light bulb with a vertical filament) along with the hologram, sold for several hundred dollars. The Multiplex Company has been producing units based on Cross's process for the past twenty years.

Variations and Refinements

Subsequently, holographers produced variations of Cross's concept. Some made stereograms with different degrees of view, commonly 60 or 90 degrees. Others began shooting the cinema by moving the camera along a track (instead of moving the stage). They went on to flattening out the cylinder, which allowed the holographic stereo- gram to be produced and handled more easily. Eventually, researchers embossed these transmission holograms onto minor backed plastics or produced reflection copies - which allowed front lighting (more practical in most situations).

LCD

A major advance in technology was the introduction of the LCD (Liquid Crystal Display). It was quickly adopted by the computer and TV manufacturers as a display de- vice. It did not take long for holographers to see the benefits of using the LCD as a source for the image being recorded. LCD origination substitutes graphics displayed on a Liquid Crystal Display screen for cinematic footage. This allows digitized images (with all their advantages) to be easily incorporated into a hologram. A wide variety of cinematic, video and still images can be scanned in a computer, manipulated, and displayed electronically using LCD technology. The hologram on the front cover was produced using a series of images that were modeled and rendered with the aid of a computer, displayed on a LCD and recorded holographically.

Inherent Problems

Before we continue, we need to discuss two major problems that confront a holographer attempting to produce a holographic stereogram that is technically proficient and aesthetically pleasing - time smear and motion blur. Both are visual effects that are related to technical and perceptual processes. Both relate to the two major components of holographic stereograms - depth and motion. They are commonly considered undesirable effects that can be controlled and manipulated with adequate forethought.

First, let's quickly review some basics about human vision. Most people with "normal" vision see with both eyes - that is they have stereo vision. Each eye collects the light from the world around us, changes this light into electro-

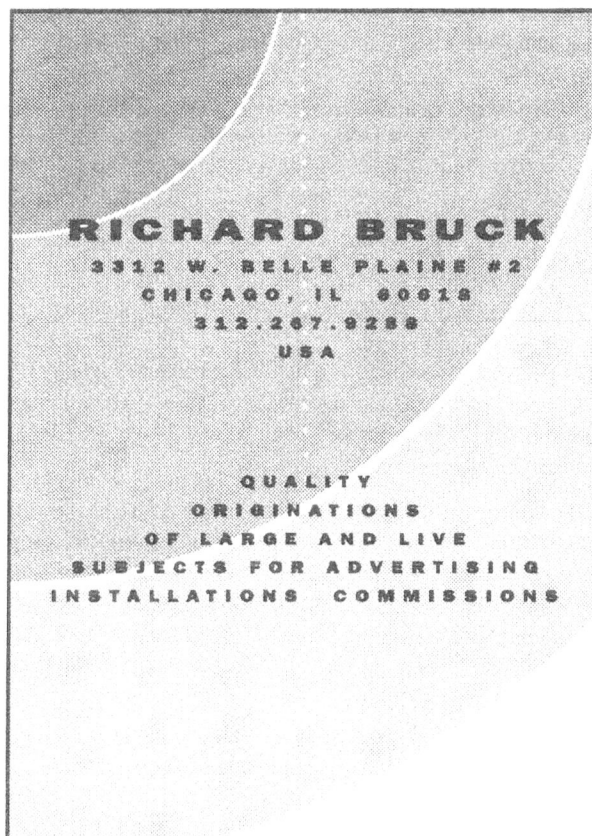

chemical impulses, and transfers these signals through a neural network that processes this visual information. Our central nervous system has evolved to combine these two visual inputs into a singular view of the world, so that we can function more efficiently in the world around us. Since our eyes are horizontally displaced by two to three inches, each eye provides us with similar, but separate views of our surroundings. This arrangement has some very important advantages. It allows us to create spatial measurements using triangulation, which in turn creates depth perception.

Time Smear

A holographic stereogram works by utilizing the concept outlined above. Each eye is fed a separate picture which is almost identical to the other, and our brain combines these into a single image with "depth". However, if each eye, instead of seeing two views of the same image, sees two views of the image taken at different times, our brain has trouble integrating both views into a coherent whole. For instance, if one eye sees a clock with the hands pointing to midnight, and the other eye sees the same clock with the hands in the one o'clock position, our brain is confused by this unnatural situation and creates an unfamiliar blurry composite. This situation is called "time smear". Each eye needs to see the same scene simultaneously. In the real world, we are not confronted by this problem, but if we look at a holographic stereogram that was not designed correctly, this problem does occur because each eye is seeing a different frame of the movie. In short, if there is little movement between the frames that the right and left eye see, the brain can combine the two images successfully. If there is significant difference between the images, you will get time smear.

Motion Blur

Another problem is images that move too fast. We have all experienced this. If you try to focus on a tennis ball or baseball as they are being hit, all you are going to see is a blur. The images simply move too fast. The brain gets a picture but before it can process it, the ball has moved and you have another picture to cope with. All you get is a blur. Motion blur can occur in stereograms if you move the seterogram too fast.

Assuming that the images used in your holographic stereogram do not have any time smear problems, the only way you will get motion blur is if you move the frames past your eyes too fast. In a hologram designed for hand-held viewing, like the one on this book's cover, the viewer has complete control over the rate they see the images. Your options range from scanning very slowly, for an almost stroboscopic ("jerky") effect, to scanning very rapidly, to create a holographic motion blur. The optimal scan rate lies in between these two extremes.

Solutions to Time Smear

There are three obvious solutions to the time smear phenomenon. One solution is to have very little motion in your cinema (or have your computer render in a bunch of images, where needed, to slow down the motion). The second solution is to have wider slits (fewer frames in the stereogram), and the third is to have the slits horizontal to the eyes.

Horizontal Slits

If you take vertical slits and turn them 90° you have horizontal slits. These slits enable both eyes to see the same image. Therefore the brain has very little, if any, compensating to do for the images presented by the two eyes.

It is still possible to get time smear with horizontal slits if you make extremely-fine slits (much finer than the diameter of the pupil of the eye). In this case, it is possible for too many frames with too much motion to be seen at once. In general, however, horizontal slits will allow many more frames of a given cinema to be seen by the eyes than vertical slits, because both eyes see the same image and the brain has fewer images to deal with. Note, however, that the viewer will have to tilt the picture up-and-down in- stead of left-to-right to see the motion and there will be very little parallax (depth perception) to the image be- cause both eyes will be seeing the same image.

The Full Parallax Grid

We have mentioned two methods for making stereograms, vertical slits and horizontal slits. A third method, used for the front cover of this book, uses both horizontal and ver- tical slits. The intent is to have movement, or parallax, both left-to-right, top-to-bottom and even corner-to-corner. In short, any way you look at their hologram will approximate a true-to-life (or full parallax) view. This procedure is best explained by discussing how the cover hologram was done.

The concept is that if you lay horizontal slits on top of vertical slits, you create a grid with many squares, which we will call cells . Each cell is a different picture seen from a different viewpoint. The center cell is a view of the picture straight on. For the picture in the next cell up and over, you have to create an image of the subject from that perspective.

This means that you either need to shoot the subject from that perspective, if you are using a real-life subject, or you have to have animation rendered so it is seen from that perspective. You then design an optical printer that prints each square in its proper position on the recording mate- rial to get an H-1.

Point Source Productions studio produced the cover image using Holo Sciences' proprietary Solid LightTM technology. The computer-rendered nature of this type of x-y perspective is a kind of Solid LightTM image the creators dubbed Cyber-sculptureTM This is the first time anyone has presented such an image in a published form.

To give you a better idea about what was involved, we have asked the creators to explain how their process works and how much it costs.

Production Team Interviews

TheAnimator

The image on the cover of this book was designed and ex- ecuted by Brian Kane of General Design, and Ben Fischler and Noah Hurwitz of Imagination Plantation, where the 3D computing was done. Kane has extensive experience in creat- ing animated images for holography. broadcast video, print, and interactive software. He has created holographic images for companies such as The Upper Deck Corporation, American Bank Note, and Live Entertainment, and his videos have been seen around the world.

Following are questions that come to mind when look- ing at the cover animation, and the designers' responses:

1) What hardware was used in creating the animation?
BK: The "TV" hologram was created with a Macintosh Quadra 650 and two Silicon Graphics Indigo2 XZ work- sta- tions.

2) What software packages were used?
BK: The video images on the screens were digitized on the Macintosh using Adobe Premiere. They were then ed- ited and saved out as separate image files . All of the mod- eling and animation was done in Alias Power Animator. We started by modeling a single TV console, then replicated it

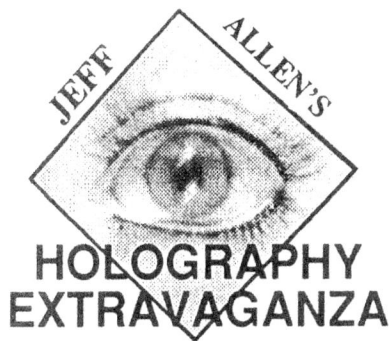

into 12 sets. The TV sets were stacked and arranged, the back wall and floor were added, and the scene was lit. We added a few extra elements, the dish antenna and cable box, to fill out the scene and make a good composition. The camera was animated by using a UNIX shell script which repeatedly invokes the Alias renderer. Some additional touch-up work was done with Adobe Photoshop on the SGI.

NH: Our tools of choice are Silicon Graphics Indigo2 workstations running Alias/Wavefront's PowerAnimator which we use exclusively for modeling, animating and rendering our 3D worlds. We have used a variety of 3D software packages on all the major computer platforms, but we have found that the combo of Alias/Wavefront's software and SGI's hardware beats any other setup around. Although very strong in all areas, Alias is an exceptionally capable modeler, which excels for producing really complex, organic shapes. More importantly, Alias gives the artist instant feedback, which means more creative control an? it really encourages experimentation. Alias not only brings the artist directly into the process, but the software almost becomes transparent to the artist and actually moves you forward, giving something back in the end. By utilizing Silicon Graphics hardware and Alias/Wavefront software we are able to apply our vision and knowledge to create superior content.

We believe in using technology as a tool to facilitate what we want to do. Technology should not exist in some rarefied atmosphere, but rather must be commandeered by artists like us if it is going to really do some good in the world. Many times in this hi-tech world we live in, people lose Sight of whether they are running the computer or if the computer is running them. If the best tool to generate 3D animation and create holograms were a rock and stick, we'd be using it! We go into PowerAnimator's OpenRender tools, and directly incorporate the creation of holographic data into the rendering process. Alias gives us an almost unlimited ability to visualize whatever we can dream up.

The advent of powerful, computer-based 3D content-creation tools has changed the landscape of holography. An animated medium like holography benefits tremendously from the major advances in high-end computer graphics. The capabilities are growing exponentially and the consumers' appetite is growing even larger. By integrating the most advanced 3D systems in conjunction with existing and new holographic techniques, we are able to break new ground in holographic imaging. The computer allows for imagery that would have been previously impossible due to time, budget, or technical constraints, to be a reality.

3) How long did it take to render the images?
BK: It took 40 hours running on two SGI machines, so about 2 days.

There are a total of 300 frames in the cover image.

4) Given a concept like the cover, how long would it take to produce the animation and how much would it cost as a rough estimate?
BK: I generally plan on between two to four weeks for a project, but it really depends on how well developed the

idea is at the beginning of a job. Also, the complexity of the modeling and animation affect how long a project will take. Characters and people tend to be the most complex, while flying logos and corporate graphics tend to be the simplest. The degree of"realism" also affects the turnaround time since the more photo-real an image is supposed to look, the longer it takes to make all of the details perfect. Between $2000-$5000 is a ballpark range.

5) What are the main design considerations that a customer needs to keep in mind when thinking about having holographic animation done?

BK: Be creative.Ho lography is a dynamic , high impact medium. It is also a unique output option for computer-generated data. Work with someone who has been through the process and understands both the technical and creative issues of holography as well as integrating foil or film into a product. A product with a hologram on it should look good even if the hologram can't be viewed properly. Use the reflective silver quality of foil or the black space of a reflection hologram as its own design element. There are some pretty advanced options available now for application and overprinting. Focus on the final product, and incorporate the hologram into the total design.

Embossed holography should be designed in color, with as much color variety and contrast as possible. Exact color matching such as Pantone color only works when the hologram is viewed properly so it shouldn't be a main design criterion. Animation should be subtle. The most important part of the image, logos or faces, for instance, should be on the image plane so they stay the sharpest.

Reflection holography should be designed in black and white, taking into consideration the playback color of the replication technique. Some are red, some are green, some are yellow. Generally speaking, you can get away with more depth in reflection holography, as opposed to embossed holography where one generally tries to keep the depth pretty shallow.

Pre-production and planning always helps a project flow smoothly. Storyboards are a good way to work out an idea. There are stock 3D data libraries available which can save time and effort. The same data can be changed to make multiple holograms, such as a character in different poses or a product in different environments. Security features can be animated into the image and easily changed for subsequent masters. I proof by sending videos of the animation at half resolution to the client before the final rendering is started. This is a good way to see what you are getting before it is mastered.

Computer hardware and graphics software have become very powerful in the last few years, and artists are now capable of creating images which could only have been previously imagined. Some of these capabilities include character animation, particle system simulations such as smoke, fire and clouds, morphing, explosions, and even hair. Live action footage on film or video can be combined with computer graphics in "virtual" space. Objects and people can be scanned in 3D with color which saves quite a bit of modeling time. As well, clients who have already developed 3D assets for film, video or

video game production can use their data to make holograms. Computer graphics give a level of control and tangibility in the design process which makes it highly flexible.

Working on the cover image has been a fun experience, since creative control was left in our hands. I'd been wanting to combine TV images with holography for a while, since it brings together the two different aspects of my career, holography and multimedia. As well, the vertical and horizontal parallax aspect of the hologram made it a new technical challenge, or at least one I hadn't done before. This was also the first time I worked on a hologram with other designers, which was a very positive experience. Working as a team, we can provide quality in quantity, which I believe is where the market is going.

The Holographers

Bob Hess of Poillt Source and Chuck Hassen of Halo Sciences collaborated all making the master for the Fan! cover project. Some of the best holograms produced over the years have been mastered by Bob Hess. Their combined years of experience in the field enabled Hess and Hassen to make the mastering device used in this project.

Looking at the project from a business point of view, it is obvious that the holographers must communicate clearly with the design team, such as General Design in this case, about what can and cannot be done. Here are some of the questions that the computer artist would need to know, with answers from Hess and Hassen.

1) What is the practical limit (upper and lower) of rows and columns that we call use? What is the optimum number and why?

First, I should say that there is no single optimum con- figuration. The appropriate choice depends on the visual goals and the budget of the creator of the image. On the one hand, live action might be the only feature of interest, in which case, we might recommend using as many as 30 or 40 frames in the vertical direction, without any horizontal parallax (one frame on the horizontal direction). This would be a format useful for converting existing film or video to 2-D animations. On the other hand, one might desire to present only horizontal parallax, but to do so with a large number of frames per degree of angle of view, to maximize the smoothness of the changes in horizontal viewpoint.

I should point out that in choosing our array parameters of 300 images, we put 20 frames of horizontal parallax one each horizontal strip, and 15 cells of animation on each vertical strip. This allowed us to produce a holographic viewing space with full parallax in the vertical and horizontal directions. A typical horizontal parallax only (HPO) integral hologram might use 180 frames to achieve the same angular spread in the horizontal direction. 15 times 180 is 2,700 frames, which would be the number recommended to produce our image according to conventional practices.

We can certainly handle such a large number of frames. In fact the hardware we built for this purpose divides the horizontal axis into more than 20,000 steps. One of our primary objectives in bringing this technology to the market-

mary objectives in bringing this technology to the marketplace, however, is to cut the costs of production for our customers, while also improving the total image quality through the elimination of time smear, and addition of vertical parallax. Holo Sciences' Solid Light™ printing apparatus is capable of using any combination of aperture width (which gives the step size on the horizontal axis) and height (which provides the step size on the vertical axis) from 1 mm by 1 mm up to 20 mm by 20 mm, or even larger by request.

The Solid Light Cybersculpture™ format we used inherently does not require as many frames of horizontal parallax as an animated HPO image to attain the same frame-to-frame smoothness, because there is virtually no difference in time between stereo pairs on any given horizontal row of cells. The high number of frames recommended in conventional stereograms are not actually necessary for a convincing illusion of three-dimensional form.

Along the vertical axis, the maximum number of rows depends on how much the subject is moving over the entire range of vertical angle of view. As one increases the number of rows per degree of view, one improves the smoothness of the vertical parallax displayed. On the downside, however, one may run into unacceptable levels of time smear if the subject is moving quickly and if the rows are so narrow that one perceives several rows simultaneously. In the general case of vertical animation, therefore, we recommend that the horizontal slits exceed the diameter of the pupil. In that way, a maximum of two rows will hit the visual field at any given time.

In the case of a stationary subject, overlap between horizontal rows is less objectionable.

For the cover image, we employed the geometry shown in the figure below. The integral hologram master is a planar grid of images, 20 cells wide by 15 vertical. We wanted to encode parallax on the horizontal and animation plus vertical parallax on the vertical axis. In the final hologram, the aperture plane is 14 inches from the image plane. Note that the view you present for "upper right" in this grid is the view from the upper left as you face the screen. Total angular displacement in the horizontal direction is 56.4 degrees (+ 28.2), and vertical is 46.4 degrees (+ 23.2).

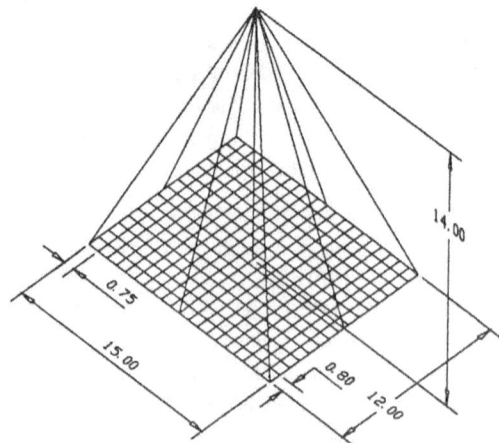

2) Since I will have to render each cell from a new perspective, how many degrees off-center is each new cell from the center square?

This is a good question. The answer is not a simple single number, but is instead a function of the geometry shown in the figure above. In Brian Kane's rendering, we chose to ignore the small errors which result from approximating the planar displacements of 0.8" and 0.75" as constant angular displacements of 3.3 and 3.1 degrees, respectively. This is computationally the equivalent of placing the center of rotation at the middle of a sphere, and moving the camera a fixed number of degrees on the surface of the sphere for each horizontal parallax step (from geometry, this is approximately $\tan^{-1} (0.75"/14.0") = 3.07$ degrees). Once a given row is complete, move the camera in the perpendicular direction by an angular displacement equivalent to one vertical step ($\tan^{-1}(0.8"/14.0") = 3.27$ degrees), and then repeat the horizontal scan. This is the equivalent of setting up our virtual object on a turntable, and mounting our camera at different elevations for each horizontal scan. In fact, the Solid Light™ integral process was de- signed to accept images created in exactly this way using a video camera. Point Source Productions even has constructed a 24" diameter stage which allows lights to rotate with the objects being filmed, so that shadows do not change with the angle of view.

Ideally, the concept is to match the changes in point of view to be rendered so they correspond to the final locations of the hologram cells (called the "aperture plane") on the master plate. This is done according to the physical size of the master and its separation from the intended final image plane. For any user who would care to be more accurate, Holo Sciences will provide the angular coordinates for every cell in an arbitrary perspective grid to be rendered.

Many software packages allow one to specify the coordinates of a "virtual camera" with respect to the rendered scene. If this capability is available, one must remember to convert the scene scale to match the final hologram geometry. For example, if your image is a vehicle concept to be rendered at 1/12 scale, then one inch in real space equals one foot in the model space, and the camera would need to be located 14 feet away from the center of the image in model space for the perspective grid to end up accurate when it is recorded at a distance of 14" from the image plane of the final hologram in real space.

3) What are the restrictions on lighting? Are dark shadows or bright highlights a problem? I know the dark image has more information for you to work with but it also tends to destroy the detail in shadow areas. A bright image is clear but you start to lose detail in the highlight areas. Infight of this, is it better to have the image I deliver a little bright, or a little on the dark side?

This sort of question has plagued people working with holographic images from the very beginning. The Solid Light™ format does release the user from many of the constraints of conventional (non-integral) holographic formats, because one may use standard lighting equipment and techniques on a real object, and the whole range of computational light modeling approaches on a virtual object in cyberspace. The degree of realism that one obtains in gen-

erating these images, and then merging them to form a hologram, depends on a number of factors:

Contrast: It is best to use a fairly narrow range of contrast within a given image. One may certainly employ clearly defined regions of high and low brightness, but the absolute difference in brightness between the highlights and the shadows should be kept fairly small.

Bright object distance from image plane. A bright object tends to cause a region of over-modulation, or burn in, relative to the rest of the image. An example of such an image component in the cover image is the violin wiping its nose in the top row. Note that when the image of the TV screen pulls away from the image plane as you swing the hologram from side to side, a ghost image remains behind in the region where the light may be seen from most angles of view. The further behind or in front of the image plane one places a bright object, the more pronounced will be the separation of the bright object from its average position, when seen from extreme angles of view. That separation magnifies the perceived level of burn-in. There are two ways around this problem. You can place the bright object far behind the image plane, so that its average exposure is smeared out over a large area, or you can place it very near to the image plane, so that the parallax-induced separation of the image from the region of over-modulation is minimized.

Desired visual effect. Remember too, that you are not necessarily attempting to match reality photon for photon. Solid Light™ is a medium that demands experimentation

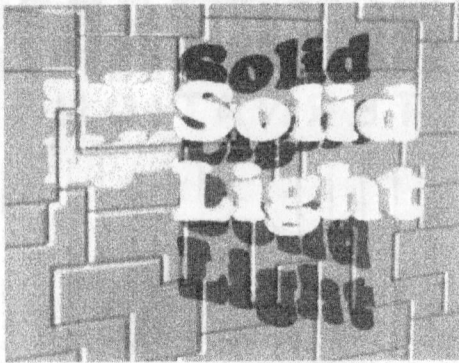

and pushing the limits. One image's flaw may well become another's special effect. This is an area where consultation with someone who has experience producing holograms, like Bob Hess or Brian Kane, can really add value at the concept stage of a project. The cover image shows many of the effects that are possible with different types of 2-D subject motion.

Attention to shadows. Remember the direction from which your hologram is designed to be lit when played back. The TV hologram is lit from the- top, so the shadows were made to appear as though cast by an overhead light source. Small deviations from the reference beam angle are acceptable, but large mismatches will confuse the viewer's eye.

Overall Brightness. Although we are reluctant to pro- vide any hard and fast rules, chances are that your holo- gram will be more visible if it is on the bright side, than on the dark side. If you feel comfortable with the issues listed above, try to remain close to your realistic vision and still stay within the guidelines. If you have any doubts, call a professional.

4) Assuming I deliver all image such as the one 011 this cover to you according to your specs, how much do you charge for mastering? Will this price drop over time or with volume?

The cost of creating a master from a set of full parallax animated images similar to the ones used for this cover would be approximately $7,500 for a single hologram contract. Simpler formats with fewer cell would cost less; more cells would cost a little more to produce. If a customer wished to produce a series of 4 or more images with similar grid geometries, all at once, that price could be cut by 30% today. This would also make sense from the point of view of creating a production tool "grid master" with more than one image at a time for replication on photopolymer. Setup costs are high in both steps, and one stands to gain very much from simultaneous production of series of images. We are still on the expensive side of the process learning curve, and hope to be able eventually to offer masters from user-provided art for only a few hundred dollars each. We hope to grow in the long run by making this format accessible to any business, small or large, that has ever toyed with the notion of making a commercial hologram.

Although space here does not allow us to list them all, an interested customer may obtain more information about other Solid LightTM formats by contacting us.

5) Is there any special file format that my animator needs produce the files in?

At this time our preferred file format is PC-based, PCX format. I am aware that many artists use other formats, and we do have the capability to perform batch conversions of most popular output formats for the PC and for the Macintosh. We use a shareware program called Graphic Workshop, which comes in a Windows compatible version, and in a DOS-only version. Another excellent graphics conversion package is called Hijack, though I have not used it personally. We will be happy to do the conversions for a flat hourly rate. We can also sample directly from videotape, but please contact us before you begin to produce the frames on video. It will save everyone time and money.

ZONE HOLOGRAFIX STUDIOS

FRED UNTERSEHER & REBECCA DEEM

• CONSULTING
• EDUCATION
• PULSE LASER HOLOGRAPHY
• CW LASER HOLOGRAPHY
• SPECIAL TECHNIQUES
• INTEGRATED TECHNIQUES

Zone Holografix Studios is devoted to pulse and CW origination and transfer holography. The facility is designed to serve both commercial clients and independent artists. Founders, Fred Unterseher and Rebecca Deem, have long and extensive careers in the field. A select ZHS team works closely with mass producers involving origination for embossed, DCG or photopolymer holograms. Contact us directly for quotations and additional information.

5338 B. VINELAND AVENUE • NORTH. HOLLYWOOD, CA, 91601, USA
TEL. 818 985 8477 • FAX. 818 549 0534

File names should follow a linear numbering convention, so that as you face the screen (looking "up" the pyramid in the grid figure), the sequence begins at X001.PCX in the upper right, and progresses to X020.PCX at the upper left corner. Numbering continues for the next row again moving right to left from X021.PCX to X040.PCX, and so on until X300.PCX in the lower left corner. In other words, if you are looking at the image, the first file should carry the viewpoint from the upper right hand corner of the viewing space, and the last from the lower left viewpoint.

6) How long does it take to make the hologram master? What kind of a proof do I get for my customer?

Right now we are set up to run our customers' work on a quarterly batch basis. As of this writing, our next session is scheduled for October 23, 1995. The actual hologram shoot requires about one day, depending on the number of frames, with two to three days of setup and disassembly time at either end of the two-week production schedule.

A reflection copy transfer of the image on silver halide film serves as a final proof for customers who request it before creation of the production tool ganged master. Additional reflection copies on silver halide plates may be added to the job at this stage. (As the price for this service continues to drop, we anticipate that some customers will want to produce small editions on silver halide, and never take the image to photopolymer replication.) Figure one additional day after the H-1 is finished before we can produce the reflection copies, at a rate of eight to ten per day.

Creation of a multi-image production tool is a separate pro-

cess. Bob Hess at Point Source Productions estimates that this step will add about $1,500 to $3,000 to the cost.

The Replicator (Mass Producer)

John Gunther, at Hughes Power Products, supervised the replication of the front cover hologram. Hughes Power Products has been producing commercial-grade holograms for a number of years. They have a lab with state-of-the-art equipment for mass replication.

Here are some of the common questions that customers might ask their replicator, along with Gunther's answers:

1) If my clients ask me what the primary steps are that the replicator has to go through, what should I tell them? Does the replicator provide me with a proof of the run before it begins?

We go through five basic steps in the hologram replica- tion process.

The first step, early in the concept development process, is to ensure that the replication customer (generally an independent holographer or holographic project coordinator) understands our capabilities and our specifications for replication masters. A description of our services and a master specification are available upon request.

The second step is to inspect the master to ensure that it is compatible with our equipment and processes. While the production master is usually provided by the customer, we are willing to assist in the preparation of the master.

The third step is to provide photopolymer proofs to the

customer to confirm the desired hologram characteristics, such as color and brightness. Note that we currently make holograms in single colors ranging from medium blue through red. The proof samples are normally hand made contact copies of the master. We will also provide twenty frames (20 square feet) of production-proof holograms for a nominal charge that can be credited, in part, against a future production order using the same master.

After the customer approves the proof samples, we en- gage our automated hologram replication line to produce the desired number of holograms, including overage to make up for attri- tion during the production processes. Replication includes the actual hologram exposure and the development processing used to set the hologram color and other characteristics.

Finally, we complete the holograms by laminating suit- able protective and/or adhesive layers on both sides of the photo- polymer film, and by die-cutting the holograms to final size. The most common form of the completed hologram has a scratch- resistant protective film on the front surface, a black backing, and a pressure-sensitive adhesive with a paper release liner. We can also provide transparent stickers with the adhesive on either surface, and a transparent semi-rigid construction.

2) From the time you have a approved master from the ho- lographer, how long does it take you to produce, for exam- ple, 5,000 copies of a hologram like the one on this cover?

While the actual production of 6,000+ holograms like the one on the cover can be completed over a period of a few days, the delivery time depends on our stock of materials and the length of the queue in our lab and at the subcontractor we use for die- cutting. Our normal business volume dictates a delivery time of three to four weeks after customer approval of the proof samples.

3) How much does it cost? Does the unit price drop with volume? Where does the unit price level off at?

Our normal minimum production order is for 430 12" x 12" frames of holograms. We do, however, currently accept orders for smaller quantities. The price for a standard mini- mum order will range from $6500 to $8500, depending on the type of fin- ishing desired. The price per frame of holo- gram will decrease with increasing quantity.

Note that the number of good holograms in a minimum run will vary with the hologram size and with the attrition or yield loss that occurs in any production process. Since the expected number of defective pieces depends on the size of the holo- grams and the quality of the master, we can provide specific yield estimates on a case by case basis only. We will be happy to provide quotations by the piece part for specific applications

Cinematography Input

The input for the cover hologram was done completely with computer animation. This is not, of course, the only way to create the input. You can shoot film and have it scanned into your computer for original input. In fact cin- ema input is still critical in cases, including celebrity shots.

In order to allow the cinemaphotographers to present their case for shooting in film we had to find someone who has ex- perience in the field and get them to explain the subtle- ties involved in the process.

Glen Gustafwn is a professional cinematographer who films mov- ies on contract through his company, Dimensional Cinematography. In additional to Hollywood cinema, Glenn has done a considerable number of films for holographic stereograms. Famous personalities are a hot ticket in the business world and Glen has participated in the film- ing of such personalities as baseball player Rickey Henderson and rockstar Prince. His corporate clients include Pizza Hut and Univer- sal Studios. Glen generally shoots in film as opposed to video, and his job ends when the desired image is on film. He was asked to give an overall picture of what equipment is needed & involved in a cinema shoot, a case history or two of problems that happen on a actual shoot, and some rough figures on costs.

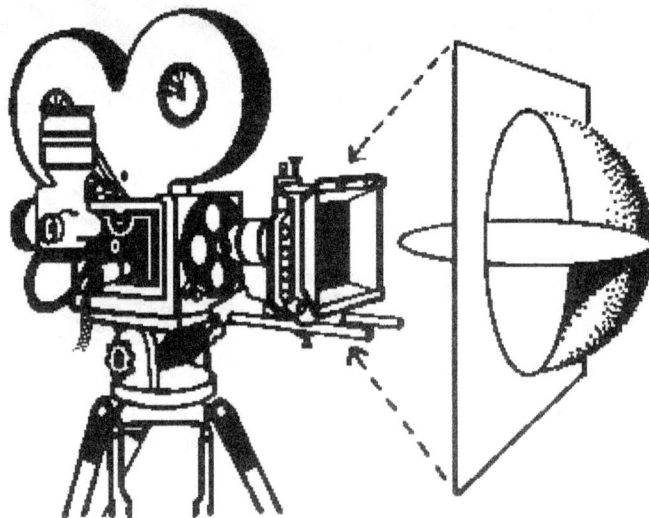

Filming For Holographic Stereograms

by Glen Gustafson

Just as a conventional hologram can be no better than the model from which it was made, a stereogram can only be as good as its film or video source. Assuming that the tech- nical requirements for producing a stereogram have been met, the factor having the greatest effect on the quality of the image is the ability of the person responsible for lighting the subject, (in film, the Director of Photography, in video the Lighting Designer).

The highest-quality moving image still comes from a mo- tion picture negative. In fact, the highest quality video comes from the direct transfer of motion picture negative, preferably exposed at 30 frames per second. Many people are surprised to learn that the majority of television production is still shot on film (aside from talk shows and news). Post production, (edit- ing, etc.), has changed dramatically with the digital revolution, but production has not changed much, yet. Because I come from a fine arts background, I tend to concentrate on results, whereas technicians focus on equipment. When video cameras enable me to obtain the results I am after in terms of drama and power of expression, I'll use video if it is the best tool for the job.

The difference between the look of film and video has as much to do with the intensity of the effort put into production as it does the medium. On a recent film shoot in L.A. I was provided with lighting and grip equipment filling a five-ton truck, a massive generator and an experienced crew willing to work like animals for 14 hours a day. On a typical video production I was usually provided with a lighting kit that fits in a single carrying case and a crew consisting of a couple of kids from a local journalism class who don't even know what grip equipment is (it is the equipment used to control the light reaching the subject). The difference between the look of the film shoot and the video owes as much to the general level of professionalism as it does the medium. The saying, "money buys the look" holds a grain of truth.

Still, there are instances when high quality video, such as Beta SP, is the best choice for shooting a stereogram. If the printed image is going to be small, the superior resolution of film will be lost in the lenticular screen used in creating the H-1 (though film still retains its advantage with regard to contrast).

Digitizing video for manipulation in Photoshop, etc. is cheaper than scanning film, but the resolution is 525 lines compared to film's 4,000. Each frame of 35mm negative contains approximately 50 megabytes of image data, allowing film to record a much more complete range of tones. This produces a smoother and more pleasing image, with detail in the lightest and darkest areas. Much of the work done in Photoshop was made necessary by shooting video in the first place, resulting in harsh contrast and blazing reds, in addition to the low resolution.

In spite of what video enthusiasts are preaching, there is far more to the "film look" than adding flicker, dirt and grain. Medium speed film, shot at 30 frames per second and properly handled, will show none of these traits, yet even the casual observer can identify the look immediately. They might not be able to explain why they prefer the image recorded on film, but nearly everyone does.

Video continues to improve, (as does film), but it will probably be some time before producers are paying stars like Jim Caney $20 million for one movie shot on video. Quality and compatibility with all formats are clearly the film advantage.

If the printing run is limited and the subject relatively unknown, the lower cost of shooting video might be the deciding factor, although stereograms are usually shot on such a small amount of film that the difference in cost is not much in terms of the entire production. The difference between what I charge to shoot a stereogram and what holographers who want to try making movies charge is due more to their failure to budget for producer's liability insurance or a proper model release than anything else. There are already clients who have been burned by such amateurism; some bargains can be expensive. The importance of the job usually dictates the level of production. Unlike tabletop holography, there is frequently no

second chance to film, and if your subject is a celebrity or of a nature where you may want to make a large stereogram of the film at some later date, the primary photography budget is not a good place to cut corners.

Background

The first stereogram I filmed was of baseball player Rickey Henderson for holographer Sharon McCormack. After that, I did a statue of the Goddess Nike, and then collaborated with special-effects cameraman Charlie Mullen on a shot of the rock star Prince for the cover of Diamonds and Pearls. When my business partner and I started Dimensional Cinematography Corporation (DCC), we traded filming service for origination at American Bank Note Holographics in order to obtain a demo. That led to our filming Universal Studios' classic movie monsters for a set of trading cards used in a Pizza Hut promotion, and an actor playing an historic figure for a security hologram to appear on bank cards starting in 1996. For that job, DCC provided all production services, as well as prosthetic make-up, custom-made wig, period costume, casting, negotiation with the actor's agent-the whole package. We can produce the entire shoot, or I can serve as Director of Photography on a "work for hire" basis, leaving production responsibilities to the client.

There are a lot of ways to shoot footage for a stereogram. I've heard of people using a non-motorized turntable, rotating it by means of a couple of ropes, and some stereograms have even been shot on consumer grade high-8 video. I think it was Henry Ford who said "Anything can be done a little bit cheaper, and a little bit worse."

I have always endeavored to produce the best footage possible for stereogram production. Because live-action filming is best suited to recording a moving portrait of a living person, that is what my work has usually involved. All of my work has been done using a turntable, because a cylindrical stereogram is best suited to show off the most interesting image in the history of art-the human face. A linear stereogram would be the natural choice for a large group or landscape.

Equipment List

HOLOGRAMAS DE MEXICO

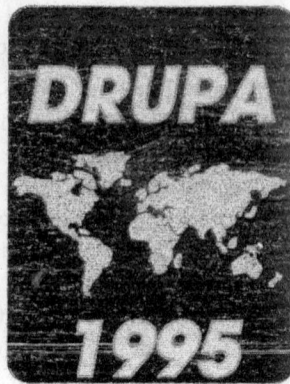

DRUPA 1995

We are a group of vertically integrated companies, something that lets us do every process in-house:

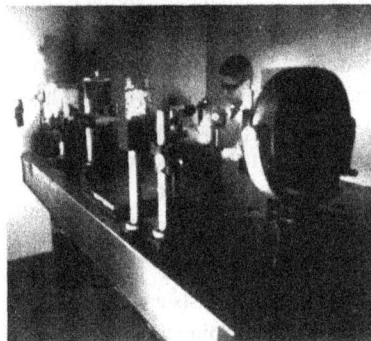

- Embossed production;
- Die-cutting and kiss-cutting;
- Hot melt and hot-stamping adhesives;
- Computerized Art Deparlament;
- Photoresist production;
- 2D/3D, 3D, Holomatrix™, and stereograms origination.

Our embossed materials are: polyester, polypropylene, PVC, paper and hot-stamping foil. Our embossed widths are between 6" to 43". Among our specialties we can have added protection to your holograms adding Bar Codes, sequential numbers, etc.

HOLOGRAMAS S.A. DE C.V.
Pino 343 Local 3 Sta. Ma. La Ribera
06400, México D.F., México.
Tels. • (525) 541 1791   • (525) 547 9046
       • (525) 541 3413   • (525) 541 1506
       • (525) 541 1696
Fax. • (525) 547 4084

Holographic Masterpieces

As the nation's leading manufacturer of embossed holography, Transfer Print Foils, Inc. offers the widest range of holographic technique. Perhaps it is the reason that everyone, from people beginning their first holography project to industry experts, relies on TPF to see their project through to completion.

Three dimensional images, 2D/3D images and TransFraction™ (prismatic) patterns are all available from TPF. We also proudly offer "pixel" images and patterns as well as an in-house Design Center to help turn your concepts into reality.

Rock Solid Security

High-tech security products such as our exclusive HoloClear™ provide the highest level of Rock Solid Security available today! Counterfeiters will run when they see a document or ID card with the HoloClear™ overlay. Call us for samples of this fraud-stopper immediately!

With a security production facility, advanced research and development and "state-of-the-art" technology, we provide unique images that are used for security on drivers licenses, international ID cards, trading cards... the list goes on and on!

Transfer Print Foils, Inc.

9 Cotters Lane, P.O. Box 538, East Brunswick, NJ 08816, USA
(908) 238-1800, (800) 235-FOIL, Fax: (908) 238-7936

A HoloPak Technologies, Inc. Company

17.04 / 432,8

4.93 / 125,2

Rubber feet
4 places

0.625 / 15,87

1.15 / 29,21

13.59 / 345,2

Power Supply: 4620P

WITH ETALON
59.50 / 1511,3

54.86 / 1393,4

TIMER

Fan

LiCONiX

BEAM OUT

4.06 / 103,1

7.34 / 186,4

7.38 / 187,4

3.67 / 93,2

TEST POINT ACCESS HOLES
.625 DIA, 3 PLACES

1/4-20 UNC-2B
4 PLACES

24.50 / 622,3

15.17 / 385,30

4.00 / 101,6

DIMENSIONS ARE IN INCHES / MM

Embosser II Series Specifications

Emb. II Model	Wavelength (nm)	Power (mW)	Mode (TEM)	Beam Diameter (1/e•, mm)	Beam Divergence (mrad, full θ)	Coherence Length (cm)	Spectral Bandwidth (GHz, FWHM)	Mode Spacing (MHz)
3620N	325	20	OO	1.2	< 0.5	10	3.0	113
3630NX	325	30	OO	1.2	< 0.5	30	1.0	113
3650N	325	50	MM	1.5	< 1.0	10	3.0	113
3660NX	325	60	MM	1.5	< 1.0	30	1.0	113
3675NX	325	75	MM	1.5	< 1.0	30	1.0	113
7625N	354	25	MM	1.5	< 1.0	10	3.0	113
7630NX	354	30	MM	1.5	< 1.0	30	1.0	113
4650E	442	50	OO	1.3	< 0.5	50	0.6	102
4660EX	442	60	OO	1.3	< 0.5	50	0.6	102
46120N	**442**	**120**	**OO**	**1.3**	**< 0.5**	**10**	**3.0**	**113**
46150NX	**442**	**150**	**OO**	**1.3**	**< 0.5**	**30**	**1.0**	**113**
46170NX	**442**	**170**	**OO**	**1.3**	**< 0.5**	**30**	**1.0**	**113**
46170N	442	170	MM	1.5	< 0.8	10	3.0	113
46215NX	442	215	MM	1.5	< 0.8	30	1.0	113

Embosser II Series Common Specifications

Vertical Polarization	> 500:1
Power Supply Voltages	Selectable 117, 100 or 220 VAC ± 10%
Weight (N style laser head)	33 lbs.
(NE style laser head)	38 lbs.
(power supply)	< 33lbs
Power Consumption	< 1000W
Power Stability (constant temp.)	< 3.0 % / hr
Pointing Stability (constant temp.)	< 50.0 μrad

Environmental Specifications

Operating Air Temp. (absolute)	10° to 30° C
Operating Air Temp. (recommended)	22° ± 3° C
Storage Temperature	-20° to 50° C
Relative Humidity	0 to 90%
Shock	20 g

DANGER

069510000

Hoizontal mounting recommended.
Specifications subject to change without notice.

Holography Presses On

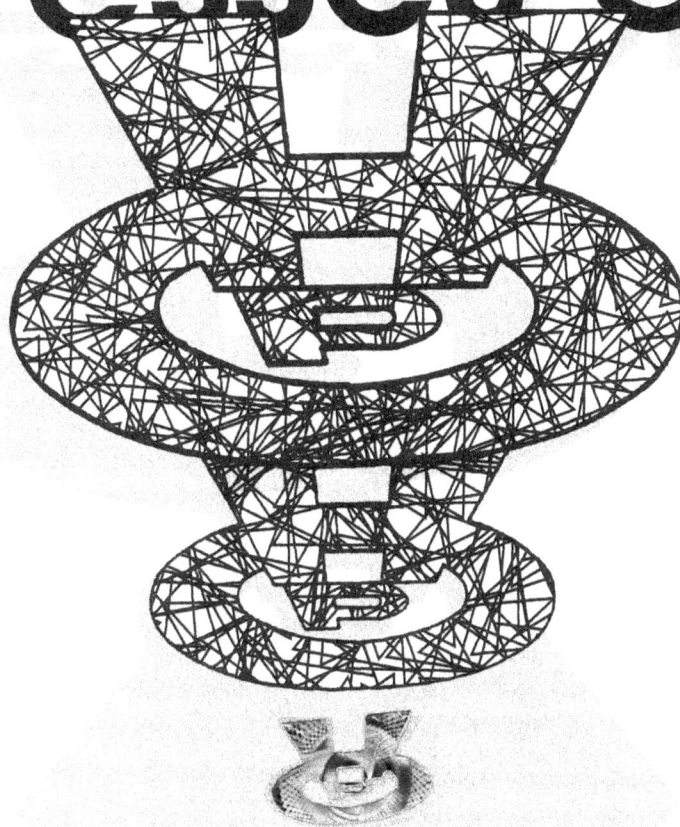

Holographic transfers applied with *heat or pressure* in stock or custom *shapes and sizes* for permanent adhesion *to all surfaces.* Sealed edges prevent delamination *in all weather...washable...dry cleanable.*

Patented product *and process.*
Distributors sought.

Holography *Presses On*
Jan Bussard
Box 193 Spring Lake, MI *49456*
Phone 616/842-5626 Fax *616/842-5653*

Camera: Mitchell GC (high-speed), 35mm, dual registration pins for excellent image steadiness even at speeds to 120 frames per second. High-speed filming makes it possible to get enough frames of subjects like athletes in action to make a stereogram or multiplex, whereas video's fixed 30 frames/second would have too much "time smear" to be usable.

Be advised that many surviving examples of these fine cameras are worn out, so there is more involved than just buying one. For conventional portraits I've adapted the camera to run with a large "stepper motor", so that camera and turntable operate via computer motion control for ab- solute accuracy. Stop-motion animation and time-lapse photography are also possible.

Turntable: Portable, designed for safety, rigidity, rapid setup and light weight. The turntables I used for earlier shoots were not designed for stereogram photography and were the biggest source of trouble and lost time. The combination of small size and massive gear reduction in this custom-made unit was difficult to come up with, but the math for calculating shutter speeds and frames per degree is simplified by having a gear ratio designed for the job. We can quickly program the turntable to yield any number of frames per degree. It will easily support three people, and is smooth enough that it was used to rotate 3-D works in the collection of the Oakland Museum for a video they were producing.

Computer: Portable, adapted to motion control of cam- era and turntable. Step and direction pulses from the computer tell the camera and turntable motors precisely how fast and how far to run, in sync with each other. The rates of acceleration and deceleration are also programmable, making the turntable a safer platform. The automatic "return to starting position" feature saves time and reduces the stress of sitting relatively still under hot lights, rather than riding the turntable the long way around after each take. Additionally, the precise camera acceleration and speed control increase exposure accuracy while reducing the amount of film needed.

Another advantage of the computer is being able to re- duce filming speed while maintaining the 3 frames/degree, in order to get the exposure needed. Increasing the

amount of light can cause trouble with the prosthetic make- up - when the actor sweats, the seams appear.

Stepper drive/DC power supply: Portable, high-power microscope drivers and supply custom made for stereo- gram filming. The computer sends instructions, but the power required by the motors is far more than it could handle directly. Also, stepper motors large enough to provide real-time motion control of large inertial loads, such as people, are capable of sending large voltage spikes back to the power supply, so they should be designed expressly for the job. Microstep drivers are preferred in special-effects photography because cheaper drivers are prone to make cameras "chatter". The drivers we use were tested and rated so highly for reliability that they are used with- out modification to power the arm in the cargo bay of the space shuttle. Reliability is an important factor when film- ing expensive talent.

Lighting & Grip: Requirements vary according to the subject. Our own lighting package includes 2000 watt fresnels (juniors), 3200w modules, 1000 watt open-face units (all with barndoors), a half-dozen C-stands, (adjust- able steel stands for positioning diffusion, flags, nets etc.), highrollers for holding up backdrop material, 4' x4' frames for diffusion, gel frames, and the unglamorous sandbags that anchor stands and help prevent on-site accidents. Dinmlable fluorescents such as "Kino-flo" are also useful, while HMIs (mercury, AC arc discharge lamps) usually are too expensive on a modest shoot, especially when a match with daylight color temperature is not needed. On a large- scale production, the efficiency of HMIs lowers operating costs enough to offset higher rental rates.

Location: While you might be able to make arrange- ments for cheap filming space through friends in your home town, when you travel to a remote location, you will be faced with the task of finding a suitable space - usually an insert stage - with lighting and grip equipment available for rent. These facilities nearly always require proof of insurance, with their company named as additional insured. In addition, they frequently accommodate crews filming commercials with budgets in the quarter-million range, so they routinely try to charge a higher "film and video rate" over what they charge for "print jobs". If you are not a seasoned producer, you will soon acquire a new respect for what the job entails.

Travel, securing a location and getting the shoot set-up are the expensive part. The best way to save money is to group shots together for a single setup. If your subject is too important to allow any sort of scheduling, his or her fee is probably so great that the cost of the shoot doesn't make much difference, anyway.

Lighting Stereogram Portraits

Photography (literally, "writing with light") is the re-cording of light reflected from the subject. The character, angle and intensity of the light falling on each part of the subject has everything to do with the light the subject re-flects, and the kind of image formed , be it on emulsion or video, etc. It is amazing how often people outside the film industry overlook the importance of controlling light, even though most have seen the generators, heavy feeder cables, large lamps and overhead silks used even for a daytime shoot. A stereogram may be only a "head shot", but in cinematography it is the close-up that receives the most attention. In fact, one of the reasons I went to study in Sweden with Sven Nykvist is the importance of the face in Swedish cinema.

Every scene has unique properties, so hard-and-fast rules for lighting do not really exist. For example, I once over- heard a client complaining that their trademark character, a stop-motion puppet, was somehow "sinister looking" in the scene being shot on a neighboring stage. It had been lit from below, casting grotesque shadows upward *a la* Boris Karloff. Yet the same lighting from below is flattering to the actors in Star Trek, the Next Generation, where the tables on the recreation deck are also a light source. The difference is the style- the higher-key lighting of the actors, with the increased fill, influences the mood. Lighting requires a degree of visual literacy, as well as the ability to actually see and analyze what you are looking at.

When color film was first introduced, technicians and engineers insisted that all lighting had to be "flat." It didn't take long for artists experimenting with the new medium to start discovering the possibilities for moody, low-key, "incorrect" lighting. We owe a debt to engineers for our tools, but the reality is that engineers usually don't produce compelling images. In most cases, holography still looks as if it was done by engineers.

When I lit Rickey Henderson for my first stereogram, I was told that lighting had to be absolutely flat. It seemed as though every time a new element was introduced into the formation of an image, engineers reasserted their need for flatness. Stereograms do have a limited range of contrast and it is possible to end up with alarming results, like a fragment of a face floating in shadow, or an area so bright it becomes a streak in the lenticular screen. Thus flat, shadowless lighting is "safest"- but the art of lighting is too subtle to lend itself to doing by rote.

A course in cinematography is beyond the scope of this article, but a description of some of the factors involved in specific shoots might be helpful.

Generally I try to pre-light for a given subject. That way, when the inevitable challenges arise, I am able to deal with them quickly and to my satisfaction.

I knew Rickey Henderson would be dressed in a light gray uniform and posed against a black background. I set up back-light to help rim the contour of his face against the dead black background, preventing any "floating half of a face" effect, and used large double nets in front of the diffused key light to reduce the light on the highly reflective uniform.

When Rickey took his place on the turntable the next day, the real surprise was his green Oakland A's hat, which was made of the "light-hungriest" green felt material I have ever seen. We were going to get a stereogram with his head lopped off at the top and a glowing "A's" floating where his forehead should be, I had to use an ellipsoidal (focus

spot) to pump in enough light to bring out the hat, using the unit's framing shutters to control spill. The "A's" logo had to be physically darkened.

Once the subject is in position, adjustments have to be made confidently and as quickly as possible. The physical strain of working in front of hot lights is the main reason for the use of "stand-ins".

Selling up for the Prince shoot.

The Prince stereogram was filmed at S.I.R. studios in Hollywood, using a rented, non-motion control turntable that arrived in sections on a flatbed truck and took the owner nearly an entire day to set up. Warner Brothers' representative wanted an image with strong shadows, which my employer felt was impossible.

To satisfy both Wamer Brothers and my employer, I prepared two lighting set-ups, one with flat, "safe" lighting and one with the key light coming from the left and a lighting ratio of 5:1, matching the still photo provided by Warner Brothers. Because the turntable had to rotate in the same direction for each take, it was not practical to mount lights on it, so units had to stand on the floor. In the test shot, the way the shadows traveled across the subjects' faces made us reject the alternate lighting arrangement idea in favor of the flat lighting, but I still felt stereograms could handle the contrast range if the moving shadow problem was resolved.

When Prince and the two dancers took their places, there were the usual surprises. Out of his incredible wardrobe of brilliant colors, the client insisted on a black suit, against a black background. We were going to get a collection of floating heads l To make matters worse, the client insisted upon a horizontal composition, a poor choice for a stereo- gram, aside from being pretty static, visually.

I rigged a "clothes light" to boost the details of the suit, flagging it off of his face and using a half-scrim (a circle of stainless steel mesh with one side cut out) to reduce the effect of the clothes light on the dancers without the visible shadow that using the fixture 's barndoors would have caused.

The turntable was the other surprise. Aside from being difficult to keep at the right speed, it ran on casters that had been chipped , causing enough rumble that the pearls hang -

ing in front of the trio danced wildly. We ended up having to tape and staple the strands to the turntable, stretching them taut and restricting Prince's planned move of pushing them aside with his hand. That was the day I resolved to build a turntable. On an important shoot, equipment problems can be a disaster. Even if you deal with them successfully, they take up valuable time. There are so many elements that have to come together for a shoot that Murphy's Law is always lurking, no matter how carefully a job is planned.

Eliminating the variable of finding turntables that could be adapted for filming stereograms has really helped speed production. Even with the motion-control turntable at the effects house where we filmed Ricky Henderson, the programming of camera and turntable was complicated by gear ratios intended for other uses. During the confusion of live-action filming, (or taping), any unnecessary complication is a mistake waiting to happen.

Frankenstein

When it came to lighting Universal's classic Frankenstein Monster, I knew it was time to put the flat lighting rule to the test. The key light coming from below has become so much a part of the image, it is almost a part of the make-up. Craig Rearden's prosthetic sculpture looked so much like Karloff in the original film that I felt it would be a shame to light it like a TV sitcom.

After shooting it flat for "safety" (you never want to leave a setup without being sure you have what you need), we mounted the key light on the turntable, casting strong shadows upwards across the face. By using a powerful lamp, we were able to get a dramatic ratio between light and shadow while keeping the exposure high for good depth-of-field and density on the slow, fine-grain negative used.

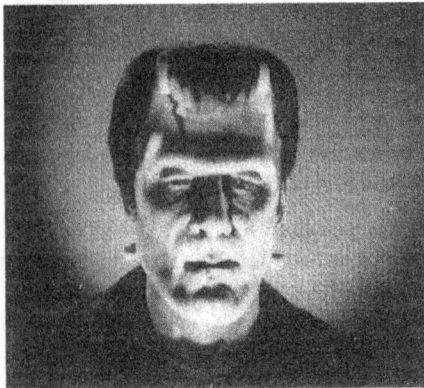

I suspect people confuse shadows with underexposure. It is the contrast ratio that gives a strong shadow side, even at an exposure level of over 600 foot-candles . The shadowy version of Frankenstein's Monster came out fine; he would have been boring with flat lighting.

Because stereogram clients usually want sharp, fine-grained images with good, solid blacks, the level of illumination required is relatively high. That's why it isn't practical to shoot in the average living room, where even if you could draw the 180 or so amps needed, the temperature would quickly soar past 120 degrees. By contrast, when I am lighting for a drama on location I typically ask the gaffer to give me a base exposure of 20 foot-candles for the background, owing partly to new, fast stock and the desirability of underexposure in many cases when shooting a story (except on video).

The difference between 20 and 600 foot-candles can be monumental. The inverse square law states that the intensity of light falling on an object drops off with the square of the distance from the lamp. Depending on the distance, getting twice the light might mean using 4, 16, or 256 times the power! Getting five times the light can be a difficult proposition. Reflectors and lenses modify the effect of the law, but generally speaking it is enough to humble most of us, given enough area to light.

Think about it the next time you see a film containing a night-time chase scene on a mile of freeway. The entire sky glows over Los Angeles when a major shoot is in progress, and you can be sure the gas & electric company notices the load. Fortunately, stereogram portraits involve lighting a small area-but the inverse square law has caught a lot of low-budget producers by surprise.

Film stocks keep improving to the point where we consider it feasible to use a faster emulsion, but continuing improvements in stereogram origination and reproduction argue for the desirability of capturing the sharpest, finest- grained image possible.

Logistics

The way to ensure the best production value - the most "bang for your buck" - is with good pre-production planning. I don't use the cheapest lab, or hire the cheapest make-up artist, but that doesn't mean I'm not stingy. I'm after the most cost-effective way to get the highest quality possible, and I've learned that a lab that gets dirt all over your negative or a make-up artist who changes his fee by a sizable amount when he thinks it's too late for you to get rid of him, (it wasn't), are not cost-effective.

I have never regretted paying for professional people. It is possible to find an occasional gem just starting out at a bargain rate, but on an important job I can't take the risk. The key word is efficiency. For example, Steven Anderson is a San Francisco Area make-up artist whose experience and cleverness have contributed to the smoothness of a number of our shoots. An individual could charge less, yet end up making the shoot cost more in terms of time or, God forbid, inability to deliver as promised. It may reduce my bottom line in some cases, but feeling free to concentrate on my job, confident that his is taken care of, is well worth it. The same goes for every service provider, and the knowledge of where to find good people is often hard-won.

Depending on the client's experience and resources, they may want to take on all aspects of production planning, have us take care of everything, or hire a consultant or middleman. The position of middleman is a legitimate one, as the field of holography is young in commercial terms and there are apparently many claiming to do things they really can't. The last thing you need is to be financing somebody else's R&D while your deadline approaches.

A consultant or middleman should know who actually has done what, and should put the client's interest first. Communication is important, just as it is in constructing a model, and in addition to sketches and other visualizing aids it is desirable for the cinematographer to be permitted to communicate with the holographer doing the origination, in part because we have learned that blind adherence to the image-height/filming-distance/diameter formula doesn't always yield the best results. We might want to shoot a black & white distortion test if it is a new system.

In my opinion, one of the things that has hindered the commercial use of holography has been the attitude of people in the field that they are "gurus". Clients are willing to pay for talent, skill and professionalism. They are not dumb, however, and people who carryon like the high priest or priestess of holography are not helping the medium. There may be new and better systems, but remarks like "I'm the only one who can do it" are asinine and make clients suspect they are being held-up. We show our best work, knowing the client has choices. If they are looking for cheap work they will go find it, and we are all happy. Hopefully they will be successful and, one day, work their way up to a point they will use us.

A typical stereogram shoot, done locally with the client providing the actor and DCC providing everything else (up to and including color separations) runs around $9,500. By contrast, the Universal Monsters shoot in Burbank came out to about $5,000 per stereogram, including travel and lodging, (but not make-up or talent). The difference is the use of the same set-up for six shots. Some cities, such as New York, are expensive places to rent even a small film stage and are often booked-up in advance.

We usually shoot 12-18 takes to ensure we have the action and expression desired, using 400 to 600 feet of film per stereogram. The actual footage used for the stereogram only measures some 12 feet, but all other things considered, it doesn't make sense to scrimp on film.

If the client has studio or industrial space available in the San Francisco or Los Angeles area that would be suit- able for filming, that would be another cost-saving option, as we have all the necessary lighting and grip equipment in those locales. We can even rent the motion control turntable for $600/day, leaving everything to the client (we do not rent the camera, however).

Another option would be to employ me via personal service contract, the way I work on films as Director of Photography, arriving with my light meters.

I hope what I have written here is helpful. The thing to bear in mind is that a stereogram is a custom job requiring highly-skilled people at each step, and it is unrealistic to compare it with a hologram of a flat, cut-out photo mass-produced on a never-changing setup by relatively unskilled labor. Stereogram portraits are especially well-suited to security applications, as they are both difficult to counterfeit and easily recognizable due to the average individual's highly-developed face-recognition skills. But stereograms' most arresting property is that of motion. It is the way the motion in the printed image stops people in their tracks that makes me believe stereograms have a bright future. Even after the novelty has worn off, I think people will be drawn to the presence of living stereograms.

The Back Cover Hologram

The most commonly seen holographic stereograms use vehicle slits and are embossed. The Chinese Lion Dancer on the back cover is an excellent example of this method. We asked Steve Smith, the creator of the image, to give his views on the stereogram process, and explain how he did his shot.

Steve Smith, president of Lasersmith Inc. has 15 years experience in professional holography, and holds three parents related to holographic stereograms and one for the replication of holograms in injection mold technology. He has been working in the process of computer modeling and animation issues related to the stereogram image for the past six year's. Lasersmith produces holographic stereográms and designs and markers complete systems for their creation.

Holographic Stereograms
by Steve Smith

The phrase "Computer Generated Hologram" has been bandied around in such a casual manner that it is now almost non-specific. If we were to give a respectable definition to the phrase, which implies a hologram totally created by the computer with no outside source, we would probably say something like: "The act of mathematically deriving the fringe structure of a diffractive surface by calculating Foire transforms of the object and reference waves. Writing these structures as surface details either by direct means via an e-beam or printing out a 2-D pattern and optically reducing this structure to the scale necessary for diffraction of light."

Holograms that fall within this domain are generally simple in design and limited in capability. These are definitely the exception to the type of hologram you see in public and not the rule.

This is not to say, however, that computers are not used. In fact, the computer workstations are commonplace among holographers, and the impact of the computer workstation on commercial holography is pervasive throughout the entire line of holographic image styles. A better phrase that describes the use of computers in generating holograms, however, might be that computers help create "Computer Aided Holographic Design."

Computer workstations have been used to create artwork for all the current styles of commercial embossed holograms for some time. 2-D 13-0 images are typically created with drawing programs such as Freehand or Illustrator. Final image layouts can be composed via Quark Express with final film RIPed to a Lino 300 image setter. The same is also true for full color 2-D images (Photo-realistic) through the additional utilization of programs such as Photoshop.

Today the new use of the computer workstation is in the creation of the Holographic Stereogram. Earlier holographic stereograms were typically created in a more direct manner, with the subject usually being a person or large object that would normally not be accessible for CW (Continuous Wave) holography because its size would make a "table-top shot" impossible. We might say that a simplified overview of the old method of doing a stereogram involved four main steps:

1) The subject is filmed.

2) The film is converted to a B&W positive film strip.

3) The film strip is sequentially shown on a "cinema screen" behind which is a slit. The recording material is behind the slit, which can mechanically move across the face of the recording material as each frame is sequentially shown. This produces your H1 master transmission hologram.

4) The Transmission Master is then optically transferred onto the white-light viewable master.

Generally, to incorporate other art or image effects onto a holographic stereogram, subsequent often cumbersome optical steps were needed, producing less-than-anticipated results.

Computer workstations, needless to say, have greatly changed the way this process is done. The film can be scanned into the computer and the computer can manipulate any individual frame; it can also render the image between any two frames to your desire. The computer can then record the final set to an optical film recorder for projection onto the "cinema screen."

Early on, pricey workstations such as the first series of Silicon Graphics were required to run expensive image modeling and rendering programs; these cost upwards of $20,000, generally out of the range of professional holographers. With the release of powerful yet lower priced graphics workstations, such as the Pentium and the PowerMac, and with more capable yet affordable image modeling and rendering software, a new realm of imaging techniques is setting the stage to present the computer aided holographic stereogram as the way to create holographic stereograms.

It is a common misconception that holographic stereo- grams are only images of people or animation. Many seemingly diverse image types can be produced in the holographic Stereogram format. 2-D, 3-D, full color 2-D, multi-layer 2-D as well as multi-channel (Lenticular) effects can easily be set up as animation. The real advantage for the computer-aided approach to holographic design is that all of the above techniques can be seamlessly intermixed, giving the holographer/designer the greatest set of tools available. It would be accurate to say that the holographic process now takes place as much in the computer station as on the holographic system.

All is not as simple as this may first seem. The image set-up and the holographic optical printer design are tightly intermeshed. In order to produce a holographic stereogram you will have three or four highly specific technologies to address.

1) Image Capture

2) Graphics/animation workstation

3) Holographic Optical Printer

4) Holographic transfer system

Let's Define Each Item More Carefully

Image Capture: Photographic or digital, resolution specified by final image size and resolution required. Linear tracking or rotation stage depending on optical printer specifications.

Graphic/animation workstation: 100mz or faster if possible, digitizing software, modeling, rendering and animation software, coupled with photo manipulation software, film recorder or LCD output.

Holographic Optical Printer: One of two main types:

Type 1:

• For high resolution, or large image size, a film-based projector type is generally best suited.

• For lower resolution or small image sizes, a LCD-based projector is acceptable.

Type 2:

• One step Optical Printer, such as MIT's system for reflection holograms or Simeon Holographic for embossed holograms.

A Modern Holographic Stereogram Shoot

To give you a detailed example, here is an overview of how the Chinese Lion Dancer on the back cover was created. First, the images of the lion dancers were recorded with the Lasersmith 36-camera imaging system. Next, the chosen animation set of 36 frames were digitized with Kodak's Photo CD. These images were then brought into a 8100 PowerMac, balanced, and an alpha channel of each lion dancer outline was created in Photoshop.

Next, a computer-modeled scene was created in the Byte by Byte software's Sculpt 4D. 36 frames were rendered to match the perspective and parallax of the filming setup; all of these files were composited in Photoshop utilizing the previously-saved alpha channels. The resulting film strip was imaged onto an RGB Transmission H1 master hologram and transferred to the final photo-resist master: a significantly more elaborate series of steps compared to the earlier four steps described above.

Hopefully, I have given a useful insight to the process of holographic stereograms. As a final thought, it would be an easy oversight to feel you could simplify the problem of generating mechanically-collect artwork for a quality holographic stereogram by simply using the typical world models that are contained in 3-D modeling, and animation programs. However, the accuracy that is required to optically match a specific holographic optical printer necessitates an extensive knowledge of holographic imaging problems as well as the optical specifications of a given optical printer. I strongly recommend that a comprehensive dialog take place between the designer and holographer before any computer or imaging work takes place, to ensure that the image parameters are set up correctly.

A Short Glossary of Stereogram Terms

Holographic Stereogram: Holographic image whose information is derived from a sequential array of 2-D images. The arrangement of the 2-D images is such that correctly spaced sets deliver stereographically correct information to the left and right eyes. Generally the array is a horizontal one. The number of 2-D images required can range from 12 to over 100 (for a single color image); for a full color RGB hologram three complementary sets of images are required.

Holographic Optical Printer (HOP): Optical printer designed to record 2-D information onto a holographic media in such a manner that the resulting image is perceived as being 3-D. Generally comprised of the Tee subunits, a projector, screen or lens device, and a slit mechanism (either optical or aperture) that permits the controlled placement of each exposure. Generally for high resolution images the input material is a high resolution film.

Film Recorder: High resolution output device used to write raster based computer imagery or data to film via a high resolution CRT, capable of writing single image resolution upward from 2K to 8K.

LCD HOP: A HOP whose projector input device has been replaced by a liquid crystal display (LCD). In this case the input is fed directly from the computer to the LCD to create H-1's for stereograms. Current resolutions of LCDs are around 680 x 440 thus limiting their application to small sizes. In their application in one step HOP printers, this small size initially confined output to small images. Clever optic solutions are slowly increasing image size without printer or segment effects.

One Step HOP: Beginning with Lloyd Cross' Multiplex printer, one step printers have offered a direct means to create a finished hologram from sequential 2-D images, bypassing the generally traveled route of the H I to H2 mastering approach. Research by Klug etal at the MJT Spatial Imaging Group led to one of the first flat output one- step printers for reflection holograms and recently a com- mercial version was developed by D. Haines and her group at Simeon Holographics specifica lly directed at embossable rainbow masters.

3-D Modeling, Rendering and Animation: Combined set of computer workstation tools used to create objects, scenes and parallax animation information applicable to stereogram images. 3-D modeling is the setting into 3-D coordinates object points and lattices to create computer versions of objects. Rendering is the process of drawing an image of these computer coordinates based on defined surface characteristics and viewer location/direction. Animation is the sequential movement of objects and/or viewer over time and subsequent renderings. Generally for holographic stereograms animation takes on two issues; object motion, and more importantly view or camera placement over the sequence. Viewer/camera placement and view direction over time is crucial in that the resulting parallax from viewing specific sets of rendered view must be stereographically correct for the final viewer aperture placement.

Image Capture: Process of recording image information over time and parallax. Image capture range must match holographic printer parallax parameters.

Parallax: Left to right stereo width that the recorded or modeled images must cover to present a "stereo pair" correctly to the viewer's eyes. These frames are not located directly side by side but rather several frames apart. Changing the parallax affects the perceived "z" axis or the apparent depth of the final scene.

Viewing Zone: This directly relates to the left to right complete extent of recorded or rendered images. The complete angle of view of the final hologram.

Digital Camera: Optical camera whose imaging media is a ccd device. Currently lower-resolution cameras are capable of the longer image sequences required of holographic stereograms. The higher resolution cameras have problems downloading the significantly larger data files in a fast enough manner, so that a long enough series of images can be continuously recorded.

3D Model Making

The final area that we need to cover is that of model making. Although some three-dimensional holographic images are being produced using compilations of a series of digitized graphics, most holograms are still created by recording the light reflected off of a physical object or a specially-sculpted scene. Therefore, the process of creating a custom designed holographic image often requires the use of an expert sculptor or modelmaker who can best translate a client's ideas (often in the form of sketches or photos) into three-dimensional artwork appropriate for holographic reproduction.

Due to the fact that most jobs call for highly-detailed miniature models to be created, and that these models are to be illuminated with laser light, and that the final image can display parallax, depth and projection - it is recommended that you utilize the services of an experienced modelmaker who is very familiar with the unique elements that go into creating a successful hologram.

George Sivy, has been a holography model maker for ten years. He worked for Polaroid during their first years in holography, making models utilized for custom and stock images (such as the popular "Brain/Skull" which appeared on the cover of the Holography MarketPlace's Third Edition). He has collaborated all numerous commercial projects, ranging from embossed security holograms to photopolymer holograms designed for the giftware market. He did the model for the 1991 Super Bowl tickets.

Reviewing our lengthy correspondence with him, we find Sivy is convinced that well-designed and well executed artwork is the basic element from which a successful hologram is created. Of the three components Sivy lists as necessary to the creation of a holographic image: the artist, the holographer, and the manufacturer, "Quality work from all three is important. Artwork, however, is key, in that good, well-conceived and carefully executed artwork will carry mediocre holography and/or manufacturing. Even the very best holography and the highest quality manufacturing can't make up for poorly-done artwork."

He has three basic suggestions for those who are working with model-makers to create a hologram that "works": plan in advance, remain in direct contact with the artist from start to finish, and be willing to pay for the artist's time and expertise.

Since the client is often unfamiliar with the technological processes that are used to create a hologram and sales-people frequently gloss over the medium's limitations, the modelmaker often assumes the role of educator as well as craftperson. This requires that modelmakers be effective communicators who can work with both conceptual and technical issues. "A hologram, under the best of circumstances, is still an interpretation of the client's 2D artwork," states Sivy.

"Nine out of ten of these people do not know what to expect in terms of time and other factors, which go into creating a piece," he states. Because of this, he says, there can be "communication gaps, misunderstandings, and deadlines which are too tight...expectations should be more appropriately formed."

Sivy advises that the client be brought into the loop even when art directors or ad agencies are involved. "It's crucial putting corporate politics aside -the guy signing the check has to be in direct contact with the artist right from the beginning."

Ideally, the client, the client's art director, the salesperson, the modelmaker, the holographer, the replicator, and the finisher will all agree on the various elements that will ensure a top quality result. All parties should agree on a realistic production schedule and should communicate regularly. Since the modelmaker is usually responsible for the first tangible output seen by the client, the credibility of the entire process is affected by the professionalism displayed by this participant.

Sivy divides the design process into two major parts - conception and execution. To best utilize the holographic medium, designers must remember that "a hologram is not merely a representation of an object in 3D, but rather it is a 'window' into a three dimensional space. The more of this space which is utilized, the greater the visual impact of the finished image."

Sivy elaborates on the "window" analogy: "Think of this space as being a room with height, depth and width. Think of the window pane as the actual holographic plate - this 'surface' is referred to as the image plane. Visualize yourself standing directly in front of the window looking into the room. Assume that your line of sight is 90 degrees to the image plane. The contents of the room are being illuminated by the sun shining from over your shoulder at a 45 degree angle to the window pane. This is referred to as the reference angle. We are now viewing the contents of the room the way in which most holograms are intended to be viewed. Standing directly in front of the window, however, permits us only a limited view of the contents. Imagine yourself moving from side to side - which allows you to see around a given object and into areas of the room which would otherwise be hidden."

This illustrates the concept of parallax view - a very unique and basic characteristic of holography which "represents considerable potential for innovative design and composition possibilities." Designers should determine viewing angles with the holographer and communicate these measurements to the model maker in order to maximize the usable image area.

Designers also need to take into account that a holographic image has "volume" - it can have depth (an object can be behind the window) and projection (an object can appear in front of the window). Images should be designed to best utilize this front-to-back dimension. It is important to note that different replication materials have different capabilities regarding the amount of depth and projection that an image can display. This "depth of field" is closely tied to how the hologram will be illuminated. For instance, embossed and dichromate holograms can only reconstruct fairly shallow images (approximately one inch), while photopolymer films can replay images of several inches or more. Silver halide glass plates can reconstruct images several feet deep under proper illumination.

A model-maker who is familiar with the physical limitations imposed by the recording materials and anticipates the conditions under which the finished hologram will be viewed can create an image that replays clearly in most situations. Therefore, many modelmakers place the most important visual components of their images on the image plane, which stays in focus under less-than-ideal lighting conditions.

After the design process comes execution, a step which utilizes a modelmaker's craftsmanship and technical proficiency. In terms of materials, Sivy uses "what will solve a given problem." Although some materials and combinations that he uses are proprietary, he did mention the readily available Sculpty, a synthetic clay that can be baked. This ensures a stable model, as most holographic mastering uses continuous wave lasers which require that absolutely no motion is present during exposure periods. Long exposure times might require even more stable materials, and Sivy may use Sculpty first, then make a mold which is used to generate an even more stable scene.

After the model is created, it is often painted to created contrasting areas of light and dark. Since most, but not all, holograms are intended to be reproduced as monochromatic images, the "coloring" process is quite different than in ordinary modelmaking. Less experienced model makers will often work under a safelight which duplicates the laser light which will illuminate the model during exposure.

Good model makers will also utilize textures, shading and special effects to maximize the hologram's visual impact. Again, this requires communication between the design team, the model maker and the holographer.

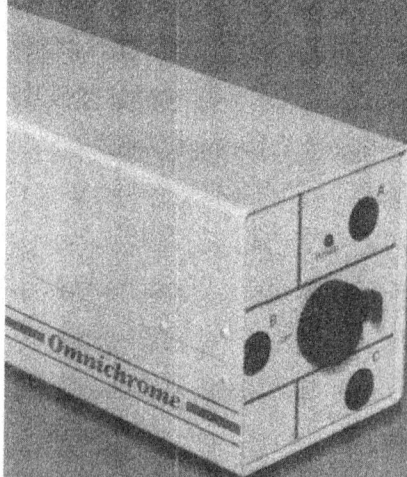

3

Embossed Holograms

This chapter covers several important issues commonly faced by clients who want to incorporate embossed holograms into their marketing promotions and ad campaigns. Peter Scheir of AD 2000, a company that supplies holograms to corporate customers, discusses the merits of utilizing existing "stock" imagery.

Introduction

Most corporate clients who want to incorporate holography into their marketing promotions, ad campaigns, and packaging will probably consider using embossed holograms, rather than other formats. The term "embossed" refers to the method by which the original hologram is replicated - a holographic image is physically stamped into a moldable substrate.

Embossed holograms are by far the most commonly used holograms due to a variety of factors, including their expense (they are relatively affordable), their availability (there are numerous suppliers), their practicality (they can be easily incorporated into existing situations) and their performance (they have proven successful in a wide range of applications).

The major drawback to using embossed holograms is that imagery must be inherently shallow due to technical considerations. Although some display more depth, most embossed holographic images tend to "smear" when image depth exceeds an inch or so. Astute designers and holographers design around this limitation. They incorporate flat graphics, photographs and even cinematic footage into the finished image. See the variety of embossed holograms displayed throughout the book to see how this hurdle has been overcome by several major producers. Of special note is the embossed hologram on the back cover, which was produced by Lasersmith.

Manufacturing

There is no question that embossing is by far the cheapest way to mass produce holograms. As further described in the appendix, these holograms are mass produced by taking a "shim" (a metal "negative" of the holographic interference pattern) and repeatedly stamping impressions onto a desired substrate as it moves rapidly through the press. Very thin foil or plastics are most commonly used as a substrate, but materials as diverse as chocolate and hard candy have been successfully embossed!

Typically, a stamping press embosses the holograms on rolls of material which measure 6 inches wide. This creates an working image area of 6" x 6". Once an impression has been made, the roll advances for the next strike. Images are usually separated by a small measure, although "continuous" patterns can be produced. To be most cost effective, producers will often "gang" several smaller images on a single shim, so each strike produces a number of images. It is quite common to sub divide a shim into nine 2" x 2" squares. Recently, embossers have constructed equipment that can handle wider rolls, thereby expanding the holographer's "canvas."

Finishing Options

Embossed holograms usually need to be backed with some type of adhesive or adhesive material so that they can be attached to a surface. Different adhesives are available depending on the final application, but the two most

If you choose hot stamping, be absolutely sure of the strength and smoothness of the material you are hot stamping onto. Since the hologram itself is very, very thin (similar to the foil used to wrap around individual sticks of chewing gum), it is imperative that the surface is clean and smooth. Textured or bumpy surfaces will distort and ruin the holographic image. It is a good idea to discuss your requirements with an experienced hot stamper who is familiar with using holographic materials. Temperature and pressure settings can differ from those used in other hot stamping processes. Beware of stamping onto coated paper and printed surfaces. These may create adhesion problems. It is best to test the materials you want to use beforehand.

Embossed holograms are the most economical way to mass produce a holographic image but. until recently, there has been resistance among many buyers because they feel that there is a large "origination fee" that they will have to face which will drive up the unit price of short runs enough to make the projects uneconomical. Fortunately, in recent years there has been significant improvement with this problem, making short run embossed holograms much more affordable. In the following article, Peter Scheir of AD 2000 reports on this topic.

Embossed Holography Made Affordable in Low Production Runs
by Peter Scheir

Over 15 years' experience in the holographic industry has helped me to identify a number of recurring themes in potential customers' objections. I have made it my goal to create a path of lesser resistance-to make holography "a solution, rather than a new set of problems." The aim of this article is to outline what many customers perceive as obstacles to using holography, and to describe at least one solution. That solution lies in the customizing of stock image embossed holograms.

common choices are stickers vs. hot stamping. Peel and stick backings are appropriate for applications that are going to be hand applied in short runs. Mechanized hot stamping is better for longer runs where accurate registration is important. Remember that in either case, the hologram itself must lay completely flat for the image to reconstruct properly. Consult your production house or an independent finisher for more specific recommendations regarding the most appropriate backing material to use.

Comparisons of Stock vs. Custom Holograms

Size / Type	Quantity	Origin-ation*	Unit Cost	Run Cost	Total Cost	Savings
3" Custom	5,000 pieces	$ 5,000	$ 0.20	$ 1,000	$ 6,000	
3" Stock	5,000 pieces	N/A	$ 0.50	$ 2,500	**$ 2,500**	**$ 3,500 (58%)**
2" Custom	10,000 pieces	$ 5,000	$ 0.08	$ 800	$ 5,800	
2" Stock	10,000 pieces	N/A	$ 0.34	$ 3,400	**$ 3,400**	**$ 2,800 (41%)**
6" Custom	2,000 pieces	$ 5,000	$ 0.70	$ 1,400	$ 6,400	
6" Stock	2,000 pieces	N/A	$ 1.25	$ 2,500	**$ 2,500**	**$ 3,900 (61%)**

* Origination is (conservatively) based upon a fully 3D image including basic model. photoresist master, electroforming. recombination. and die work.

A Powerful Promotional Tool

Impact is the word, in my mind, which best befits graphic holography. It is the visual impact of the medium which gives it such great power and generates so much interest. Companies of all sizes are constantly searching for means of promotion which will set them apart from competitors and get them noticed. Holography adds three distinct advantages to any promotion:

1) Impact: Holograms are eye-catching. People will look at holograms for significantly longer periods of time than other graphic mediums. If a customer's name or slogan is on the hologram, that information is enforced!

2) Pass-Around Value: people are often impressed with good holograms, and they are likely to bring them to the attention of colleagues and associates.

3) Retention: people tend to keep holograms. We receive new leads from potential customers who have holograms on their desk which were created more than 10 years ago!

Holographic Misperceptions

Given these three factors, it may be hard to understand why holography is not found in much wider use in the promotional marketplace. I feel that it is largely due to false perceptions about holography and its costs.

Unfortunately, in the minds of many, affordability has never been a word associated with holography. It would be impossible for me to count the times that I have heard potential customers say "holography is great but too expensive," "we can't afford to use holography for our project" (can they really afford to ignore the impact of holography?) or, "holograms are only for large runs." These decisions are made before a price quote has even been made!

Objections such as these are sometimes based upon past experience; but more often than not, they come from an incomplete or incorrect conception of the holography industry and what we have to offer. Mainly, objections of this sort stem from experiences in which an inappropriate solution was applied to a project.

For instance, a customer may have been looking for 5,000 hologram labels to give away at a trade show. Calling a company found in the phone directory, they were told that there would be a set-up cost of $8,000; or, that the mini- mum run had to be 50,000 pieces. This type of incomplete information creates obstacles in customers' minds. After such an experience, further exploration into the use of holography would be limited due to the customer's presumption that holography could not be cost-effective.

It is true, however, that fully-custom embossed holography carries set-up (origination/tooling) costs which are many thousands of dollars. Making embossed holograms

affordable in low production runs (runs too short to amortize the tooling costs associated with fully custom-embossed holography) is the key to tapping this marketplace. I have found that the best solution to this problem is the use of stock images.

There are a great number of projects which are defined as "not cost effective", given fully custom holography's high set-up costs. Stock image holograms have just as much impact as fully custom holograms, without the high set-up costs and stock image embossed holograms can, with great ease, add huge impact to any promotion.

Direct mail can be enhanced with a stock hologram label. Memorable premium and specialty items such as magnets, calculators, and t-shirts can be created from stock holograms. Powerful brochure covers, pocket folders, annual reports, magazine covers, advertisements, etc. can be produced and packaging can benefit from the draw and appeal of a stock hologram. New life can, in fact, be breathed into almost any promotion with a stock image hologram.

Many hologram production companies have existing stock images. Sometimes images are created 'on spec' for projects which do not come to fruition, and the originator retains the rights of 'replication' (embossing the image). In other circumstances, originators create holograms for self-promotion or for retail markets that can then be utilized as stock images for advertising and promotional projects. These images can often be customized to suit individual customers' needs.

Customizing a stock image can take a number of forms. A simple post-printing stage adds a customer's logo, ad copy, or slogan to a stock image, making it (for all practical purposes) a custom image! Stock images can also be die-cut (cropped) to specific shapes and/or sizes which give them a custom appearance. Placing ad copy around the image on the literature or product can further enhance the design and make a stock image unique.

Generally, stock images are more expensive per piece than custom images due to 'short run costs' and/or royal- ties (paid to the owner of the image) which are included in the price per piece. This higher cost per piece is deceptive. Since no custom tooling costs need to be amortized, the total cost for the production run is low enough to make the project viable where a fully custom hologram would have been out of the question.

Though the average "total dollar value" of stock image jobs is lower than that for fully custom projects, the lack of high set-up costs and the availability of low minimum runs create a much greater conversion rate, so there are more jobs. Just as important, customers see the image before the project begins, eliminating the long, potentially problematic proof stages and chances for image rejection.

Origination is (conservatively) based upon a fully 3D image, including basic model, photoresist master, electroforming, recombination, and die work.

Qualification of customers is often complicated in fully custom holography. It can take a long conversation and an expensive packet of samples just to determine that a cus-

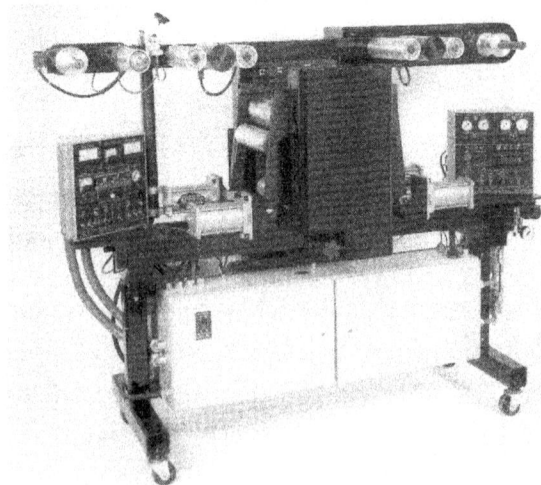

tomer does not have the budget for a custom image. Stock images offer an excellent back-up option. If the budget does not allow for custom work, the project may often be salvaged with a stock image hologram. Furthermore, some of our sales reps have found that offering an affordable alternative in stock imagery can remove the obstacles in a customer's mind. As the customer looks further into the project, he or she finds that a custom piece is really more attractive , and a budget is ' created' which includes custom tooling!

Stock Images Work in other Industries

In the photographic industry, stock images have long been a well-used alternative to custom photography. Various collections of stock photos offer a dazzling array of images, so that one can be found to suit even the most obscure project parameters.

Holography has not yet developed into such a serviceable industry as photography. If you believe that a stock image may work for your particular project, request a list of available images from your favorite hologram supplier; or, look to one of the recently-developing compilations of stock images [NOTE: If you are shopping for a stock embossed hologram, be sure to see if it is available in the form you require (i.e.- hot stamp foil vs. pressure sensitive stickers, logo add-ons, custom cut sizing or cropping, roll vs. sheet, etc.).

Some Stock Image Suppliers

AD 2000 INC. of New Haven, CT has created a stock hologram collection named HOLOBANK. We have assembled a large selection of stock images by licensing available images from many of the world's finest origination facilities. This growing compendium has over 400 stock images ranging in usable size from 3/4" to 6" square, and customizing is supported in the form of add-on printing as well as custom die shapes, all for runs as low as 500 pieces.

You can also find limited parts of the HOLOBANK collection available in various forms from companies including Applied Holographics, Astor Universal, Bridgestone Technologies, and Light Impressions Europe.

Smith and McKay Printing in San Jose, CA is one of the country's leading hot stampers of holograms. They have taken a unique approach to the sale of stock image holograms. Purchasing large rolls of stock images in hot stamp foil and producing a number of promotional products using these images allow them to keep products on the shelf ready to customize as required. Their line of business cards, note cards, stationery, and presentation folders is available in volumes as low as 100 pieces, and can be customized through hot stamping.

American Banknote Holographics offers a large variety of stock images ranging in size from under an inch to postcard size. Many of their images are available in both hot-stamp foil and pressure-sensitive formats, though custom die shapes and add-on printing are only available in larger runs.

Holographic marketing companies can benefit from stock image holograms. Pre-cataloged assortments of stock images allow easy use and access to a wide variety of images. Besides easy access and low minimum runs, a stock image project can run smoothly on the side, leaving important man hours to be spent on larger and more pressing projects.

It is important to recognize that short runs and low budgets are not restricted to small companies. Many large corporations make use of short runs for specific projects which have relatively small budgets but are extremely important nevertheless. These projects can be high profile; and, they can create an 'in' because your company was there for them on the smaller project. The customizing of stock images ultimately makes them custom images. The majority of those viewing the images will never know that they were not created specifically for the company using them, particularly if holographic diffraction imprinting is employed.

In summary, customized stock embossed holograms add a powerful new dimension to the sales of holograms and related products. Consider the benefits:

• The visual impact of custom holography without the high tooling costs.

• Affordability in low production runs.

• Multiple applications, from print advertising to premium items.

• Additional sales to customers who are presently turned away.

• Ease of presentation and higher lead conversion.

All in all, the best solution I have found to the issue of "affordable impact embossed holography in low production runs" is the customizing of stock images. More and more customers are turning to stock images to enhance their promotions. In the future, the use of their images will be wide spread. Companies dedicated to the promotion of stock holography will continually increase their image selection, and begin targeting images to particular industries. Eventually, stock image holography could become as inexpensive and common a graphic element as stock image photography is today.

'Comment made by Doug Miller while president of Holographic Design.

4

Lasers

This chapter is devoted to the lasers used in holography. Michael Fisk of Liconix researched much of the information contained in this chapter. Our chapter starts by carefully defining the fundamentals of electromagnetic radiation and then proceeds to a detailed explanation of what types of lasers are best for holography and why.

This chapter concludes with a article from Ron Olson of Laser Reflections who makes the case for using pulsed lasers in holography.

1. Electromagnetic Radiation

To sufficiently understand the operation of lasers, their many advantages and their necessity in the production of holograms, one must first comprehend certain properties of our physical world.

The entire universe consists of only two things: matter and energy. Matter is all things that have physical substance; energy is the mover, or potential mover, of physical substance. Matter is the stuff we see, smell and feel. It has mass and occupies space.

Energy, on the other hand, is more abstract. It is most often invisible, though sometimes not. Yet, it is everywhere. It lurks in the crevices of every molecule and sweeps the skies with its magnificence. A master of transformation, energy facilely converts itself from one of its many forms to another, all without sacrifice.

Energy is the driving force behind all forms of motion: the motion of our car, the motion of planets, the motion of atoms. Nothing moves without it. Matter, without energy, is reduced to a dark, frozen lump of nothingness. In a dynamic universe, matter both possesses energy and is affected by it.

Energy not only changes form, it is easily passed from one object to another. Interestingly enough, no matter how many times it transforms or transfers, the amount of energy involved in any given transaction never changes. The law of conservation of energy, one of the most important laws in the universe, dictates that energy is never created or

destroyed; it can only be transferred to another object or converted into a different form of energy.

Because the amount of energy in the universe remains fixed, phrases such as "energy shortage" and "depleted energy" are misnomers. You can not lose energy, nor can you be in short supply. The amount of energy in our environment is so great that it is beyond our comprehension. The discomforts in past decades from "energy shortages" were created only by our inability to either convert energy to a usable form or distribute usable energy to where it was needed.

Energy is measured in Joules, in honor of the British scientist James Joule. One joule is roughly the amount of energy required to lift an apple from your kneecap over your head.

A glass of apple cider has 502,092 joules (equal to 120 Calories-the Calorie is another unit of energy often used when referring to the content of food) of food energy. A gallon of gasoline has over 200 million joules of energy.

The process of applying energy to matter is called work (also measured in joules). Work is the mechanism that transfers energy through a system. It is produced by applying a force on an object such that motion occurs over a distance. For example, when you pick up a book, you have performed work on the book. The heavier the book, the greater the force that is necessary to raise it-therefore, the more work done. The farther you raise the book, the greater the distance in which the force must be applied. Again, more work is accomplished.

Energy is formally defined as the ability to do work. It can be classified in two categories: stored energy and motion energy. Stored energy is more commonly called **potential energy.** When raising a book in the air, work must be performed on it. While suspended in air, however, the book has the ability to perform work in the opposite direction-courtesy of the earth's gravitational field. This "stored" energy may be released simply by dropping the book.

The potential energy an object possesses due to its position in a gravitational field is called, predictably enough, **gravitational potential energy** (or GPE). Any object raised above the earth's surface has gravitation potential energy. Water behind a dam has a significant amount of OPE that may be converted (as mandated by the law of conservation of energy) into electrical energy.

Other forms of potential energy are also commonplace. A coiled spring is the good example of **mechanical potential energy.** By performing work on the head of a Jack-in-the-box, one can push it into the box. With the lid secured, the box has stored energy (mechanical potential energy) hence, the ability to do work. By unlatching the lid, the stored energy is released and work is performed in the reverse direction on our friend Jack.

The axiom that "opposites attract" is especially true for electric charges. Electric charges that are positive attract those that are negative, and vice-versa. Equally, two electric charges of the same type (both positive or both negative) repel each other. The attractions and repulsion of electric charges are caused by invisible **electric fields** produced by each charge. An electric field permeates the territory around each charge, affecting all other charges that occupy its space. The larger the charge, the more influence its electric field exerts on its occupants. We encounter electric fields to varying degrees throughout a typical day. We witness them when we use our dryer, for it is electric fields that cause static cling.

To separate two opposite charges- or unite two like charges-requires work. Like the Jack-in-the box, when work is applied to bring like charges together (or to separate opposite charges), **electric potential energy** is created. Remove whatever constraint that holds the stored energy (in the Jack-in-the-box the constraint was the lid, in electric charges it is usually a non-conductive material like air or plastic) and work is performed in the opposite direction.

A familiar device utilizing electric potential energy is the battery. The stored energy in a battery can be released by placing a conductive path between the positive and negative terminals. The performance of battery is stated in terms of **voltage** (also called **electric potential,** abbreviated with a V and measured in **Volts**). Voltage, an important element in laser operation, is the ratio of electric potential energy (**EPE**, not to be confused with electric potential) to the amount of charge (abbreviated in equations with a **q**, measured in **coulombs**). In equation form:

$$V=(EPE)/q$$

Another common source of voltage is the generator. Generators are machines that convert various types of energy into electric potential energy. Generators in dams convert gravitational potential energy from elevated water. The energy is transported to homes and businesses, readily available for those who wish to do a little work. Since generators produce higher voltages than batteries, they are used to supply power to all gas lasers and most others.

Matter is the greatest repository of energy. Atoms arranged together have binding electrical forces (called bonds) that act much like infinitesimal coiled springs. When bonds are broken, stored energy is released. This stored energy in molecules is called **chemical potential energy**. The gas we pump into our cars and the food nourishing our bodies are two common forms of chemical potential energy being utilized in our lives. Forces holding the nucleus of an atom together store an astounding amount of **nuclear potential energy**, as witnessed on July 16, 1945, when the Manhattan Project unveiled the atomic bomb.

Matter in motion possesses **kinetic energy**. An object will gain kinetic energy when work is done to it. An object will lose kinetic energy when work is done against it. The amount of kinetic energy an object gains or loses is exactly the same as the amount of work done on or against it.

For example, the engine of a train converts chemical potential energy into kinetic energy and performs work on the train. The train will move; it now has the amount of kinetic energy equal to the net gain of work done on it. If you turn off the engine, the train eventually stops, even if it is riding on a perfectly level set of tracks. This is because friction (between the wheels and the track; between the wheels and their axels; and between the air molecules and the front of the train) is performing work against the train.

When a train in motion hits a stationary object in its path, work is performed on the object. The object will move. Some of the train's kinetic energy is transferred to the stationary object. If one removes all sources working on and against a moving train on perfectly flat tracks-engine, friction and objects in its path- the kinetic energy of the train will never change. The train will continue to move forever at a constant velocity.

Other kinds of motion energy include heat, sound and electromagnetic radiation.

Heat occurs from the motion of molecules. The faster the molecules move, the more heat generated. A common source of heat is friction. In our previous example, friction performed work against the train. The kinetic energy of the train transformed itself to frictional heat in its wheels, axles and tracks (to a lesser degree, the air molecules). The train eventually stopped because its kinetic energy was entirely transformed into frictional heat. In most energy exchanges in nature, heat is part of the transaction.

Sound is another form of motion energy that occurs when a disturbance in a medium (commonly air) produces molecules to vibrate back and forth creating "sound waves". Each molecule receives the wave, vibrates back and forth and returns to its original position, but not before imposing a similar disturbance on its neighbor. The neighboring molecule repeats the same maneuver, as does each successor, thus creating a chain of disturbances that allows sound energy to propagate through the medium. Eventually, the sound waves hit an eardrum causing it to vibrate. The vibrating eardrum creates signals to the brain that enable us to "hear" the sound energy.

One of the most important forms of motion energy is **electromagnetic radiation.** It exists everywhere throughout the universe and comes in many forms. Radio and television waves can be transmitted hundreds of miles through the air enabling music, images and conversation to magically appear in our homes. Microwaves, used in radar and modern cooking devices, ensure safe travel and a fast meal. Infrared radiation warms our skin and other vital regions of the universe. Visible light, the only form of electromagnetic radiation that we can see, enables our world to have definition and beauty. Ultraviolet radiation burns our skin and cures our plastics. X-rays help doctors diagnose problems in our bodies while gamma rays are found in many forms of radioactive decay.

Although we perceive and apply them differently, all forms of electromagnetic radiation are essentially the same phenomenon. Only the amount of energy per fundamental unit distinguishes a microwave from a beam of light.

The fundamental unit of electromagnetic radiation is called the **photon**, an infinitesimally small "packet" of energy. Radio waves have relatively low energy per photon. Microwaves have more energy per photon than radio waves but not as much as infrared radiation. A photon of visible light has more energy than a photon of infrared radiation but less than ultraviolet radiation. X-rays and gamma rays carry the most energy of all.

Electromagnetic radiation is created by accelerating or decelerating an electric charge. The greater the acceleration (or deceleration) of an electric charge, the more energy it will produce. It would take a much greater deceleration of an electric charge to create an x-ray photon than a microwave photon. Electric charges that are stationary, or those moving at a constant velocity, do not create electromagnetic radiation.

An electron, the most fundamental unit of negative charge, is the most common vehicle for creating electromagnetic (radiation. For example, a radio station produces radio waves by accelerating electrons up and down a transmission antenna in a process called oscillation. An antenna is limited in its ability to rapidly accelerate and decelerate electrons, however. This is why antennas do not create visible light. Electron activity in atoms is the most prolific manufacturer of visible light. As explained later, this activity will be the basis from which lasers are created.

2. Properties of Electromagnetic Waves

Water waves make a good model for the study of electromagnetic waves because they are commonplace, exhibit comparable properties, and move slowly enough to care-of a wave. The unit of one cycle per second is more commonly called a **Hertz** (abbreviated **Hz**) in commemoration of German physicist Heinrich Hertz. Because the frequencies of electromagnetic waves are quite high, larger units such as **megahertz** (one million hertz, abbreviated **MHz**) and **gigahertz** (one billion hertz, abbreviated **GHz**) are commonly used.

The velocity (v-measured in meters per second) of a wave is directly related to the medium in which the wave is travelling. If the pool was drained and filled with molasses, the velocity of the waves would be less than those moving in water. For the remainder of this section, it will be assumed that all waves generated in our fictitious pool are travelling at identical velocities.

The distance between two successive crests (or two troughs) on a wave is called the **wavelength** (λ) which is measured in meters or subunits of meters. The most common subunit for wavelength measurement of electromagnetic waves is the **nanometer** (abbreviated **nm**) which is one-billionth (1×10^{-9}) of a meter.

When the swimmer slaps the water with slow intervals, the distances between the crests and troughs of the waves are large, hence the wavelength is long. The sunbather times very long cycles per second on such waves. The frequency is small. However, when the swimmer rapidly slaps the water, the crest and troughs seem to bunch together. The wavelengths are short. The sunbather counts many cycles per second. Waves with higher frequencies, therefore, have shorter wavelengths and waves with lower frequencies have longer wavelengths. The relationship between the two can be summarized in the equation:

$$f = v / \lambda \text{ or } \lambda = v/f$$

It is important to remember that, as long as you adjust their numerical values per the above equation, frequency and wavelength are interchangeable. In the study of light it is common to use either term .

The swimmer may also notice that when he slaps the water with more force, the crests become taller and the troughs become deeper. The height of the crests (or in many cases, the depth of the trough) is called the **amplitude** of the wave. If the swimmer continues producing waves, one long, unbroken string of crests and troughs will span the entire length of the pool. This is called **continuous wave** transmission (often called just **CW** transmission).

But, the swimmer could decide to produce one wave, rest (thus saving his energy), and then produce another wave-followed by another rest period. By saving his energy between waves, the swimmer could slap the water harder and produce waves of greater amplitude. This is called **pulse** transmission. Lasers transmit in a similar manner; they are either continuous wave or pulse.

In a V-shaped swimming pool with two shallow ends converging to one deep end, two swimmers at rest (swimmer A and swimmer B, equal distance from the deep end) start slapping the water at exactly the same time and with exactly the same intervals. Not only would both waves (wave A and wave B) have the same frequency but, as they passed

fully observe. There are a few profound differences be- tween the two (for example, water waves must propagate in a medium where electromagnetic waves need not), but not enough to impugn our comparison.

A wave is created by a disturbance in a medium (for electromagnetic waves, a disturbance may be created in empty space). In a swimming pool, a swimmer resting in the shallow end of the pool slaps his hand on the water. The disturbance creates a wave that moves from the shallow end to the deep end (for this example and all fictitious pools in this section, allow the edges to absorb all waves that hit it, thus eliminating the effect of reflected waves). The wave has a "crest" (high point) and a "trough" (low point).

A closer look at the wave would reveal that the water molecules do not travel with the wave. In fact, if you measured the net movement of all the water molecules due to the wave, it would total zero. One may ask, "If the water molecules aren't moving in the direction of the wave, what is?" The answer is energy.

If the swimmer slaps his hand many times in regular intervals, a wave with a series of crests and troughs is created. Each pair of one crest and one trough is called a cycle due to their tendency to repeat. If the intervals are fast, the crests and troughs (or cycles) will appear to be closer together. If a sunbather sitting halfway between the deep and shallow end of the pool had a watch, she could count the number of cycles that pass by her each second. This value, the number of cycles per second, is called the **frequency** (for space economy, we use the letter **f** in equations)

the sunbather, all the crests of wave A would pass exactly at the same time as all the crests of wave B. Similarly, all the troughs of wave A would pass at exactly the same time as the troughs of wave B. The two waves are said to be **in phase.**

Two waves being **in-phase** or out **of phase** refers to a comparison of the two waves at a given point. In the example above, the given point is the exact spot where the two waves pass the sunbather. Two waves can be out of phase at a given point for a variety of reasons. The disturbances could have started at different times. The frequencies of the two waves could be different, or the distance travelled to the given point (called the path length) could be more for one wave than the other. If the two waves were in different mediums, they could have different velocities.

Two waves having identical frequencies and velocities that are out of phase, will be so to the same degree at all points. For example, two waves at 60 Hz that are 80° out of phase at the sunbather will be 80° out of phase at the deep end of the pool and all points in between.

Two waves with equal frequencies and velocities starting at the same time can be out of phase if their path lengths are different, e.g., if swimmer A was slightly further from the sunbather than swimmer B. Holographers use this fact to create their holograms.

When wave A and wave B meet at the deep end, they will join together and this phenomenon is called **interference.** How they interfere depends on what degree the two waves are in or out of phase (called their **phase relationship**). If wave A and wave B are in phase , the crests of the two waves will meet and combine to form one large crest for each cycle whose amplitude is the sum of the two individual crests. The troughs of wave A and wave B would also combine to form one large trough in each cycle whose amplitude is the sum of the two individual troughs. This is called constructive interference (see diagram

Constructive Interference
(In phase)

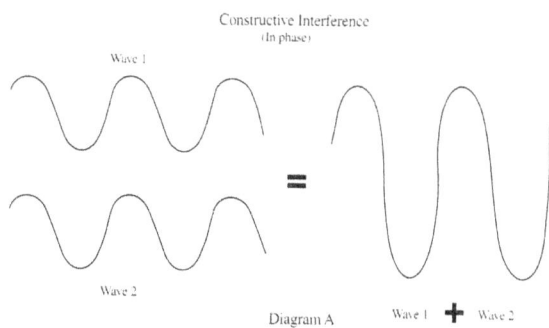

Diagram A   Wave 1 **+** Wave 2

A).

If the two waves are 180° out of phase, the crests of wave A would merge with the troughs of wave B (and vice versa), cancelling each others' amplitudes. The result is a combined wave with little or no amplitude. This is called destructive interference (see diagram B).

Because the phase relationship of two waves can change from one point to another, the two waves can be in phase

when they pass the sunbather, but out-of-phase when they hit the deep end of the pool. The ability of the two waves to stay in phase while they travel the length of the pool is called

Destructive Interference
(180° out of phase)

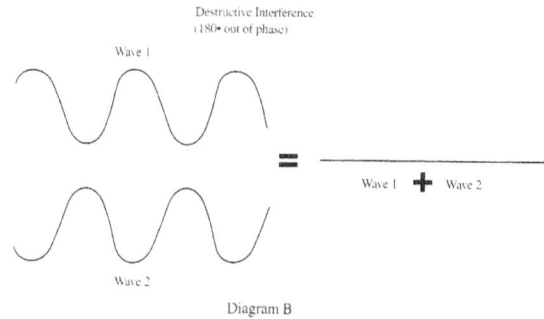

Diagram B

coherence. Waves that stay in phase for a long time are said to be very coherent.

Suppose the two swimmers agreed to create continual constructive interference at the deep end of the pool. They would start slapping the pool at the same time while trying to maintain exactly the same frequency. If both swimmers have extremely good timing, they can keep the two waves coherent for a long period. But the swimmers, like other producers of waves (lasers, for example) aren't perfect. Their frequency may be slightly off.

The difference in frequency of any source (source is a common term for a device or system that produces waves) is called its bandwidth (abbreviated **f**, and is measured in Hertz). The coherence of a source can be determined by measuring how long the waves stay in phase. Be it swimmers or a laser, the distance that the source can guarantee the waves will stay in phase is called its coherence length (abbreviated **L**; it is measured in meters or subunits of meters). The coherence length of a source is of great importance to holographers. It is directly related to bandwidth by the equation:

$$L = v/\Delta f$$

Lasers are used to produce holograms primarily because no other source offers enough coherence.

Although electromagnetic waves exhibit exactly the same wave properties as the water waves described above, there (are some notable differences between the two. Water waves propagate in two dimensions on a plane. Electromagnetic waves tend to propagate in three dimensions. As mentioned earlier, electromagnetic waves can propagate with or with- out a medium. Electromagnetic waves move extremely fast; water waves move relatively slowly. The interference of a water wave is determined by its amplitude-with electromagnetic waves, it is a function of its intensity.

In empty space, electromagnetic waves move at a velocity of 300 million meters per second (3×10^8 m/S, also known as the **speed of light**-it is abbreviated in equations with the letter c). In his 1905 paper on special relativity, Albert Einstein correctly defined the speed of light as the absolute fastest velocity possible-a cosmic speed limit, so to speak. In air, the velocity of electromagnetic waves is just slightly less than " c " . In most applications where electromagnetic waves are travelling through air, it is acceptable to use "c" as the velocity of the wave. Therefore, for electromagnetic waves travelling in either free space or air:

$f = c/\lambda$ or $\lambda = c/f$

and

$L = c/\Delta f$

Electromagnetic waves are a union of electric and magnetic fields that are at right angles (90°) to both each other and the direction of their movement (see diagram C). When electromagnetic waves propagate, there are infinite amount of directions they can travel. A laser is designed to channel waves such that their propagation is substantially in one direction.

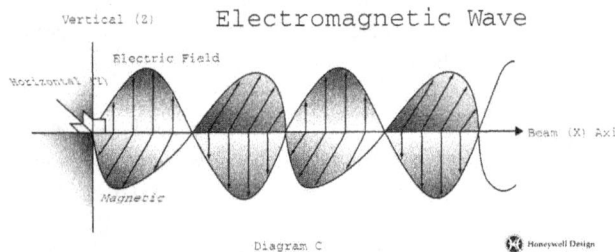

Diagram C

This unidirectional propagation of electromagnetic radiation is generally referred to as a **laser beam**.

Even while moving in a common direction, each wave can have its electric field and magnetic field oriented differently. The electric field can point straight down on the first wave, sideways on the second. There are an infinite amount of directions the electric field (or magnetic field, which stays exactly at a right angle to the electric field) can be pointing on the beam "axis" for each wave moving in the same direction.

Many properties of waves are more consistent if their electric and magnetic fields are properly aligned. The ability of electromagnetic waves to be aligned in the same orientation on the beam axis is called **polarization**. The human eye is not sensitive to polarization and can not distinguish between polarized or unpolarized light waves. Some insects, like bees, are more sensitive and use polarization to determine direction. In holography, where consistency of the source's waves is critical, polarization is essential.

Because a reference point is needed in defining polarization, the electrical field is used to identify the position. If the electric field is travelling directly on the xz plane, the wave is defined as **vertically polarized**. How close the beam is to being polarized in the vertical position can be described by its **polarization ratio**. A laser beam with a 100:1 polarization ratio is very close to being polarized in the vertical plane- a 500:1 ratio is closer still.

Polarization can be achieved by several means, including reflection, transmission, scattering and birefringence. **Birefringence** is a phenomenon that occurs in certain crystals- such as calcite-and other materials. Such materials limit the absorption of waves to those with specific electric field orientations. Sunglasses use this effect reducing the amount of glare received by the eyes. Crystals with maximum birefringence allow only one orientation to be absorbed and transmitted. The beam exiting the crystal is polarized.

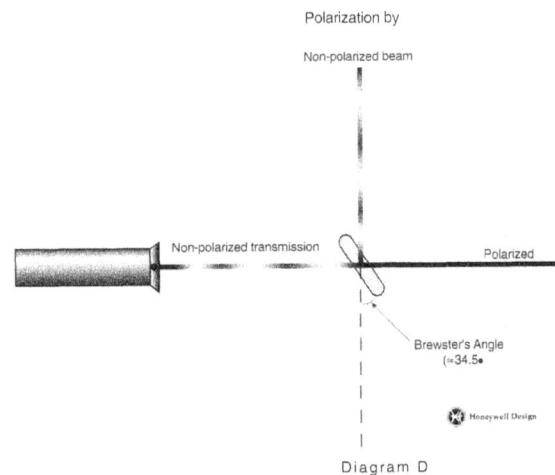

Diagram D

Scattering occurs when an atom deflects a photon away from it. Scattering of electromagnetic radiation in the earth's atmosphere produces partial polarization. Bees use this type of polarization for navigation. Polarization by scattering is not an effective source for most photonic applications. For lasers, polarization by transmission is more relevant.

If an unpolarized beam strikes a non-conductive, transparent target at a specific angle (called **Brewster's** angle, see diagram D), a polarized beam will pass through the target. This process is called polarization by transmission. Waves in the beam that do not have the selected electric field orientation are reflected from the target. Brewster's angle varies for different wavelengths and materials, but for most lasers it is in the neighborhood of 34.5 degrees.

3. Particle Properties of Light

On October 19, 1900, Max Planck introduced a concept that was to revolutionize science. In an effort to resolve conflicts between scientific theory and experimental evidence, Planck suggested that **energy** (abbreviated in equations with the letter **E**) is not continuous, but instead comes in discrete little "packets" called **quanta**. Further, an "energy packet" of light was directly related to its frequency by the equation:

$E = h \bullet f$

where **h** is defined as **Planck's Constant** and is equal to 6.63×10^{-34} Iouie Seconds.

Although the mathematics seemed to work, and it did resolve current conflicts between theory and experimental data, the implications in Planck's hypothesis were rather hard to accept. Clearly, light exhibited wave properties such as **diffraction** (bending of a wave around a comer), interference and polarization. Waves are inherently continuous and not discrete. Frequency, for example, used in Planck's equation is a wave phenomenon. Yet the energy in the same equation describes light as discrete packages of $h \bullet f$. How could light possibly consist of particles and demonstrate properties of waves?

The numerous experiments and the profound mathematics that followed are extremely significant and detailed. The final result of two decades of scientific fervor was a new definition of the laws of physics now known as **quantum mechanics**.

At the core of quantum theory is the concept of **duality** which states that light, electromagnetic radiation, energy and even matter is both a wave and a particle. Electromagnetic radiation itself is composed of minute "wave packets" called **photons** that demonstrate properties of both continuous waves and discrete particles.

In terms of wavelengths, the energy of one photon is expressed as:

$E = h \bullet c / \lambda$ or $\lambda = h \bullet c / E$

As stated earlier, all electromagnetic radiation is the same phenomenon. Only the energy per photon is different. The fundamental unit of electromagnetic radiation is the photon. One can classify all forms of electromagnetic radiation by the wavelength (or frequency) of the photon. Because the wavelength involved in the common forms of electromagnetic radiation is small, it is usually measured in nanometers (1×10^{-9} meters, abbreviated **nm**). In visible light, the wavelength of a photon determines its color. Red had the longest wavelength (740-622 nm), followed

by orange, yellow, green (577-490 nm) blue (489-430 nm) and violet (429-390 run). White light is a mixture of all colors . Photons with wavelengths greater than 740 nm pro- duce infrared (below red) radiation. Photons with wave- lengths less than 390 nm create ultraviolet (beyond violet) radiation.

Electron Jumping to a Higher Energy State

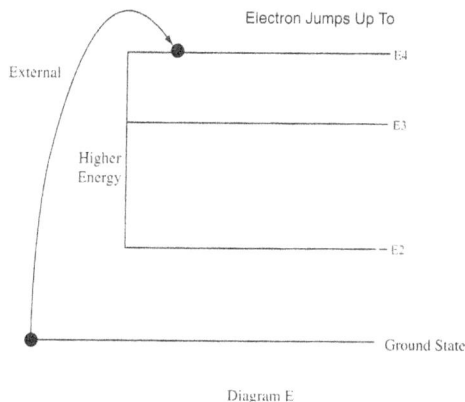

Diagram E

Photons can be created by transferring energy to an atom. In an atom, electrons reside in various positions and energy states. If the atom is stable, the electrons are defined as being in their ground (lowest level of energy) state. When external energy is transferred to the atom, the electrons get "excited" and respond by jumping up to higher, or excited **energy states** (see diagram E).

Quantum mechanics dictate that an electron cannot reside between two energy states. It has to jump up all the way to a higher state or not jump at all. The electron will stay in the excited state for a very short period and then spontaneously drop back down to a lower energy state.

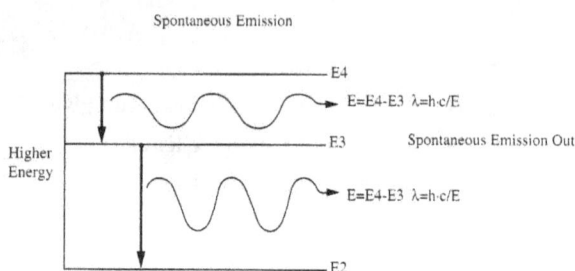

Diagram F

When the electron drops one or more energy steps (also called **transitions**), it releases the amount of energy difference between the two states (see diagram F) in the form of a photon. The wavelength of the photon is determined by the amount of energy released- -the energy difference of the two energy states-   through the formula A= h.c/E. If the energy released is small, the wavelength of the photon will be large and vice versa. This process is called **spontaneous emission.**

There are two principal mechanisms that enable energy to be transferred to an atom: absorption and collision. **Absorption** occurs when a photon bumps into an atom. If the photon's energy (determined by the wavelength of the photon using $E=h \bullet c/\lambda$) matches the energy difference between a lower energy state where an electron resides and a higher energy state (called an energy band gap), the atom will absorb the photon. The photon energy is transferred to the atom, kicking the electron up to the higher energy state. If the photon 's energy does not match any of the energy differences in two excited states of the atom, scattering occurs, redirecting the photon without otherwise altering it.

Collision occurs when a moving particle (an electron, ion, atom or molecule) smashes into the atom with the proper amount of momentum. Some or all of the particle's kinetic energy is transferred to the atom, again raising the electrons of the atom to a higher energy state.

4. Laser Theory

Albert Einstein was the first to recognize the significance of Planck's concept of quantized energy. He used Planck's $E=h \bullet f$ equation to derive his explanation of the photoelectric effect in 1905, for which he later received the Nobel Prize in Physics. In 1916, Einstein predicted another phenomenon now known

as **stimulated emission-**the basis for all laser technology. The same year, Einstein also released his most prized work, the general theory of relativity. The theory of stimulated emission-went unnoticed until the late 1940's.

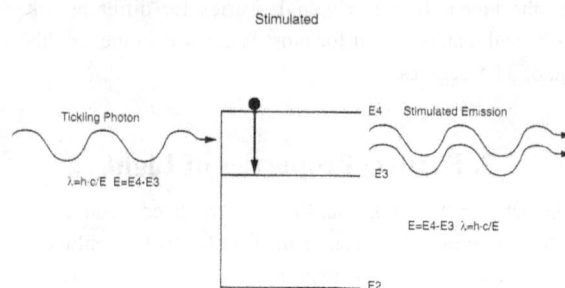

An excited electron in a higher energy state will spontaneously drop to a lower energy state and emit a photon. The wavelength of that photon is determined by the difference of the energy between the two energy states ($\lambda = h \bullet c/E$). If, before the electron drops to the lower energy state, a photon with a wavelength identical to that which is about be produced by spontaneous emission passes by the exited atom, it will stimulate the electron to drop. This stimulation (also called "tickling"--see diagram G) will force an emission of a photon that is identical to one passing. This process is stimulated emission. Both photons have the same wavelength (and therefore the same frequency), are in phase, coherent, and travelling in the same direction. It is important to note that the "tickling" photon is not absorbed by the atom; it must only pass closely.

In an environment with many identical excited atoms, photons can multiply rapidly through stimulated emission. Two photons quickly become four which quickly become eight. Since atoms and photons are extremely small, eight photons can become billions in a reasonably short distance. This process of multiplying photons is called **amplification**; it is the essence of a laser. The term LASER itself is an acronym for **L**ight **A**mplification from **S**timulated **E**mission of (electromagnetic) **R**adiation. Billions of photons, with the same wavelength, in phase, coherent and travelling in the same direction can be a very useful tool.

For lasing to occur, it is essential that there are more at- oms with electrons in a higher energy state than those at the lower energy state- a condition called a population inversion. An atom with its electrons in lower energy states will absorb a passing photon instead of duplicating it. If there is more absorption than stimulated emission, amplification will not occur. How can a baker increase his inventory if he has five hungry children and only four cookies in the oven? A population inversion assures continuous multiplication of photons.

To create a laser (see diagram H), many atoms of one type must be contained in a given space. The atoms used for lasing are called the active medium and are housed in a container. Since the energy states of the active medium create most or all of the stimulated emissions, it is the active medium that determines the possible wavelengths produced by the laser ($\lambda = h \bullet c/E$). **Solid state** lasers have active mediums made from matter that is solid at room temperature. Neodymium and chromium ions (an ion is just an atom with either an excess or deficit of electrons) are common active mediums used in solid state lasers. The active medium in **liquid lasers** are, of course, liquid and those in **gas lasers** are gas. Common active mediums used in gas lasers are neon, argon and ionized cadmium.

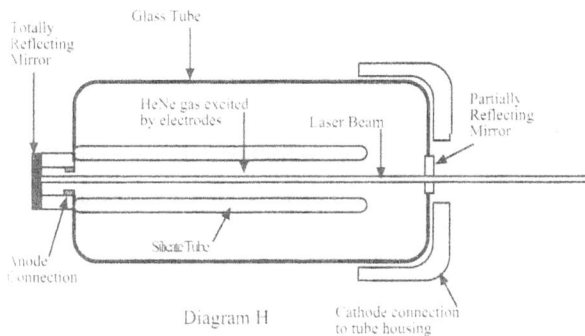

Totally Reflecting Mirror

Glass Tube

HeNe gas excited by electrodes

Laser Beam

Partially Reflecting Mirror

Anode Connection

Silicate Tube

Cathode connection to tube housing

Diagram H

A **pump** transfers energy to the active medium, raising their electrons to an excited (higher energy) state. Brisk and continuous pumping will create and maintain a population inversion. As stated earlier, energy can be transferred by either absorption or collisions. In solid state lasers, an optical pump produces a flood of photons with energies that are easily absorbed by the active medium. Common optical pumps are flash lamps, laser diode arrays and other gas, liquid and solid state lasers.

The vast majority of gas lasers use electron collisions to pump energy to the active medium. Normally, electrons will not flow through a neutral gas. If a large voltage (from 1,000 to 20,000 Volts, depending on the gas) is applied to the gas, the gas will "break down" and allow a **discharge** of electrons to rush through it. This initial voltage is called a **spike**, and is applied to the gas through two electrodes (an **anode** and a **cathode**) on opposite ends of the container. Once electron current is flowing through the tube, the voltage is automatically reduced (by 90 to 4,000 Volts, depending on the gas). The discharge of electrons and other charged particles collide with the atoms in the active medium, enabling energy to be transferred. Both the spike voltage and the operating voltage necessary to operate the laser are furnished by a power supply, usually an external box that transforms either 117VAC or 220VAC to the required voltage.

Many active mediums are not efficient at receiving energy from the pump . In such cases, the active medium must be combined with a transfer medium- a substance that compensates by efficiently collecting energy from the pump then passing it on to the active medium. Solid state transfer mediums should be good absorbers; gas transfer mediums require the ability to efficiently receive energy from collisions.

The transfer medium must be compatible with the active medium in two ways. First, it must have some common higher energy states that provide a channel to efficiently pass energy to the active medium. Second, it must be chemically compatible, allowing peaceful coexistence with the active medium as well as the other components inside the tube.

There are typically 5-20 transfer medium atoms for every active medium atom. Inert helium makes a good transfer medium and is used in helium-neon (**HeNe**, pronounced Hee-Nee) and helium-cadmium (**Heed**, pronounced Hee- Cad) gas lasers. Common transfer mediums in solid state lasers are yittrium atrium garnet (called "**YAG**"), yittrium lithium fluoride (**YLF**, pronounced "Yelf") sapphire and ruby.

The container that holds the active medium (and transfer medium, if applicable) must protect it from elements that may interfere with the lasing process. In solid state lasers, the crystalline transfer medium encompasses the active medium atoms, forming a strong durable solid structure that serves as the container. Because the transfer media in solid state lasers house and transfer energy to the active medium, they are called **hosts**.

For gas lasers, the container is almost always a long cylindrical **tube**. Tubes are made of various materials; ceramic and glass are the most common. Inside the tube is an equally-long, yet very narrow (<3 millimeters) bore that allows lasing to occur in a straight, usually horizontal path.

During lasing, a wide variety of spontaneous and stimulated emissions occur throughout the tube with photons propagating in every conceivable direction. All emissions except those travelling down the bore exit from all sides of the tube with limited or no amplification. Those photons travelling down the length of the bore continue to multi- ply through stimulated emission. By making the bore long enough, one could continue the lasing process until ad- equate amplification was achieved. At this point, a billion or so photons exit the bore in the form of a laser beam.

Unfortunately, in order to achieve sufficient amplification, the bore would have to be forty feet long. A forty foot laser would be extremely awkward to both transport and operate. A shorter bore is needed to make the device more practical.

By passing photons back and forth (called **optical feedback**) several times through a shorter bore, adequate amplification can be attained. This is accomplished simply by placing mirrors on both sides of the bore. If both mirrors are 100% reflective, extremely high amplification is achieved. However, no photons exit the laser (photon exiting the laser is called **transmission**). This, of course, has no value whatsoever.

However, if one mirror could reflect some of the photons back into the bore while allowing the remainder to exit, both amplification and transmission could occur. Such a mirror found on lasers is called the **output coupler**, or **OC**. The second mirror is known as the **high-reflecto**, or **HR**. A perfect high-reflector will provide 100% reflection. In actual lasers, there is a small amount of light transmitted from the HR, referred to as **leakage**.

Power is the amount of energy a source produces each

second. It is measured in units of Joules per second, more commonly called **Watts**-in honor of the British scientist James Watt. The power in a laser reflects the amount of photons per second exiting the laser.

Laser designers strive to maximize the power produced by the laser. However, if the OC allows too many photons to exit, the number of photons returning to the bore may not be adequate to provide significant amplification. This, in turn, limits the amount of transmission. Therefore, the amount of transmission and reflection provided by the OC must be properly balanced to compliment both the amplification process in the laser **cavity** (area between the two mirrors) and the transmission from the cavity. In most lasers, output couplers allow 1-3% of the photons to be transmitted .

It is common to have multiple wavelengths lasing inside of the cavity. Because there are a variety of paths of energy states for which an excited electron can travel back to the ground, there are a variety of energies (and therefore, wavelengths) that can be emitted throughout its descent. The electron can also bounce down two steps at a time-maybe three each time emitting photons of higher energy and lower wavelengths.

Certain transitions are more dominant, however. The more spontaneous emissions produced at a given wavelength, the greater probability of stimulated emission. The more stimulated emissions, the more amplification. The most dominant wavelength in a given laser is called the **primary wavelength.** The second most dominant wavelength is called the **secondary wavelength.**

Photons with undesirable wavelengths can be eliminated from the cavity by putting special thin film coatings on the mirrors. Such coatings only allow a specific range of wavelengths (for example 430-460 nm) to reflect back into the cavity. Thus photons with undesirable wavelengths will not multiply.

The lasing process described above is continuous wave transmission. Lasers that produce this kind of transmission are called, expectably, **continuous wave lasers** (or just **CW lasers**). Lasers that provide pulse transmission are called **pulse lasers.**

A continuous wave laser can be converted to a high energy pulse laser by installing a Q-switch in the laser cavity. **Q-switches** are devices that enable the active or transfer medium in the cavity to collect maximum pump energy before beginning the process of stimulated emissions. The Q in Q-switches is an abbreviation for "quality factor" to represent the quality of a feedback system. In the "low Q mode", a Q-switch limits optical feedback in the cavity for a very short period. If optical feedback is blocked, continuous stimulated emission will not occur.

During this period, the active medium continues to collect energy from either the pump or the transfer medium. A high percentage of electrons are elevated to higher energy states creating a large population inversion. When the Q-switch opens (high Q mode), a rapid and powerful episode of stimulated emission occurs in the cavity until the Q-switch is again closed. The active medium begins receiving energy, and again its electrons begin to elevate in preparation for the next high Q mode.

The repetitive bursts of lasing in the cavity result in a string of powerful energy pulses departing from the output coupler. The average length of time of a pulse is referred to as its pulse length (measured in seconds and subunits of seconds). The number of pulses per second is called its repetition rate (or more commonly, rep rate, measured in Hertz) .

There are four types of Q-switches: chemical, electro-optical, acousto-optical and mechanical. The first three are common and found in a variety of applications. Mechanical Q-switches are seldom used because they tend to be slow, noisy and produce unwanted vibrations.

Although the overwhelming majority of holographic applications use CW lasers, a new branch of holography called pulse holography is emerging. This kind of holography uses high-energy pulse lasers to expose the recording medium.

The high energy pulses produced by a pulse laser significantly reduces the exposure time of the recording medium (usually silver halide) to less than one hundred nanoseconds Reduced exposure times yield some significant benefits.

Perhaps the most important benefit of pulse holography is that vibration is no longer a factor. In CW holography, one or two minute exposure times require the interference patterns on the medium to stay constant until the medium is exposed. Vibrations in the optical train or table can cause the interference fringes to overlap, which distort the image of the hologram. Vibrations are a primary enemy to a CW holographer. Expensive vibration -dampening optical benches are needed to keep vibration to a minimum.

In pulse holography, vibration is of little concern. Not much can happen to their interference patterns in less than one hundred nanoseconds. Because of this, motion can be captured in a hologram. This enables the holographer to produce holographic images of real people, pets and other animated subjects.

Pulse lasers capable of sufficient energy are extremely expensive, in excess of $50,000. The most common pulse laser used today is a Q-switched, frequency doubled Nd: Yag (neodymium YAG) which produce high energy pulses (0.5-1joules per pulse, with a pulse width of 15 nano seconds) at 532-nm.

Pulse holography has developed a cult following in the 1990's. It is used mainly in artistic holography. High cost and other production limitations has prevented pulse holography from being adopted in commercial applications (ED: for another opinion on this topic, please see Ron Olson's article, "Arguments for Pulsed Holography", immediately following this section).

The laser tube and mirrors are held by a mechanical support structure almost unanimously referred to as a resonator. The title, nevertheless, is wrong. A resonator is a system consisting of a laser cavity and mirrors that enables rapid bi-directional optical feedback. It oscillates. Most physicists will readily admit that the term is incorrect; however, there is no other term available other than "mechanical support structure". Conforming to the majority, I will use the term "resonator" in this paper with the knowledge it is incorrect.

The resonator has the task of keeping the bore straight and aligned with the mirrors. This is not an easy assignment. The lasing process produces an ample amount of heat. The heat creates thermal expansion, which tends to shift the mechanisms that hold the mirrors (mirror mounts) and bore. The bore itself, being long and quite narrow, is extremely susceptible to thermal distortions.

When the laser cools down, the components of the laser tend to contract. Even the best resonators will sometime fail to keep the bore and mirrors in line. Because it is easier to align the mirrors than the bore, alignment devices are placed on the mirror mounts.

Such devices, called **tilt plates** or **xy plates**, enable the operator to change the positioning of the mirrors either sideways (horizontal, or "x" position) or up/down (vertical, or"y" position) without changing the z position (frontwards and backwards - this would change the cavity length which can only hurt the laser's performance-see section V). Proper adjustment of the tilt plates enables maximum amplification inside the cavity, producing maximum laser power. The resonator must also provide mechanical protection from the routine bumps and bruises that may occur.

The tilt plates and mirror mounts are secured in the **resonator** by three or four resonator rods, which span the length of the laser. Resonator rods are the backbone of the resonator. They give mechanical support to the entire laser head and the hardware that holds the tube.

On large lasers, the resonator rods are made from carbon graphite and are generally one to two inches in diameter. Carbon graphite has a very low **coefficient of expansion**, which means it will have minimal movement (expansions and contractions) when temperatures fluctuate. In smaller lasers, invar is generally used. Invar has a larger coefficient of expansion than carbon graphite, but provides equal strength at one-fourth the thickness.

In polarized lasers, **Brewster windows** are attached to the ends of the bore. The windows, non-conductive and transparent panels placed at Brewster's angle, seal the bore and polarize the photons that pass through it. One of three methods may be used for sealing the windows to the bore: epoxy, frit and optical contact. Sealing the bore and securing the Brewster window by epoxy was one of the first methods used on gas lasers. A space-grade epoxy glue is evenly distributed on a qualiz **Brewster stub** and meticulously fastened on the bore. Frit sealing involves heating glass between the Brewster stub and the bore.

Perhaps the most effective method of sealing the windows, yet hardest to do properly, is optical contact. Optical contact requires a precise mechanical fit between the Brewster stub and the bore. The bore is heated, microscopically melting the Brewster stub directly to the bore.

In low-powered lasers, such as air-cooled ion lasers, HeNe and HeCd lasers, the Brewster windows are almost always made from fused silica. Fused silica is preferred due to its low absorption of electromagnetic radiation, which enables the highest possible transmission. In higher powered lasers, such as large frame ion lasers, fused silica is susceptible to solarization. Solarization occurs when an excessive amount of the transmitted photon energy is absorbed by a Brewster window, changing its optical properties. Two of the more common effects of solarization are thermal lensing and color centering.

Thermal lensing is caused when photon energy absorbed by the Brewster window is converted into heat. Heat circulating in the window changes its optical properties. It also warms the air surrounding the Brewster window, distorting the optical properties of the air. Both the window and the surrounding air act as a randomly shifting lens that causes a slight variation the direction of the beam. When the shifted beam hits the mirror, it may not reflect precisely down the center of the bore. Part of the beam may "clip" upon entering the bore, causing a significant reduction in power.

The effects of thermal lensing are very similar to those of a misaligned mirror. It is common for an unaware operator to try to correct the malfunction by adjusting the tilt plates. Often, it will work-temporarily.

Unfortunately, heat energy is not stationary. The distortions in the lens and its surrounding air can change, causing the beam to shift again. Or, if the operator turns the laser off, thermal lensing may not occur again until hours after restarting it. All previous adjustment of the tilt plates are no longer valid. The laser is now legitimately misaligned.

Thermal lensing is extremely frustrating if not detected. An operator who finds his large frame ion laser constantly out of alignment may find it necessary to inspect his Brewster windows.

Color centering is the result of extreme solarization. In color centering, the Brewster window absorbs enough energy from the laser beam to change its molecular structure. When this occurs, the Brewster window will lose its transparentness.

Because most of the energy of a beam is in its core, the molecular restructuring is generally restricted to the center of the Brewster window. Photons will no longer pass through the damaged region, producing a beam that has no light in its center. This "donut" shaped beam is unusable in most applications. To help reduce the effects of solarization, thermallensing and color centering, Brewster windows on large frame ion lasers are generally made from crystal quartz.

5. Additional Properties of Lasers

Any light source that delivers exactly one wavelength is said to be **monochromatic**.

Ideally, stimulated emissions from a group of identical fuel atoms should produce very distinct wavelengths (or frequencies) lasing within the cavity; for example, a HeCd laser would produce amplification at 325.0 and 441.6 nm. By using proper coatings on the mirrors, the 325.0 nm wavelength can be removed from the lasing process, thus producing a monochromatic beam at 441.6 nm. In actuality, however, this is not the case.

In the same manner the two swimmers discussed earlier produced slightly different wave frequencies, the lasing inside of the cavity produces minute variations of photon frequencies (or wavelengths) in the transmitted beam de- fined as the laser's **bandwidth** (Δf).

Several factors contribute to variation of photon frequencies inside the cavity, including the motion of an atom at the time it emits a photon.

Variation of photon wavelengths in a laser, called its **line-width** ($\Delta\lambda$, measured in nanometers) is exactly the same phenomenon as a lasers bandwidth (Δf). Quantitatively, however, the two will not have the same numerical values. Linewidth can be numerically converted to bandwidth (and vice versa) by the following equations:

$$\Delta\lambda/\lambda = \Delta f/f \text{ or } \Delta\lambda = \Delta f \bullet \lambda/f \text{ or } \Delta f = \Delta\lambda \bullet f/\lambda$$

Often, it is helpful to eliminate frequency entirely from our equations. This may be done by inserting the expression $f = c/\lambda$ (see section II) into the above equations and applying some basic algebra. The results are:

$$\Delta\lambda = \Delta f \bullet \lambda^2/c \text{ and } \Delta f = \Delta\lambda \bullet c/\lambda.^2$$

Coherence length (L) can also be expressed in terms of wavelength and linewidth.

$$L = c/\Delta f = \lambda^2/\Delta\lambda$$

Waves moving back and forth in a confined region create constructive and destructive interference similar to that in Section 2. If the frequency of the waves matches the resonant frequency of the region, constructive interference will occur throughout the length of the region. A set of non- moving waves, complete with crests and troughs, form in the region. These "standing" waves now dominate.

By changing the length of the region (or the frequency of the waves), the two frequencies no longer match. Destructive interference is introduced, and the standing waves disappear. If you continue to change the length, you will find other discrete distances (called harmonics) that will enable the frequency of the region to match those of the waves. Destructive interference is again replaced by constructive interference, and the standing waves reappear.

By increasing the length of the region, you enable more standing waves to exist within its boundaries. By decreasing the length of the region, less standing waves exist.

A laser has waves (photons are waves) travelling back and forth in a confined region (the laser cavity). Because many frequencies exist within the lasers bandwidth (Δf), some of them will match the resonant frequency of the structure. Standing waves will form within the cavity. Only those frequencies creating standing waves will continue to lase. These frequencies are called longitudinal modes.

Within the bandwidth are a set of distinct longitudinal modes spaced equally apart. There no frequencies lasing in between them. The distance between each mode is called the longitudinal mode spacing (abbreviated m in terms of frequency , measured in Hertz~see Diagram I).

Longitudinal

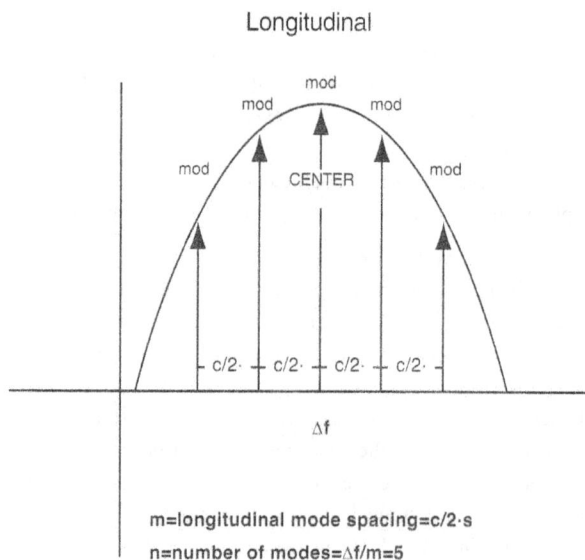

m=longitudinal mode spacing=c/2·s
n=number of modes=Δf/m=5

Diagram I

The longitudinal mode spacing is detem1ined by the

separation of the cavity mirrors (abbreviated **S**, also called the cavity length, it is measured in meters) and can be calculated with the following equation:.

$$m = c/2 \bullet S$$

In terms of wavelengths, longitudinal mode spacing (**M**, in nanometers) can be calculated:

$$M = m \cdot \lambda^2/c = \lambda^2 / 2 \bullet S$$

The number of longitudinal modes (n) in the bandwidth can be determined by the equation:

$$n = \Delta f/m$$

In terms of bandwidth or linewidth, the number of longitudinal modes should be identical. The equation in terms of linewidth is:

$$n = (2 \bullet S \bullet \Delta\lambda)/ \lambda^2$$

Because the mode spacings are very close together (generally less than 111000 of a nm), very small changes in the cavity length can cause the modes to move. In argon ion and other lasers that generate a significant amount of heat, the cavity can expand or contract enough to cause the longitudinal modes to literally jump over each other. This phenomenon is called **mode hopping**. In many laser applications, such as holography, mode hopping can cause undesirable effects.

Another mode that manifests itself inside the cavity is the transverse electric and magnetic mode, more commonly called the TEM mode. Light will propagate with distinct and defined geometrical paths. The most fundamental path will produce a clear, uninterrupted spot when projected on a target. Other paths, or "modes" will have dark irregularities (called nodal lines) separating the spot. The nodal lines can be either vertical or horizontal (see Diagram J).

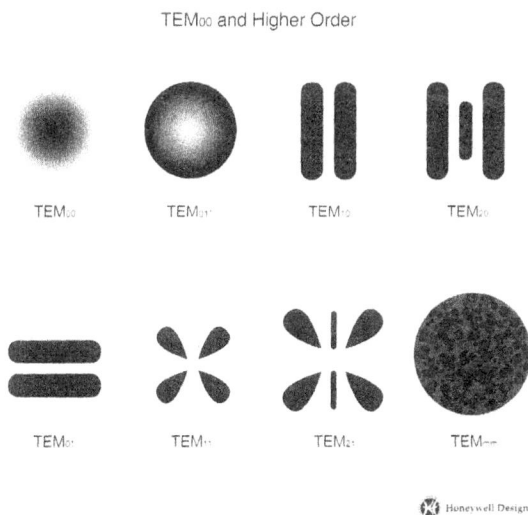

TEM$_{00}$ and Higher Order

These patterns are classified with subscript using the form TEM$_{vh}$ (where "v" designates the number of nodal lines in the vertical direction and "h" designates the number of nodal lines in the horizontal direction)- for example TEM$_{10}$, TEM$_{20}$, TEM$_{11}$.. One of the more interesting patterns is the TEM$_{01*}$ a mode that produces a large circular node in the middle of the beam. Because of its distinct pattern, TEM$_{01*}$ mode is often referred to as the "**donut mode**."

Lasers can be built to produce only the fundamental mode TEM$_{00}$ by reducing the ratio of bore diameter to the diameter of the TEM$_{00}$ beam, usually to less than 3:1. This can be achieved by properly selecting mirror combinations that encourage TEM$_{00}$ transmission.

Generally, long thin bores produce TEM$_{00}$ mode more readily than fatter ones. In holographic applications, TEM$_{00}$ transmission is essential.

One primary advantage of TEM$_{00}$ is that it has a symmetrical energy distribution (see Diagram K) in the cross section of the beam. The energy in the center of the beam spot is greatest, symmetrically lessening as you measure further towards its outer edges. This is called a **Gaussian energy distribution**; it is very desirable in holography and many other laser applications. A TEM$_{00}$ beam will not exhibit phase shifts in the electric field of the wave, as higher modes will. For holography, where phase relation- ships are critical, this is extremely important.

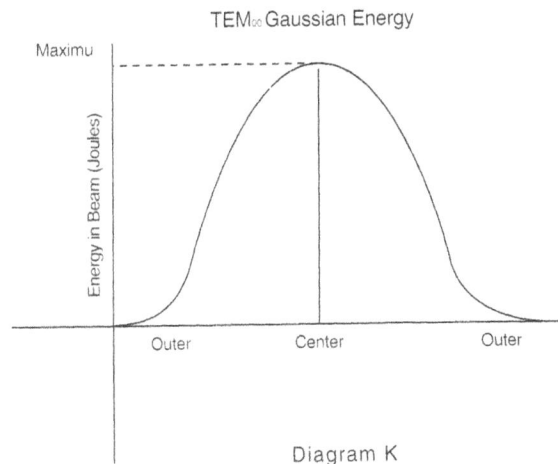

Diagram K

TEM$_{00}$ beams can be focused down to the smallest possible size spot known as the **diffraction limit**. In theory, the diffraction limit of a beam is its wavelength. In real applications, the smallest possible size of a focused spot is slightly higher. In holography, the ability to focus down to a "tight" spot is not an advantage since the beam is expanded.

The beam of a **multimode laser** has a combination of many modes resulting in an uneven energy distribution in its cross section. Multimode beams (commonly written TEM$_{mm}$) de- liver more energy than, but have none of the advantages of those modes listed above. A TEM$_{mm}$ beam can also be focused down, but not nearly as tight as TEM$_{00}$' Although multimode lasers are used in a variety of applications where laser power is the primary concern, they can not be used for holography.

Mechanical instabilities in the optics and tube can cause the beam to wander. A laser's beam-pointing stability (also called angular drift, measured in microradians) measures how much the laser beam drifts from the beam axis. Beam pointing stabilities of 60 microradians or less are considered good.

6. Types of Lasers Used in Holography

The three primary types of lasers used in holography are argon ion, HeCd, and HeNe lasers. Each has distinct advantages that are related to the holographer's needs. Typically, the type of recording material, size of the hologram, and operator budget determines which laser is best suited.

In embossed holography, photoresist is the primarily medium used for recording images. The use of photoresist enables mass-produced holograms. Photoresist chemically etches the holographic image onto a glass plate. The optically-engraved glass plate (called a master) is made conductive, and then electroplated- which produces a shim. The shim is placed on an embossing machine for mass stamping of embossed holograms

Because photoresist is extremely sensitive to wavelengths between 420-450 nm, HeCd lasers (which lase at 441.6 nm) are most often chosen. HeCd lasers are much more cost-effective to operate than comparable lasers, enabling originators of holograms to keep their operating costs down.

Artistic holography isn't constrained by the necessity to mass produce. This gives the holographer freedom to choose from a variety of emulsions to produce their holograms. In such cases, most holographers prefer to use emulsions sensitive to the primary wavelength (514.5 nm) of an argon laser. Argon lasers provide an attractive combination of high power and long coherence length, enabling holo- grams that are both large and visually striking.

In most forms of holography, the coherence length of the laser determines the size of the hologram. Generally, a 10 cm coherence length laser produces holograms that are 10 cmx 10cm(4"x4", I inch = 2.54 cm). Seasoned holographers routinely extend the size of their holograms without increasing their coherence length, some being able to shoot a 6" x 6" while using a source with 10 cm coherence length.

Novice holographers typically can't justify the hefty expense of an argon laser. A HeCd laser, though much more attractively priced, still costs over ten thousand dollars. For "weekend shooters" on a budget, the HeNe laser provides a cost-effective alternative to the higher-priced argon and HeCd lasers.

Certain laser parameters are essential to all forms of continuous wave holography. The beam must be TEM_{00} mode. A polarized beam (at least 100: 1) is necessary in all forms of holography except dot matrix; the low exposure times of dot matrix holography, typically 5 milliseconds per dot, reduce the need for polarization. The laser can not generate excessive heat or vibration, two mortal enemies of the holographer.

With the exception of dot-matrix holography, high power is very desirable. Less photon power requires longer expo- sure times. Not only do longer exposure times raise the cost of labor (professional holographers do not work for free), but they increase the odds of something going wrong. As every holographer will gladly tell you, things do go wrong.

Argon Ion Lasers

Before the advancement of the helium-cadmium laser, argon ion lasers were the preferred choice for almost all forms CW holography. The argon laser provides a generous amount of power, polarization and coherence- three of the more prized parameters in holography.

Argon ion lasers produce lasing at many wavelengths between 454 .5 -528.7 nm- usually eight to ten-and can be equipped with a prism wavelength selector in the cavity to allow the operator to select a specific wavelength. Of primary interest in holography are the 514.5, 488.0 and 457.9 nm wavelengths. The 514-nm line is the most powerful, followed closely by the 488.0 nm line. Large-frame argon lasers can provide nearly 10 Watts of power at 514.5 nm and over 6W at 488.0 nm. Most argon lasers have a polarized beam with a polarization ratio of 100: 1 or greater. The combination of high power and high coherence length make argon ion lasers the best laser for the serious artist who can afford its substantial price.

In embossed holography, the powerful 514.5 and 488.0 nm wavelengths of an argon laser have little effect on the photoresist used to record the holograms. Because photoresist is extremely sensitive to wavelengths between 420 and 450-nm, only the 457.9 nm line can effectively expose it. Unfortunately, the 457.9 nm wavelength is relatively weak in comparison to the more powerful 514.5 and 488.0 nm lines-less than 20% of the relative power. Large frame ion lasers produce approximately 1W of power at 457.9 nm.

Argon ion lasers can be equipped with another intracavity device called an etalon. An etalon is a wedge-shaped piece of high quality optical glass which, by means of constructive and destructive interference, can eliminate longitudinal modes from the laser cavity. The etalon acts as a separate laser cavity inside of the main laser cavity. When the beam enters the etalon, it is reflected inside the wedge to form its own optical feedback before exiting. By adjusting the angle where the beam enters and the temperature (which changes its index of refraction and cavity length) of the glass, the etalon can choose which longitudinal modes will survive (through constructive interference) and which ones will not (through destructive interference). The beam exits the etalon, plus or minus a few longitudinal modes, to resume amplification in the main cavity.

By reducing the number of longitudinal modes, the etalon reduces the bandwidth. Lowering the bandwidth raises the coherence length. In essence, the etalon allows the holographer to control how much coherence length the laser beam will have. While most lasers refer to their coherence length in centimeters, an argon laser with an etalon can attain coherence lengths of many meters. The trade-off, however, is power. An etalon will produce losses of 30% or more, depending on degree of bandwidth reduction.

The active medium in an argon laser is a pure argon gas. Because argon is a good energy absorber, it needs no transfer medium. Argon atoms are excited by passing a high-current density discharge through a ceramic tube. An initial spike of a few thousand volts breaks down the low pressure gas (approximately 1 torr), then the voltage drops to 90-400 volts while the current jumps to 10-70 amps (dc). The discharge current is concentrated in a small-diameter bore in which stimulated emission takes place and must be high enough to ionize the argon gas (hence, the title argon ion laser). An external magnet placed immediately outside the tube produces a magnetic field parallel to the bore axis that helps confine the discharge to the bore.

The conditions inside an ion-laser plasma tube are extremely harsh . Highly charged electrons and ions violently collide with the tube and bore eroding the surfaces, and contaminating the gas. A sizeable amount of UV radiation is generated in the cavity which, over time, tends to damage the Brewster windows and other optical surfaces.

Many times, microscopic particles from the walls of the bore can peel off and fall into the beam path. These pesky flakes can cause minute losses in power called drop offs. More advanced materials used in the inner cavity of argon ion laser tubes have reduced the amount of lasers experiencing drop off problems.

The average lifetime of an argon ion laser tube is directly related to the how high the operator sets his tube current. Those trying to achieve maximum power will find an average tube lifetime of around 2400 hours. More prudent operators can expect approximately 3000 hours per tube.

Although argon lasers are capable of producing a significant amount of optical power, the energy efficiency of the argon laser is actually quite poor. Because stimulated emission takes place in energy transitions far above the ground state, much energy is required to pump the electrons up to an excited state.

The inefficient conversion of energy creates a substantial amount of heat. To remove this heat, metal disks are brazen inside the tube to allow it to transfer from the bore to the outside of the tube. A metal duct outside of the tube circulates water which transports excessive heat away from the tube. To keep the temperature of the tube low enough to operate, large-frame ion lasers require a water flow rate greater than three gallons per minute.

Once the water exits the laser head, it must either be recirculated or dumped down the drain. To recirculate water, a heat exchanger is placed in the water flow loop. The exchanger works much like a radiator, extracting heat from water, exiting the head and transferring it to a different medium (usually air or other water). Cooler water exits the heat exchanger and is sent back to the tube.

Although heat exchangers can be relatively effective, their high vibration makes them unappealing in sensitive applications like holography. Most holographers prefer an open-cycle system in which the water flows from the tap to the laser, and then down the drain.

The power supply must also be water cooled, although water circulating in the middle of high-voltage circuitry is not a favorable combination. Condensation and leakage jeopardize the performance and safety of the unit.

Due to the high-current requirements needed to produce stimulated emissions, argon lasers operate from 208VAC, three-phase line voltage. The extreme conditions of the laser tube tend to induce nontrivial expansions and contractions in the laser cavity, making argon lasers susceptible to mode hopping and poor beam pointing stability.

Though not perfect, argon ion lasers offer the best combination of power and coherence length for artistic holography. The price of a large-frame ion laser generally starts at $30,000. Re-tubing costs range from $13,000 to $)5,000.

Helium Cadmium (HeCd) Lasers

The 441.6 nm line of a HeCd laser exposes photoresist

HELIUM-CADMIUM LASER

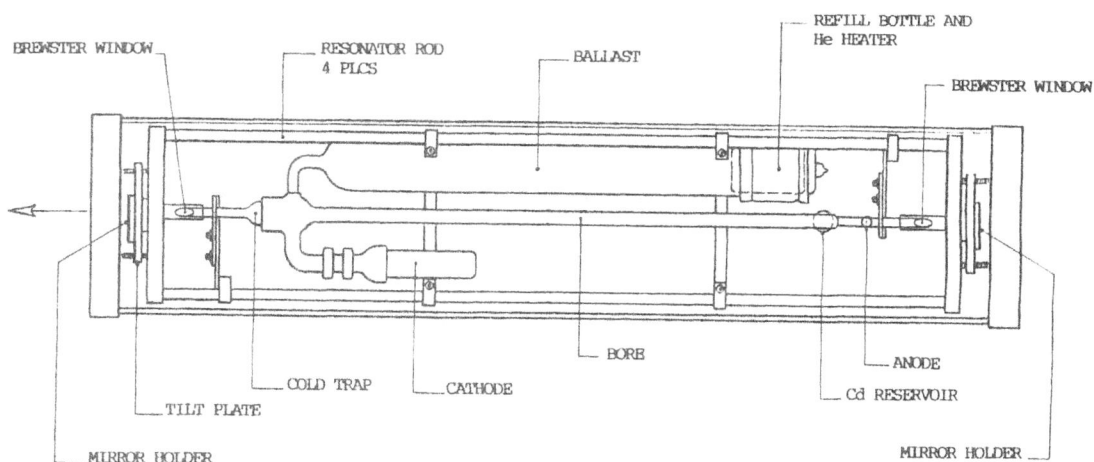

Diagram M

CROWN
MASTERS
SPACE
&
TIME

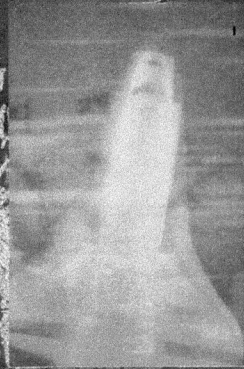

CROWN HOLO-GRAFX™ IS FULLY INTEGRATED TO:

- **SAVE YOU TIME**
- **REDUCE COSTS**
- **MAINTAIN PROJECT SECRECY**
- **LET YOUR IMAGINATION EXPLODE**

Crown Roll Leaf, Inc. was founded in 1971. Since then we've produced over a billion holograms embossed on roll leaf and pressure sensitive label stock for packaging, publishing, promotion and other purposes.

In 1991, Crown added state-of-the-art holographic origination capabilities to become the first fully integrated manufacturer of embossed holograms of all types.

Our secured Crown Holo-Grafx origination facilities are served by four fully equipped origination labs and an expert holographic art department that works with customers from concept through application to the finished product.

Today, with more than 180 employees, we help clients transform their ideas into cost-efficient custom holograms for:

- **Value-enhancing decoration**
- **Product identification and anti-counterfeiting purposes**
- **Security applications.**

Everything from collectible comic books and baseball cards to CD covers and tickets for sporting events are adorned with Crown custom holograms.

For holographic excitement at a reduced cost, Crown offers stock images that can be turned into 2-D/3-D images with a company or product name or logo. Using advanced computer imaging, Crown can produce a Holoproof in 48 hours, up to seven times faster than competitors.

Crown Roll Leaf produces over 50 million feet of product each month and ships products to 49 countries around the world.

No matter what you decorate, Crown provides FREE IN-HOUSE PRODUCTION AND ART CONSULTATION. Call for samples and details today! Crown's capabilities let your imagination explode!

1-800-631-3831

ABOUT THE FRONT OF THIS INSERT...

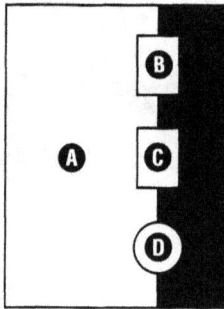

A *FIREWORKS (HOT STAMP)* - This surface diffraction grating utilizes one and two-directional diffraction gratings and Pixel-Grafx™. Designed to be run on a web press without concern for registration, the random stamping of the pattern creates one-of-a-kind collectibility.

B *SPACE SHUTTLE (HOT STAMP)* - The main image is a model of a space shuttle. The secondary image (also called a flip image) of the ship is a photograph. The background of stars is line art.

C *FLAG (HOT STAMP)* - This 2-D/3-D pseudocolor image utilizes line art fireworks behind the flag photograph which has been separated into red, green and blue (light separation colors).

D *EAGLE (HOT STAMP)* - This hologram combines a model of the eagle with a rainbow diffracting background.

America's #1 Integrated Source for Holograms, Diffraction Gratings, and Foils

CROWN ROLL LEAF, INC.® & HOLO-GRAFX™

Paterson, NJ • Elk Grove Village, IL • Burlingame, CA • Torrance, CA • Stone Mountain, GA • Toronto, Canada • Montreal, Canada   (201) 742-4000 ■ FAX (201) 742-0219

1 Holography MarketPlace - 5th Edition

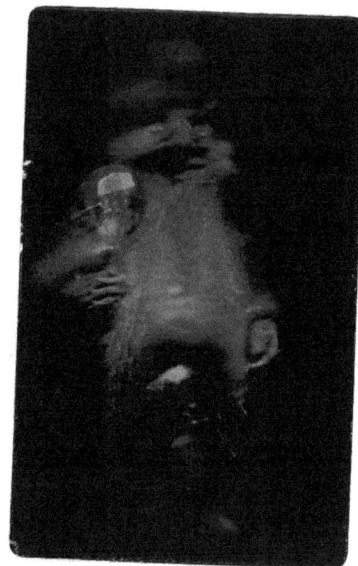

used in embossed holography ten times more effectively than the 457.9 nm line of an argon laser. Unfortunately, for many years HeCd laser were only capable of delivering a maximum of 40 mW TEM_{00}. The marginal power resulted in unusually long exposure times for originators of 2D/3D (also called multiple plane), 3D and composite holograms. A 1W argon laser at 457.9 run, comparable to a 100 mW HeCd laser in its effective exposure rate, could expose photoresist 2 112 times faster than the most powerful HeCd laser. Further, a HeCd laser could only provide 10 cm of coherence length, adequate enough for a standard 4" x 4" master. As noted earlier, only the more resourceful holographers could use a HeCd for the more illustrious 6" x 6" image. In comparison, Argon lasers can provide an unlimited amount of coherence length-albeit at the expense of power- by having an etalon installed.

As for originators of dot matrix holograms, in which the exposure time is trivial, HeCd lasers were readily adopted. A HeCd laser costs less to buy and operate, is easier to use, has lower maintenance, and lasts longer than a comparable argon laser. HeCd lasers do not suffer from common argon laser malfunctions such as thermal lensing, color centering, power supply leakage and mode hopping. Further, HeCd lasers operate off of standard 117VAC, need no water to cool the tube and, therefore, no special plumbing is required in the facility.

In the last four years, HeCd lasers have quadrupled their power output. A typical HeCd laser can deliver more than

150 mW TEM_{00} and provide 30 cm coherence length-more than enough for a 6" x 6" hologram. The effective exposure on photoresist for HeCd lasers now meet or exceed large-frame argon lasers, while saving the average holographer in excess of $800 per month per laser on their electricity and water bills. This improvement has created a profound change in embossed holography marketplace. Today, seven out of every eight lasers bought for commercial embossed holography is a HeCd.

The active medium used in a HeCd laser is cadmium. Standard HeCd lasers use naturally reoccurring cadmium that consists of a blend of three isotopes: Cd_{112}, Cd_{114} and Cd_{116}. It is abundant in nature, and therefore, costs pennies per gram. The stimulated emissions of cadmium provide a primary lasing wavelength of 441.6 nm, with a secondary wavelength of 325.0 nm. One manufacturer has patented a 353.6 nm HeCd laser. With naturally reoccurring cadmium, the spectral bandwidth is 3.0 Gigahertz, which translates into 10 cm coherence length. Standard HeCd lasers can produce up to 120 mW TEM_{00} at 441.6 nm.

By processing naturally reoccurring cadmium in a manner similar to processing nuclear grade uranium, anyone of the three isotopes can be isolated, producing **isotopically enriched cadmium**- more commonly called **single isotope cadmium**. Because the technology and equipment required to produce single isotope cadmium is restricted, the processing of it is extremely expensive. This significantly raises its price. Currently, single isotope cadmium sells for over $1 ,600 per gram.

An average HeCd laser consumes from five to eight grams of cadmium. Using single isotope cadmium can increase the price of the laser many thousands of dollars. The highest powered HeCd laser with naturally occurring cadmium is priced around $18,000. An equivalent laser with isotopically enriched cadmium sells at $25,000.

Having only one isotope lasing in the cavity produces a third of the spectral bandwidth (l.0 GHz). One third of the bandwidth nets three times the coherence length. In essence, the use of single isotope cadmium raises the coherence length from 10 to 30 cm. As an added benefit, single isotope cadmium also produces more power-about 30% more at 441.6 nm. Single isotope HeCd lasers deliver up to 170 mW, TEM_{00} at 441.6 nm. This translates into 1.7W of equivalent power when compared to the 457.9 nm line of a large frame argon ion laser in effectively exposing photo- resist.

Similar to argon ion lasers, the pumping process in HeCd lasers (see diagram M) is accomplished by a high density discharge that breaks down helium gas (the transfer medium). The discharge is produced by an anode and a cathode and confined to a long, narrow glass bore. The spike voltage is typically 13,000 volts and after breaking down the gas (ionization), is reduced to approximately 2,000 volts.

After breakdown of the helium gas, cadmium placed near the anode in a cadmium reservoir is heated to about 250 C by a heater wrapped around the reservoir. In approximately five minutes, the cadmium starts to vaporize. Through a natural process called catephoresis, the cadmium vapor migrates uniformly from the anode, through the bore, and towards to the cathode. Because catephoresis exists, a uniform distribution through the bore is possible. Otherwise, the lasing process would be impossible to con- trol.

It is critical that the cadmium is properly "trapped" before reaching the cathode. Cadmium ions being deposited on the cathode would drastically alter the electrical properties of the laser, affecting the laser's performance. A cold trap is placed centimeters from the cathode to stop the advancing cadmium. Another cold trap is placed in front of the Brewster window, to eliminate the deposition of cadmium on the window.

The strict regulation of both the helium and cadmium vapor pressures are vital to the performance of the tube. The amount of cadmium in the bore can be easily detected by measuring the tube's voltage. A feedback circuit is placed in the head that adjusts the cadmium heater should the tube voltage read too high or too low.

Helium must also be kept at a proper pressure. Helium, being a very small atom, can diffuse out of the tube; the amount is seldom of consequence. During operation, though, the cold trapping process tends to trap the smaller helium atoms under the larger and more massive cadmium atoms. The helium atoms get buried, and the tube pressure eventually lowers. To replenish the lost helium, a refill bottle is placed in the tube with a supply of high pressure helium . When the tube senses low helium pressure , a heater wrapped around the refill bottle is switched on. The elevated temperature raises the kinetic energy in the bottle, creating even higher helium pressure in the refill bottle. The higher temperature also widens the atomic spacing of the container. The combined effect creates an increased diffusion rate of helium from the refill bottle to the tube.

Helium-cadmium lasers that are stored risk subtle migration of helium atoms from refill bottle into the tube. This creates an excess of pressure in the tube, which can make the laser difficult to start. HeCd lasers, when stored, should be started at least once a month and operated from 2-4 hours to stabilize the system.

A typical HeCd laser tube typically lasts 4000 hours. Single isotope cadmium laser tubes, because manufacturers tend to limit the amount of cadmium in the reservoir, last about 3500 hours.

HeCd lasers have quickly reached a position of dominance in the embossed holography marketplace. The combination of high power, effective exposure rate, low cost of operation, and ease of use make it the laser of choice for embossed holography.

Helium Neon (HeNe) Lasers

The HeNe laser was the first gas laser to be commercially available, brought to market in 1961. Over 30 years later, the HeNe laser is still the most commonly used laser today. Supermarkets use HeNe lasers to scan the bar codes on packages for quick and efficient customer check out. High schools and universities find their low price, ease-of use, good beam pointing stability and long tube lifetimes extremely attractive. HeNe lasers operate from a 117VAC source and are air-cooled.

An average HeNe laser costs a few hundred dollars, making it an affordable tool for those who normally could not afford the expense of an argon ion or HeCd laser. HeNe lasers are low powered, typically delivering between 0.5 and 1mW, TEM$_{00}$ at 632.8 nm. More expensive models are available, delivering up to 35 mW, TEMoo at 632.8 nm. The beam is generally polarized with a coherence length between 20 and 30 cm. An intracavity etalon may be installed for greater coherence, but the corresponding loss of power tends to create extremely long exposure times. The average lifetime of a HeNe laser tube is about 15,000 hours.

The first HeNe laser ever demonstrated emitted an 1153- nm wavelength, but almost all HeNe lasers are utilized at the 632.8 nm line. HeNe lasers emitting green, yellow and orange wavelengths are also available, but their low power makes them ineffective in most commercial applications.

The active medium in a HeNe laser is neon. As in HeCd lasers, helium is the transfer medium. The laser tube (see Diagram N) consists of a glass envelope (bulb containing the cathode) with a narrow bore through its center. The bore can be anywhere from 10 to 100 cm in length, depending on how powerful the laser is.

A 10,000 volt (dc) spike breaks down the two gasses in a narrow capillary tube. The voltage drops to between 1,000 and 2,000V with a current of a few milliamperes. Electrons in the discharge pump both helium and neon atoms to excited states. The more abundant helium atoms collect most of the energy, then transfer it to lower energy or ground-state neon atoms through a series of inelastic collisions.

This transfer of energy is very efficient for two reasons. First, both the helium and the neon gasses have two higher energy states with comparable energy values. Second, both pairs of matching higher energy states are characterized by prolonged delays before allowing their electrons to drop. Energy states that hold their electrons longer (up to many milliseconds) are called metastable states.

When the abundant helium atoms in the tube are excited by the discharge current, more of them will have electrons in one of the two metastable states than in other higher energy states. A leaky bucket in a rain storm that holds water longer than a comparably sized peach basket, will more likely have rainwater in it the next day.

A sizeable population of helium atoms with electrons in the metastable state is built up. The excited helium atoms wander in the tube, collide with non-excited (ground state) neon atoms and transfer their energy to the them. Since the higher energy levels of the two gases match, the amount of energy transferred to the neon atom is just enough to raise its electron to a metastable state. Soon, the majority of neon atoms have their electrons in metastable state. A population inversion soon develops and lasing begins.

HeNe lasers are being challenged by low-powered laser diodes that emit at the 650 and 670-nm wavelengths. Laser diodes are extremely small, light, use very little current, need extremely low voltages, and are less than one fourth the price of a HeNe laser. Laser diodes have inherent

Laser Type	λ (in nm)	Power (in mW)	Effective Exposure for Photoresist	Coherence Length (in cm)	Price (in US$)	Average Tube Life (in hours)	Linear Polarization
Argon Ion	514.0	3,000*	------	adjustable*	$50,000*	2,800	100:1
Argon Ion	488.0	3,000*	0.1	adjustable*	$50,000*	2,800	100:1
Argon Ion	457.9	1,000*	1	adjustable*	$60,000*	2,800	100:1
HeCd**	441.6	150-170	1.5 - 1.7	30**	$25,000**	3,500	500:1
HeCd	441.6	120	1.2	10	$18,000	4,000	500:1
HeNe	632.8	2	----	20-30	$725	15,000	500:1

*The values for the argon ion lasers are derived from a large frame system, water-cooled and incorporating an etalon that allows >2 meters coherence length.

** The values for the extended coherence length of HeCd are derived from a large frame system using isotopically enriched cadmium.

properties that are detrimental to holography, however. Low coherence lengths (less than 2 cm), and wavelength instability make the use of laser diodes in holography unfeasible.

The low cost BeNe will always be the perfect laser for novice holographers. The combination of affordability, high-reliability and ease-of-use make this laser perfect for the production of budget holograms.

Ron Olson of Laser Reflections (formerly Positive Light) has been shooting pulse holographic portraits with a customized set-up in his lab in the Santa Cruz Mountains of California for the past several years. Here he presents an argument for the Pulse Laser as the laser of the future for holographic applications.

Arguments for Pulsed Holography
by Ron Olson

In photography, the flash is taken for granted. Few photographers would take the camera seriously if limited to shutter speeds >30 seconds. Fewer still would even consider their profession if relegated to imaging collections of plastic fruit, bathtub toys and taxidermy. Yet this seems the contented lot of many of those who practice conventional holography. As for the consumer, holography has been presented as a thing in and of itself, that which is imaged being for the most part arbitrary: "you should buy it because it's a hologram."

This was true of photography in its formative years. Eventually, however, camera technology matured and camera operators proliferated to the point that generic photography soon became worthless. Holography, unfortunately, has not been so lucky. With relatively few techno-artists contributing in the seventies and eighties, and the technology limited to unrealistic and inanimate subjects, the necessary critical mass was never met. The technique, splendid in its potential, has not truly risen above the level of curiosity and its nearly-forgotten promise remains as ripples on the pond from stones dropped all the way back in the sixties.

In 1990 we designed an Nd: YAG/Glass laser specific to the task of imaging live subjects. Our Q-switched, greenlight laser allows us to make highly accurate holographic recordings of nearly anything which can be placed in a roughly $1m^3$ imaging area. We work primarily with human subjects but also do considerable work with animals, rang- ing in size from race horses to sea horses.

Yet when I speak to other people within holography regarding our work, I often hear things like, "Oh, yeah.....the pulsed guys" almost as if to cast us apart from conventional holography. While I believe there will always be a place for tedious exposure holography, as it might better be tagged, I see little evidence that it can successfully compete on the same platform with other, increasingly advanced, visual display techniques - e.g. lenticular photography. This having been said, the unparalleled presence of high-resolution, deep-image holography remains unique - but there must be something with presence in

front of the film in the first place. Flash holography and pulsed-laser systems provide the necessary keys.

To make matters worse for pulse holography, the laser manufacturing community seems to have conspired with the uninspired in offering as a solution: egads! the ruby laser. This laser is an absolutely splendid light source for interferometrically recording tires under pressure and guitars in resonance - and absolutely useless for imaging living subjects. In the current literature, two of the more popular holographic references talk about the advantages of visible light over near-infrared light (clean surface reflections vs. penetration/water absorbtion), but then go on to say that difficulties with visible light systems preclude them from serious consideration. Some ruby laser user! advocates even state that YAG lasers are so difficult to maintain that only a handful of people (they're usually good enough to include me) could hope to operate them, or -and I'm not making this up - visible light lasers "give you a headache ."

I feel compelled to advise the holographic market to reconsider these opinions, and to those contemplating holography as a serious pursuit I offer this opinion regarding pulsed vs. CW holography: there can be no argument as to the superiority of a pulsed-light source for holographic originations - you can speak in terms of subject matter, set- up times, yield - it makes no difference, a pulsed recording system wins hands down. The primary argument against pulsed lasers is their cost, but I don't see people pounding away on toy pianos for very long if they're truly serious about music or driving GEO Metros if they're really enamored with motorsports.

For those with reservations about Neodymium based lasers (YAG, YLF and glass): yes, they are more difficult to operate than ruby lasers - cameras with interchangeable lenses are more difficult to use than disposable cameras - but the issue is image quality, and there really is no comparison. For those who talk about the higher energy or longer coherence length in ruby lasers, yes, if the intended live subject is bigger than a horse then you may need more energy than 1J and more coherence length than ~3 meters.

We have been working with our Neodymium YAG! Glass laser for more than three years now - a laser system which I designed given available commercial components and my background in custom, high-energy lasers. Like most solid-state lasers, it is extremely reliable - we have not lost a day in production since moving to our former studio in January of 1993. During that time we have produced more than 200 master transmission pieces and roughly the same number of reflection transfers - all on 30cm x 40cm format or larger.

Recently we have switched to 36cm x 60cm for the majority of our masters and transfers. We have produced as many as ten different masters in a day; and as many as six copies - one each from six different masters - within that same (approx) twelve-hour time frame.

SetUp

Our diagnostics package consists of a laser calorimeter (to measure single pulse energy), a Fabry-Perot etalon (to monitor the laser linewidth) and linagraph paper (to monitor alignment and beam spatial profile). We spend approximately 15 minutes in laser system preparation before each shoot, less time than it takes to prepare the processing chemistry.

We align our system and light subjects with the low energy output of the oscillator/ preamplifier (E~2ml) at 0.5Hz; the eyes are extremely sensitive to light at 532nm and the 0.5Hz repetition rate is ample for image composition/subject lighting. Unlike our ruby counterparts' reports, we do not see any fall-off in image quality as a result of delayed film processing - there are occasions where we shoot a number of plates one day and process them the following day after the model (or the bald eagle) leaves the studio.

We recently finished a series of human form images which we mastered on 24" x 20" format plates and are transferring onto our new standard: 24" x 14" plates. We use the same Q-switched laser for producing the H2 reflection copies on silver halide, but the next iteration of our studio (this summer) will include the possibility of CW green (YAG or Argon) transfers as well as allowing production of H2s on DuPont photopolymer. The incompatability of Q-switched lasers with photopolymer is a legitimate gripe - however, silver halide is at least three orders of magnitude more sensitive than photopolymers (μJ vs. mJ) and should therefore be the chosen material for mastering in any case. We are currently offering to customers a color-tuned 24" x 14" H2 portrait in five working days.

I know there are many people who believe that true color holography is a requirement for portraiture - I am in strong disagreement. It has been our reaction that Disney and Pepsi are much more concerned with the issue of color than anyone seriously involved in portraiture. Furthermore, I was asked by a concern that is spending considerable effort in promoting color holography to propose a design for a synchronized three-pulse (red, green and blue) Q- switched laser. While marginally feasible, it would cost more than $250,000, and this truly would be a laser that few people in the world could hope to operate. Trying to balance a three-color recording is by no means trivial for people in photolabs. To consider it seriously for holography of live subjects is a bit premature and much less important, in our opinion, than progress in black & white.

In any case, too much continues to be said regarding "color", "first-order diffraction efficiencies", "viewing angles", etc. and not nearly enough about the deplorable images offered up ad nauseum by a holographic community seemingly preoccupied with cool dudes on skateboards . There is a generation gap which should be of much more concern to serious holographers - that of laser hardware past and present. Holography which continues to SCOI11 the living, aesthetically blessed world is doomed to reside pre-packaged in hologram stores alongside other dubious visual gimmicks.

The reaction to our recent work within the art and image market has been consistently good and we have gained the attention, support and patronage of many prominent people within the fine art, advertising and museum display communities. The new hardware costs for our recording/transfer system which we call nsLooks™ are roughly $150,000 and this is not likely to change for the better in the near future. For those who wish to believe differently, the laser we are using costs a little more today than it did 20 years ago.

Occasionally, used laser systems do become available which are good candidates for reconfiguration as holographic systems (E at 532nm ~0.5-1J, pw~15ns, coherence ~3m). I am aware of several such systems advertised within the past year for a cost $30k> <$50k - not unlike what many people wind-up spending on a CW laser plus isolation table. It has been almost three years since we published a paper describing our work and our system, yet in that time we have received very few inquiries into our Q- switched YAG/glass system.

Ilford Plates and Film for Holography

A Secret Cache Located!

The last remaining plates and film from Ilford have been discovered in Ilford's Binghamton warehouse and are now available from Hologram Research of Stony Brook, New York.

These superb materials for holography, the highest quality silver-halide emulsions ever designed and produced by a major film manufacturer _specifically for_ holography, are being stored at 45°F in their original unopened boxes, and are available in very limited quantities (see chart below).

If you want the **absolute best** plates and film for your next museum piece, limited edition, or trade show display, call Hologram Research at: (516) 444-8839

fax us at: (516) 444-8825

email us at: hologram@lihti.org.

Once these are gone, that's it!

	Emulsion	Size	Pieces per Box	Available Boxes
RED SENSITIVE				
SHEET FILM	SP673	8"x10"	25	15
	SP737T	4"x 5"	50	12
	SP737T	8"x 10"	25	2
	SP737T	12"x 16"	25	3
	HOTEC R	4"x 5"	25	5
ROLL FILM	SP673	9.5"x 200+'	1	4
GLASS PLATES	SP696T	4"x 5"	20	40
	SP696T	8"x 10"	10	13
	SP696T	30cm x 40cm	10	39
BLUE/GREEN SENSITIVE				
ROLL FILM	HOTEC G	110cm x10m	1	2
GLASS PLATES	SP695T	4"x 5"	20	100
	SP695	8"x 10"	10	27
	SP695T	30cm x 40cm	10	32

Processing information available in Silver-Halide Recording Materials for Holography and Their Processing by Hans J. Bjelkhagen, Springer-Verlag Publishers, Springer Series in Optical Sciences, Volume 66, pages 99-102.

Pricing, detailed Ilford specifications, and processing information available on request.

HOLOGRAM RESEARCH, 25 EAST LOOP ROAD, STONY BROOK, NEW YORK 11790-3350

Phone: (516) 444-8839, Fax: (516) 444-8825, Email: hologram@lihti.org, CompuServe: 73363,3106

5

Recording Materials

In this chapter, we discuss the four major types of recording materials: Silver Halide, Dichromate, Photopolymer and Photoresist.

Photopolymer

There are two major producers of photopolymer materials, DuPont and Polaroid. Following is a description of their products, based on each company's literature.

Polaroid

Polaroid continues to supply the industry with photopolymer holograms. There are two ways you can obtain a Polaroid photopolymer hologram.

Stock holograms: Choose from selection of already-created holograms. These holograms are available on panels of pressure-sensitive (self-adhesive) labels.

Custom holograms: Hologram images that are determined by the customer. Choose the image or model you would like to capture in a hologram, and Polaroid will create it.

Polaroid's film is not for sale to the public; they offer complete production services from the start of a project to its completion. For more information call 1-617-386 8676.

DuPont

DuPont's line of holographic photopolymer films and replication equipment features OmniDex-352, a blue-green reflection material, currently available in 500' x 12" rolls.

Two high-performance OmniDex films are available:

HRF-700 and HRF-600. HRF-700 is a high Δn blue-green sensitive reflection film and HRF-600 is a blue-green sensitive high Δn film for recording transmission or reflection holograms.

In addition DuPont also has full-color and red reflection films. HRF-705X film has a film speed of 30 mJ/cm2 in the red and 20- mJ/cm 2 in the green. Films are available in film thicknesses of 10 and 20 microns. (see chart for details).

Processing of DuPont's photopolymer consists of a few steps: holographic exposure followed by UV or white-light (exposure, optional lamination with a color-tuning film, and heating to brighten the image. The finished hologram is stable to environmental conditions such as heat and moisture.

High-volume replication and processing equipment is also available from DuPont for use with OmniDex films. The DuPont OmniDex Replicator operates much like a printing press, using a cylindrical scanning geometry for continuous production of holographic images typically at five square feet per minute.

The DuPont Omnidex Laminator is available with the replicator for color-tuning of rolls or sheets of images, and for downstream conversion operations such as application of pressure-sensitive adhesives or anti-scratch cover films to rolls or individual sheets of holograms. Equipment is also available for heating of large rolls of holograms.

For further information on DuPont materials or replica-

Table 1: DuPont Photopolymer Products

Film	Thickness (µm)	Type of Film	Film Speed* (mJ/cm^2)
HRF-150X001	38	Blue/Green Transmission	514nm: 100
HRF-600X001	10 or 20	Blue/Green Transimiison high Δn	514nm: 50
HRF-700X001	10 or 20	Blue/Green Reflection high Δn	514nm: 15
HRF-700X071	20	Full Color Reflection high Δn	476nm: 11 532nm: 7 633nm: 10 647nm: 12
HEF-750X179	20	Full Color Mastering Film** (removable Mylar base)	476nm: 13 532nm: 9 647nm: 17
OmniDex 706	20	Blue/Green Reflection	514nm: 25
OmniDex 352	25	Blue/Green Reflection	514nm: 30

OmniDex 706 comes with equal amount of GA2-RED Color Tuning Film (CTF)

OmniDex 352 comes with equal amount of GA1-GOLD Color Tuning Film (CTF)

For Quantity purchases we also offer the following 500 ft rolls :(500 ft roll is 72,000 sq. inches with 12 inch wide imaging area):

500 ft. roll OmniDex 706 with 500 ft roll GA2-RED CTF

500 ft roll OnmiDex 352 with 500 ft roll GA1-GOLD CTF

* Single beam exposure with front surface mirror for reflection films, or single beam copy of 50% Diffraction Efficiency master for transmission films

** Film has enhanced glass adhesion premitting removal of cover sheet and Mylar base, leaving only photopolymer layer and nonbirefringent protective layer

tion equipment, contact the DuPont Company, Optical Element Venture, Experimental Station, P.O. Box 80352, Wilmington, DE 19880-0352 USA. Tel. 302-695 2838. Fax: 302-695 9631.

Dichromated Gelatin (DCG)

You can get dichromate from any major chemical manufacturer, by the barrel. As for previous statements about its short shelf life, at least one person in the field claims he's had batches around his facility for over a decade.

The biggest improvement in the recent past for dichromates is the ability to do mass-produced color holograms.

While dichromates used to be considered expensive, the supplier problem (Ilford dropped out) in the silver halide field, along with the relative expense of photopolymer, has made dichromates more accessible. As for working with the DCG, a person who's done it for two decades says "DCG is one of the easiest materials in terms of processing. There is the fix bath, water, and then drying in isopropyl alcohol. It's really quick and doesn't involve any bleaching like silver halide."

There are problems of atmospheric differences and moisture exposure to DCGs. There's nothing that can really be done about it, although some research is underway. There also is a "spectral smear" which can cause a little blurriness in playback of certain frequencies of DCG holograms.

Mastering time for DCG holograms depends on the size of the project and specifications. It can vary from "60 minutes to 6 weeks," says one supplier. Some people have said that mastering charges for dichromates were "very expensive." A spokesperson for a major player in the dichromate business disagrees. "Depending on the size of the order and the type of master being produced," she says, "the fees can be as low as $200.00 or waived entirely."

The most popular dichromate products continue to be watches as well as 30mm x 38mm discs, and 4" x 5", 5" x 7" and 8" x 10" frames.

The laser power necessary to make hologram depends on what the dichromate itself is mixed with. You usually need blue or green (wavelength), but it's not very sensitive to green light. In addition to the 5w argon, holographers use both 40nw argon and Helium Cadmium laser, depending on the size of the hologram.

One of the advantages of DCG is that there is a lack of scatter in blue light. This makes DCG ideal for making Holographic Optical Elements.

Silver Halide Emulsion

Photography has utilized silver halide emulsions for decades. As you would expect, some major manufacturers of photographic film have adapted their formulas to provide film and plates for holographers.

Years ago, there were three primary manufacturers of silver halide films and plates: Ilford, Agfa-Gevaert, and Kodak.

Ilford decided to drop out of the market and no longer supplies holographic emulsions. Agfa-Gevaert now has the lion's share of this market.

Silver halide requires less light for exposure than most of the other recording materials, so the primary reason for its popularity is its speed. Fast emulsions reduce stability problems. This allows people to buy less powerful, hence less expensive lasers and still make good holograms. It is also convenient to buy silver halide emulsions in precoated film and plates that have a reasonable shelf life, something which cannot always be said of the other recording materials.

The silver halide emulsion is granular, so it scatters more light than the other recording materials.

Although most of the mass manufacturing of silver halide holograms on film and plates is not fully automated, there are fully automated machines for silver halide film reproduction which are capable of runs in the tens of thousands. The mass manufacturing quality, which at first was poor, has improved considerably.

Emulsion Content

There are five atoms which, because of their atomic similarity, are called the halides. They are chlorine, bromine, iodine, fluorine and astatine. Silver halide emulsions are made using either silver chloride, silver bromide, or silver iodide. The other two halides are not used because silver fluoride is insoluble in water and astatine is radioactive.

A typical silver halide emulsion is made by adding a solution of silver nitrate to a solution of potassium bromide and gelatin. Silver bromide crystals form in the emulsion. The emulsion is heated for a certain amount of time, which is called the ripening process.

During the ripening process, the grain size increases and the speed of the emulsion is increased. Some doping agents may be added to the emulsion at this time to foster proper

crystal growth. Afterwards, the gelatin is allowed to cool. It is then shredded, and the soluble potassium nitrate is washed out of the emulsion.

The emulsion is heated again, with more gelatin added; then it is cooled and applied to a base. The thickness and hardness of the emulsion is important in holography because emulsions too thick tend to deform during development. Emulsions that are too hard can either retard chemical reactions or create vacuoles in the emulsion left by migrating atoms. These vacuoles tend to scatter light.

Let's assume the emulsion is made and we now want to expose it to light. It sounds surprising, but a perfectly structured crystal of silver bromide does not react to light in any appreciable way. A crystal with defects, however, does react with light. Fortunately, most silver bromide crystals will have defects which consist of some interstitial (out of order) silver ions displaced in the crystal structure.

The process of the photochemical reaction is not known in exact detail, but it is believed that when light strikes a silver bromide crystal, enough energy is available to remove an electron from an occasional bromide ion. The electron produced is able to migrate through the crystal until it comes in contact with an interstitial silver ion. The silver ion takes the electron and becomes silver metal. Silver atoms formed by this mechanism apparently act as a nucleus for the formation of aggregates of 10 to 500 silver atoms, known as **latent images** because they are too small to be seen by the naked eye.

After exposure, the emulsion is developed. The developer goes to the site of any silver bromide crystal with a latent image and causes all the silver in that particular silver bromide crystal to be reduced to silver metal and deposited on the already-existing latent image of silver metal. This causes a worm-like grain of silver metal to form which is limited in size by the amount of silver available in the silver bromide crystal. This growth is considerable, amplifying the size of the latent image silver metal by a factor on the order of 10^6.

Common Silver Halide Emulsions					
Plates	Film	Spectral Sensitivity	Emulsion Thickness Plate	Film	Grain Size
Agfa-Gevaert					
10E75	E75	10Red	7 μm	5 μm	.090 μm
8E56	8E56	Blue-Green	7 μm	5 μm	.044 μm
8E75HD	8E75HD	Red	7 μm	5 μm	.044
Eastman Kodak					
649F	649F	All	15 μm	6 μm	.060 μm
120-02	SO-173	Red	6 μm	6 μm	.050 μm
	SO-253	Red	9 μm	9 μm	.070 μm
	SO-424	Blue-Green	7 μm	3 μm	.065 μm
Note: There exist Russian and Bulgarian emulsions with grain size of about .015 μm - see directory for suppliers.					

If the developer is left in contact with the emulsion long enough it eventually attacks all the silver in the emulsion. The speed of development is slow enough, though, that you can use a timer to take the emulsion from the developer just after the latent image, but not the unexposed silver bromide crystals, have been developed. At this point the developer has converted silver ions to silver metal if and only if they belong to a silver bromide crystal that was exposed to light.

The emulsion is then placed in a fixer solution which attacks all silver bromide crystals that were not exposed to light. The fixer makes these silver bromide ions soluble and removes them from the emulsion.

The result is an emulsion with black spots where light has struck, and clear spots where no light struck.

An ideal silver halide emulsion depends somewhat on its use but there are three main factors to consider in any emulsion: thickness of emulsion, grain size of silver halide crystals, and sensitivity (or density of silver halide crystals) in the emulsion. We can generally state the following:

It is agreed that emulsions of more than 10).tm are neither practical or theoretically necessary to produce most volume holograms. Thicknesses above this size cause problems in development.

Grain size becomes an important issue in holography because it involves recording fringe patterns that are wavelengths apart. Too large a grain size may create excessive scatter, which may fog or destroy your hologram, and too small a grain size makes the emulsion have no usable sensitivity. It is generally agreed that the most ideal grain size is in the range of .0lmm to .035).tm.

The ideal exposure would probably be 100 - 300) $\mu J/cm2$ to give a useful density (D=2-3). If exposures are much longer than this, the main attraction of silver halide emul- sion, its speed, comes into question and other emulsions become more attractive.

Some of the common brands of silver halide emulsions used in holography are listed next, and then we present recommendations by the manufacturers.

Notes from Manufacturers of Silver Halide Emulsions

Agfa-Gevaert

Transmission holography: "While theoretically the maxi- mum diffraction efficiency that may be achieved with an amplitude hologram will be 6.25% at the utmost, theoretically 100% may be achieved with phase holograms.

Technical literature describes a great number of processing systems that, with the highest possible diffraction efficiency enable the noise to be kept as low as possible.

The exposure doses that are required for making a good phase hologram amount to ~50 mJ/cm² for the emulsion 8 E 56 HD, and to ~25mJ/cm² for the emulsion 8 E 75 HD, as a relatively high density (between D= 1.5 and D=2.5) proves to be necessary. As the HOLOTEST 10 E emulsions after bleaching produce more noise than the 8 E types, they are not recommended for phase holography."

Theoretically emulsion layers of a thickness of 20 mm are necessary for reflection holography so as to achieve reflection holograms of top quality, but it is still recommended to use the materials 8 E56 HD and 8E75 HD with thickness of the emulsion layer of 7 mm; with these materials the distortion of the Bragg planes after processing will be smaller. This is why it is also possible to achieve high-quality reflection holograms on thinner emulsion layers.

Kodak

"When selecting a silver halide material, it is important to match peak sensitivity as closely as possible to the wavelength emission of the laser being used. Early attempts to produce holograms also demonstrated the need for emulsions with lowest possible graininess characteristics, and highest possible resolution.

"Another factor to consider is latent image fading, a propensity for microfine-grain emulsions to lose density after exposure; that is, for a given exposure, developed density is likely to decrease with increased time between exposure and processing. Storage conditions between exposure and processing can also contribute to the rate of fading. When conditions do not permit prompt processing, adoption of a uniform protocol for exposure and processing is necessary, especially when each hologram in a series must have the same density."

Photoresist

Photoresist has long been used in the microcomputer industry to stamp out circuit boards. Holographers found out that because of its ability to make an imprint in relief, photoresist was good for creating the shim for stamping out embossed holograms.

There are two options with photoresist: you can buy all the raw materials and coat your own plates, or you can buy pre-made plates. Following is information about the coating process, and then recommendations from manufacturers about using their pre-made plates.

Coating Your Own Plates

Several artists and technicians have successfully coated their own photoresist plates. However, it is an exacting and time consuming process. Please see earlier editions of the Holography MarketPlace for a complete and detailed discussion of this topic.

Emulsion

Most holograms to be embossed are recorded in Shipley AZ135OJ-emulsion because it is a commercially available positive photoresist emulsion with sufficient resolving power for holography. This photoresist emulsion is designed to be used for the microcomputer industry, however, and consequently does not have the sensitivity that one would wish for making holograms. Hopefully Shipley or a competitor will supply an emulsion more suitable to the holography industry.

Some individuals have tried to make their own resist coating. It is not recommended as resist plates of excellent quality are readily obtainable up to 7 x 7 inches. Emulsions are generally applied by either spin coating or dipping. Spin coating is the preferred method: you can find suppliers that will coat plates for you but you should be very cautious about the results. Your coating should be mirror-smooth, free of defects and striations when viewed in yellow safelight.

Exposure

A frequently used exposure is at 457.9 nm using an argon laser. This is the argon wavelength to which this emulsion is most sensitive but, unfortunately, it is also one of the weakest argon wavelengths. If you use an etalon the available power is reduced by about 50%. At our desired wavelength of 457.9nm, an 18 watt argon with an etalon can put out about 700mw and a 5 watt argon with an etalon produces 120mw. Bums reports the Shipley emulsion is a factor of about ten times more sensitive to the 441.6nm wavelength, which is available from Heed lasers, but Heed lasers are not readily available with more than 40mw of power. In addition, the coherence length of HeCd lasers is

Popular Methods for Coating Photoresist w/ Electrically-Conductive Silver			
METHOD	COST	SPEED	QUALITY PROBLEMS?
Silver Spray	inexpensive	fast	possible quality problems
Vacuum-deposited silver	expensive	moderate	excellent quality
Electroless nickel	inexpensive	moderate	depends on operator

poor relative to argon lasers equipped with etalons. Therefore, an argon laser with an etalon is preferred for HIs and H2s; whereas HeCd lasers are preferred only for contact copies.

To give you an idea of the sensitivity, Burns reports that the Shipley emulsion requires a threshold energy to be effectively exposed. At 457.9nm, it generally requires a total exposure energy of 250,000 $ergs/cm^2$ to 2,500,000 $ergs/cm^2$ depending on the thickness of the emulsion. Burns found he could achieve this exposure by delivering 150 $ergs/cm^2$ per second for a 30-minute exposure or 420 $ergs/cm^2$ per second for ten minutes. Either way it is a long exposure and system stability could be a problem.

The exposure Burns finally used delivered 1,000 ergs/ cm2 per second and the exposure time was 4 minutes 38 seconds. Development was for 40 seconds with tray agita- tion at a temperature of 53+ F using Shipley 303A developer diluted 6: I with distilled water.

We now coat the photoresist with a layer of electrically conductive silver. Three popular methods for coating and their respective benefits and disadvantages are described in the following table. Regardless of the method, the thickness of the conductive layer needs to be only several hundred angstroms. You want a nice, even thickness with no pinholes or cracks. Watch the temperature to make sure it does not exceed the melting point of the photoresist.

Silver Spray

This method offers high production rates with low start up costs. It requires only a spray-gun system, spray booth, and solutions. You can use a two-nozzle spray gun to mix the two reagents with a commercially-available, two-part silver solution or mix your own solution from existing formulas.

Vacuum-Deposited Silver

To proceed with this method, you need to buy vacuum metallization equipment. In the procedure, the photoresist is affixed at the top of a vacuum bell jar and a small quantity of silver is evaporated from a hot filament in the bottom of the jar. The benefit of this method is that it requires little skill on the part of the operator, and therefore leaves less room for error.

Electroless Nickel

This is a three-step immersion process that is inexpensive. It requires dip tanks (one of which must be heated) and solutions that may be mixed in house or purchased commercially. The steps are first to sensitize the photoresist with stannous chloride solution, then dip it in palladium chloride. The third and final step is immersion in electroless nickel deposit, which takes place in a heated tank. There is a wash step between each of the three solutions. The electroless nickel process does take some operator experience since the type of agitation, temperature of solution, immersion time, etc. can all change the outcome.

BUYING PRE-MADE PLATES
Towne Labs

Towne Labs of Somerville, NJ (USA) is a producer of photoresist plates for use in holography and other fields. Towne supplied the following description of how its plates are manufactured.

The materials that are used to produce the large Iron- Oxide coated holographic plates are purchased to specification requirements of the micro-electronic, semiconductor and printed circuit board industries for which it was originally designed.

For example, an optical grade polished (both sides) soda lime, float glass substrate 24" x 32" x 0.190" (609.6 x 812.8 x 4.83 mm) has a non-flatness specification not to exceed 150 x $10^{(-6)}$" per linear inch (1.5 microns per linear cm). Before it is acceptable for FeO_2 coating, each piece of glass is cleaned and surface-polished to ensure that the slightest surface imperfections and even micro-dust particles are removed.

The pure Iron-Pentacarbonyl used has a controlled specific gravity of 1.44-1.47 @ 20°C and its deposition is carried out in class 100 clean room conditions. After the FeO_2 deposition 16.5 x 10 (-6) inches (0.42 micrometers) thick, the plate is inspected for integrity of coating. In an area 19" x 27" (482 x 685 mm) no pinholes or imperfections greater than .006 inches (.152) are permitted and only 6 maximum .003 to .006 inches (.076 to .152 mm) are permitted.

The plates are dried in a thermostatically-controlled Class 100 environment, (then) cleaned and inspected again. From that time to the deposition of the photosensitive coating, nothing is permitted to contact the surface of the plate.

The non-carcinogenic Microposit-S-1400-30 highly sensitive photoresist used for coating is specifically formulated to be striation-free. On plates up to 18" X 18", (457 mm square) the photoresist is applied by a spinning process to a final standard thickness of 1.5 +/- 10% micrometers subsequent to a 0.2 micrometer filtration process. This entire procedure is doomed to failure immediately if any portion of the image area separates from the glass substrate (during exposure, developing, etching, electroplating or when the mother shim is removed from the imaged surface.

The final results are that unless specific attention is devoted to very controlled conditions and to all phases of materials and processes used to produce plates, there is a significant margin for error and the times, effort and expense expended on producing unacceptable material will greatly outweigh the initial investment required to produce quality plates.

COMMENT: The success of the iron-oxide coating is owed primarily to two inherent characteristics of FeO_2 coating, e.g. the iron oxide coating effectively absorbs any laser light that may be transmitted through the photosensitive coating. This virtually eliminates light backscatter and the possibility of damage to the primary image. Second, and possibly more important, the iron-oxide coating greatly increases the adhesive quality of the photosensitive coating, thus ensuring the integrity of the imaging and electroplating processes to follow.

Charlie Ondrejik, the marketing manager of Towne, says that of the people who buy their plates, about 20-25% use them for holography, and "the number is growing." A particular area of expansion has been the Far East and China, which Ondrejik called" the most active area at the moment"; they are involved mostly in artistic applications.

Towne has not made any specific modifications to its plates for holographers. "It's a by-product of what we're doing for the semiconductor industry," says Ondrejik. The

semiconductor industry typically uses a plate made with Shipley AZ-1350-J photoresist in small plate sizes up to 9" x 9", whereas holographers use plates of Shipley S-1400- 30 photoresist and start at larger sizes of 8" x 10", working up to larger sizes of 24" x 32". The plates used by both industries are coated with the same Iron-Oxide materials, which is vapor-deposited to 3300 angstroms; the difference is in the photoresist material on the plate itself.

6

Business Listings

In this chapter we list businesses involved in the field of holography. First we have a general lising and then we cross index the addresses according to a number of categories. NOTE: Due to a programming error, some city names do not appear in the general listings. See page 123 and 124 for any missing address information.

Master Listing

21st Century Finishing Inc.
215 Pennsylvania Avenue
Paterson, NJ
Postal Code: 07503
Country: United States of America
Contact: Anthony Olmo
Voice Phone: 1 20 1 279 21 00
FAX Phone: 1 201 279 5659
Business Description: Converting company that specializes in holographic foil stamping, embossing, etc.
****

3 Deep Hologram Company
PO Box 410027
Cambridge, MA
Postal Code: 02141
Country: United States of America
Contact: Alex Cheimets
Voice Phone: I 7 I 4 969 5354
FAX Phone: I 7 14 9695354
e-mail address:acheimets@aol.com
Business Description: Supplies Russian silver halide emulsion on plates. Manufactures and distributes retail and promotional holograms.
****

3-D Worldwide Holograms, Inc
7503 N.W. 36 Street
Miami, FL
Postal Code: 33 166
Country: United States of America
Contact: Jeffery Schwartz
Voice Phone: I 305 994 7577
FAX Phone: I 305 994 7702
e-ma il address:hologram@shadow. net
Business Description: We are speciali sts in full color animated holographic stereograms. Our fully computer controlled laser labs and origination studios are supported by our 20 million dollar S.G.I. digita l production facility.
****

3 D Holographics
31 Willow Park Ave
Glasvenin
Country: Ireland
Voice Phone: 353 1 843 6200
****

3D Im ages
3 I The Chine
Grange Park
London, England
Posta l Code: N2 1 2EA
Country: United Kingdom
Voice Phone: 44 181 364 0022
FAX Phone: 44 181 364 1828
Business Description: Manufacturer and distributor of 3D images.
****

3D Vision
Hologramme-Laserprodukle
Ostertorsteinweg 1-2
Postal Code: 28203
Country: Germany
Contact: Uwe Reichert
Voice Phone: 49 42 I 767 97
FAX Phone: 49 421 767 97
Business Description: Holograms , Holographic projects. General commercial holograms for sale and distribution
****

3M
3M Center, Bldg 225-4N- 14
St Paul , MN
Postal Code: 55144- I 000
Country: Un ited States of Ameri ca
Contact: Morien Tholen
Voice Phone: I 800 328 7098
FAX Phone: I 6 I 2 736 2298
Business Descri ption: Supplier of authenticating labels, including some holographic applications.
****

3M Optics Technology Center
13 31 Commerce Street
Petaluma, CA
Postal Code: 94954
Country: Un ited States of America
Contact: Dr. Willima French
Voice Phone: I 707 765 3240
Business Desc ription: HOE projects for 3M Company.
****

A. H. Prismati c, Inc
1175 Chess Dri ve, Suite D
Foster City, CA
Postal Code: 94404-1 108
Country: United States of America
Contact: Shiela Bagley
Voice Phone: I 415 345 4855
FAX Phone: I 415 345 4854
Business Description: Manufacturer and distributor of exclusive ranges of holographic gifts, toys, Jewelry, Photopolymers. Licensed products: Star Trek, Deep Space Nine. and Star Wars.
****

A.H. Pri smatic, Ltd.
New England House
New England Street
Brighton, Eng land

Postal Code: BN I 4GH
Country: United Kingdom
Contact: Ian Dayus
Voice Phone: 44 1273 686 966
FAX Phone: 44 1273 676 692.
Business Description: Manufacturers of exclusive ranges of holographic gifts, toys, jewelry, photopolymers, and film holograms. Licensed products available: Star Trek; Star Trek: The Next Generation; Star Trek: Deep Space Nine; Star Wars.
****

Academy of Media Arts Cologne
Peter-Welter Platz 2
Postal Code: 50676
Country: Germany
Contact: Prof. Dieter lung
Voice Phone: 49 221 20 I 89 I 15
FAX Phone: 49 221 20 I 89 124
Business Description: International Academy for Media Arts. Extensive holography lab and teaching.
****

ACT Systems
7002 South Revere Parkway
Suite 40
Englewood, CO
Postal Code: 80 11 2
Country: United States of America
Contact: Patrick Townsend
Voice Phone: I 303 790 7565
Voice Phone: I 800 797 7565
FAX Phone: I 303 790 8845
Business Description: US Agent for Kimmon Electric (Japan), Maker of special HeCd and HeNe lasers for holography.
****

Acme Holography
12 Sunset Road
West Somerville, MA
Postal Code: 02144
Country: United States of America
Contact: Betsy Connors
Voice Phone: 1 617 623 0578
e-mail address: bconn@media.mit.edu
Business Description: Acme Holography is Boston's first private holography lab. We offer full service in reflection, transmission and computer generated holography, including design consutation and large-scale environmental holography.
****

AD 2000, Inc.
948 State Street
New Haven, CT
Postal Code: 06511
Country: United States of America
Contact: Peter Scheir
Voice Phone: 1 203 624 6405
Voice Phone: 1 800 334 4633
FAX Phone: 1 203 624 1780
e-mail address:pscheir@ad2000.com
Business Description: Fully custom & customized stock image embossed and photopolymer holograms. Our HoloBank conta ins the world's largest selection of stock image embossed holograms, ava ilable plain, as labels, foil, magnets, pins, roll stock, etc. LOW RUN SECURITY OUR SPECIALITY'
****

Adlas G.M.B.H. & Co Kg.
Seeland Strasse 9
Postal Code: 23569
Country: Germany
Voice Phone: 49 451 390 9300

FAX Phone: 49 451 390 9399
Business Description: Established 1986. Manufact urer of diode laser-pumped solid state lasers which opera te in CW and pulsed mode with wavelengths in IR, visible and Uv. Branch office: 636 Great Road, Stow, MA 0 1775, USA. Owned by Coherent, Santa Clara, CA, USA
****

Advance Photonics
A-147 Ghatkopar Industrial Est
Ghatkopar
Postal Code: 400 086
Country: India
Voice Phone: 91 22 582 204
FAX Phone: 9 1 22 202 4202
Business Description: Branch office of Newport Corporation .
****

Advanced Optics, Inc .
5 East Old Shakopee Road
Minneapol is, MN
Postal Code: 55420
Country: United States of America
Contact: Sharilyn Loushin
Voice Phone: 1 6 12 888 0868
Business Description: Manufactures of custom and precision optics, first surface minors for use in holographic equipment.
****

Advanced Technology Program
Bldg 101--Room A430
Gaithersburg, MD
Postal Code: 20899
Country: United States of America
Voice Phone: 1 800 287 3863
FAX Phone: 1 303 926 9525
Business Description: United States Govern ment program providing funding for R&D proj- ects.
****

Aerospatiale
Ets D'Aquitaine
Saint -Medard- En-J al les
Postal Code: F-33165
Country: France
Contact: C. Lefloc'H
Voice Phone: 33 56 57 34 80
FAX Phone: 33 56 57 30 70
Business Description: Scientific and industrial research, NDT testing
****

Aerotech Inc.
Electro Optical Division
101 Zeta Drive
Pittsburgh, PA
Postal Code: 15238
Country: United States of America
Contact: Steve A. Botos
Voice Phone: 1 412 963 7470
FAX Phone: 1 412 963 7459
Business Description: Manufacturers of helium neon tubes, power supp lies and complete systems for OEM and end users. Other product lines include optical table positioners and precision rotary and linear positioning systems. Subsidiary Companies: Aerotech Lt- dEngland, Aerotech GMBH-Germany, Aero- tech Australia
****

Ag Electro-Optics Ltd.
Tarporley Business Centre
Tarporley, England
Postal Code: CW6 9UY

Country: United Kingdom
Contact: Dr. J.A. Gibson
Voice Phone: 44 1829 733 305
FAX Phone: 44 1829 733 679
Business Description: Distributor of lasers, op- tics, lab equipment.
****

Agfa Div of Bayer Corp.
US Headquarters
100 Challenger Road
Ridgefield Park, NJ
Postal Code: 07660
Country: United States of America
Contact: Mark Redzikowski
Voice Phone: I 201 440 2500
FAX Phone: I 201 440 5733
Business Description: Manufactures of film , plates and recording materials
****

Agfa Gevaert
Septestraat 27
Postal Code: 2640
Country: Belgium
Voice Phone: 32 3 4448242
Business Description: Supplier of holographic silver halide materials.
****

Agfa Ltd
27 Great West Road
Brentford, England
Postal Code: TW8 9AX
Country: United Kingdom
Business Description: Manufactures of film, plates and recording materials
****

Agfa N.V.
Holography Film Dept
Septestraat 27
Postal Code: B 2640
Country: Belgium
Contact: Mr. Frank Morthier
Voice Phone: 32 3 444 8251
FAX Phone: 32 3 444 8243
Business Description: manufacturers of film, plates and recording materials
****

AKS Holographie-Galerie GmbH
Potsdamer Strasse 10
Postal Code: 45145
Country: Germany
Contact: Gudrun Sot!
Voice Phone: 49 20 I 756455
FAX Phone: 49 201 753582
Business Description: Embossed hologram man- ufacturer.
****

Amazing World Of Holograms.
Corrigan's Arcade Foreshore Road
South Bay
Scarborough, England
Postal Code: YOII I PB
Country: United Kingdom
Contact: Carl Racey
Voice Phone: 44 1723 500 696
FAX Phone: 44 1482 492 286
Business Description: Exhibitors and retail- ers of film , glass, embossed, dichromate and related products. Permanent display of 200 holograms updated and changed regularly. Main season May-October. Distributors of film & glass.
****

American Bank Note Holographics
51 West 52nd Street
New York, NY
Postal Code: 10019
Country: United States of America
Contact: Gilbert Colgate
Voice Phone: 1 2 12 582 9200
FAX Phone: 1 212 582 9331
Business Description: World leader in develop-
ment of embossed holography for security and
commercial applications. Produces embossed
holograms in a range of formats: foil. pressures-
sens itive & tamper-evident labels, clear &
wide-web laminates for packaging.
****

American Holographic In c.
60 I River Street
Fitchburg, MA
Postal Code: 01420
Country: United States of America
Contact: Orick Kelley
Voice Phone: I 508 343 0096
FAX Phone: I 508 348 1864
Business Description: Design, develop &
manufacture optical components & instruments for use in industrial & medical measurements. Using holographic diffraction
grating design & manufacture capability to
produce components for unique measurement
instruments.
****

American Laser Corporation
1832 South 3850 West
Salt Lake City, UT
Postal Code: 84104
Country: United States of America
Contact: Dan Hoefer
Voice Phone: I 80 1 9721311
FAX Phone: I 801 972 5251
Business Description: Established 1970. Manu-
facturer Argon, Krypton and Mixed gas laser
systems from 3 mw to 10W in air or water con-
figuration.
****

American Propylaea Corporation
555 South Woodward
Suite 1109
Birmingham , MI
Postal Code: 48009-6626
Country: United States of America
Contact: Hans Bjelkhagen
Voice Phone: I 810 642 7000
FAX Phone: I 810 642 9886
e-mail address:dddimage@aol.com
Business Description: Color holograms, holo-
graphic optical elements (HOE) autostereoscop-
ic real-time HOE based display device for CAD,
Medicine, education and entertainment.
****

Amity Photonics Co.
26 Gibbs Road
Amity Harbor, NY
Postal Code: 1170 I
Country: United States of America
Contact: Paul Westphal
Voice Phone: I 516789 1099
FAX Phone: I 5162268701
Business Description: Manufacturers and dis-
tributors of optical len ses, prisms, filters,
reticles, mirror and optical sub-assemblies for
all purposes. Scientific, medical and industrial.
Consulting services ava ilable. Established
1985 3 Employees at this address
****

Ana MacArthur
P.O. Box 15234
Santa Fe, NM
Postal Code: 87506
Country: United States of America
Contact: Anna MacArthur
Voice Phone: 1 505 986 9220
FAX Phone: 1 438 8224
Business Description: Holographic instilla-
tion artist - unique in sta llations. Also produce
limited edition dichromate holograms and holo-
graphic sculptures.
****

Another Dimension
637 NW 12th ave
Deerfield Beach, FL
Postal Code: 33442
Country: United States of America
Contact: Mark Anoff
Voice Phone: I 305 429 1017
FAX Phone: I 305 421 2391
Business Description: Distributes holograms of
all varieties.
****

Ap Holografika Studio
(Ga lvanart Bt)
PO Box 113
Postal Code: H-1701
Country: Hungary
Contact: Tibor Balogh
Voice Phone: 36 1 282 4921
FAX Phone: 36 1 282 4921
Business Description: The AP Holographic
Studio provides mastering, whole process for
custom embossed holograms (mainly in secu-
rity applications), optical design using HOEs.
Wholesaler of holographic novelties and diffrac-
tion foils.
****

APA Optics Inc.
2950 Northeast 84th Lane
Blaine, MN
Postal Code: 55449
Country: United States of America
Contact: Anil Jain
Voice Phone: I 612 784 4995
FAX Phone: I 612 784 2038
Business Description: Design and manufacture
of custom computer generated holograms and
binary optical elements. Manufacture IAT, an in-
terferometer using CGH's to test aspherie optics.
****

Appl ied Holographics, PI c.
40 Phoenix Road
Crowther, Washington , England
Postal Code: WE38 OAD
Country: United Kingdom
Contact: David Tidmarsh
Voice Phone: 44 191 417 5434
FAX Phone: 44 191 4163292
e-mail address: 100407 .1630@compuserve
Business Description: Holography, stereo
grams, narrow and wide web embossing, hot-
stamp foil, pressure sensitive, polyester, OPP,
acetate, large format holograms, compact discs,
paper.
****

Applied Physics Research
1050 Northfi eld Court
Suite 300
Roswell, GA
Postal Code : 30076
Country: United States of America
Contact: Betty Woodman

Voice Phone : 1 404 751 0704
FAX Phone: 1 404 751 0806
Business Description: Manufacturer of high
quality, wide web, holographically embossed
film for packaging applications.
****

Arbeitskreis Holografie B. V.
Boeckelter Weg 47
Postal Code: 47608
Country: Germany
Contact: Herman-Josef Bianchi
Voice Phone : 49 2831 3034
Business Description: Artistic holography
****

Archeozo ic Incorporated
777 Gravel Hill Road
South hampton, PA
Postal Code: 18966
Country: United States of America
Contact: Paul E. Yannuzzi
Voice Phone: 1 215 322 4915
FAX Phone: 1 215 862 0528
Business Description: From product de-
velopment to custom work. We present the
most start ling images of earth's past, pres-
ent and future through a unique fusi on of
model making and the highest quality ho-
lography.
****

Armin Klix Holographie
A-Z Schnelldruck
Postfach 260218
Postal Code: 40095
Country : Germany
Contact: Armin Klix
Voice Phone: 49 211 3 17 775
FAX Phone: 49 211 3 17 749
Business Description: Anfertigung von Display-
hologranunen im Auf trag von Werbung und
Tndu strie und Einzel--und Grosshandel. Kata-
log von Li efe rb aren Hologrammen vorhanden.
Einzelstucke und Grosserien.
****

Art Institute Of Chicago
Holography Department
112 South Michigan Ave
Chicago, IL
Postal Code: 60603
Country: United States of America
Contact: Ed Wesley
Voice Phone: 1 312 345 3998
Business Description: The SATC offers an
MFA degree with concentration in holography,
and is equipped with three tabl es, wi th one
containing a stereogram printer for computer
generated imagery or life. 1 instructor, 3 grad
assistants.
****

Art Lab
1000 Ri chmond Terrace
Staten Island, NY
Posta l Code: 1030 I
Country: United States of America
Voice Phone: I 718 447 8667
Business Description: The Art Lab is an an
school which offers classes and workshops in
the art of holography. Free brochure available.
****

Art, Science & Technology Inst
Holography Coll ection
2018 R Street, NW
Washington, DC
Postal Code: 20009
Country: United States of America

Contact: Odile Meulien-Ohlmann
Voice Phone: 1 202 667 6322
FAX Phone: 1 202 265 8563
Business Description: Research & educational organization for advancement of the art, science and technology of holography. Research focus: 3-D imagery in art and portrait. Permanent Art-work: Holo graphy collection. Training sessions.
****

Artbridge Light Studies
Mergesstr 16
Postal Code: 38108
Country: Germany
Contact: Odi le Meulien
Voice Phone: 49 531 352 816
FAX Phone: 49 531352816
Business Description: Agency for design & pro-duction of holographic products, services, & commercial art. Custom-made holograms; ap-plications in architecture; in-house innovative products. Exhibit & educational services includ-ing lectures and training. Wholesale/ distribu-tion Private collection.
****

Artigliography Co.
7130 Mohawk West Drive
Indianapolis, IN
Postal Code: 46236
Country: United States of America
Contact: Kerry 1. Brown
Voice Phone: I 3 17 823 0069
Business Description: Distribute/display for sale al1work by: Ed Wesly, James MacShane, Dick Dellsterberg, T. All en Black, Kerry J. Brown, Mike Crawford. Custom designer of holographic displays for all uses. Estab li shed 1987. Gallery in Indianapolis.
****

Asahi Glass Co.
R&D General Divis ion,
2-1-2 Marunouch i Chiyoda-Ku
Posta l Code: 100
Country: Japan
Contact: Fumihiko Koizumi
Voice Phone: 81 3 3218 5825
FAX Phone: 81 3 32 14 5060
Business Description: R&D on holography and single copy holograms.
****

Association fo r Hologram Techniques (AHT)
N iederesch 28
Postal Code: D-49214
Country: Germany
Contact: Gunter Deutschmann
Vo ice Phone: 49 5424 5365
FAX Phone: 49 5424 5359
Business Description: Association of German hologram manufacturers.
****

Astor Universal - AHL Holographics Divis ion
14400 West 96th Terrace
Lenexa, KS
Postal Code: 662 15
Country: United States of America
Contact: John Thomas
Voice Phone: I 800 255 4605
FAX Phone: I 9 13 338 2037
Business Description: Worldwide designers, originators and manufacturers of holographic pattern and image foils and films. See Astor Universal advertisement for further details. Additi onal offices in North Carolina and the UK.

****
Astor Universal - AHL Holographics Division
Loughborough Universi ty 040
Ashby Rd
Loughborough, England
Postal Code: LE II 3TU
Country: United Kingdom
Contact: Dr. Laurence Holden
Voice Phone: 44016 1 789 8 131
FAX Phone: 44 1509 232 772
Business Description: Worldwide design-ers, originators and manufacturers of holographic pattern and image foils and films. See Astor Universal advertisement for further details. Additional offices in USA.
****

Astor Universal - AHL Holographics Division
10440 Ontiveros Place, Unit
Santa Fe Springs, CA
Postal Code: 90670
Country: United States of Ameri ca
Contact: Norman McGahan
Voice Phone: 1 800 255 4605
FAX Phone: 1 310 946 5743
Business Description: Worldwide designers, originators and manufacturers of holographic pattern and image foils and films. See Astor Universal advertisement for further details. Additional offices in North Carolina and the UK.
****

Atelier Holographique De Paris
13, Passage Courtois
Postal Code: F-75011
Country: France
Contact: Pascal Gauchet
Voice Phone: 33 1 43 79 69 18
FAX Phone: 33 1 40 09 05 20
Business Description: Artistic holography; Buying & Selling; Consulting
****

Austral ian Holographics Pty Ltd.
PO Box 160
Kangarilla, South Austral ia
Postal Code: 5157
Country: Australia
Contact: David Ratcliffe
Voice Phone: 6 1 08 383 7255
FAX Phone: 61 08 383 7244
e-mail address:austholo@camtech.com.au
Business Description: Manufacturers of white light transmission holograms up to 2.2 x 1.1 metres and reflection ho-lograms up to 1 x 1 metres. Pulse and CW mastering. Thematic exhibitions of holograms are a specialty, also pulse la-ser sales.
****

Avant-Garde Studio
34 North Roachdale Ave.
Roosevelt, NJ
Postal Code: 08555-0296
Country: United States of America
Voice Phone: 1 609 448 6433
FAX Phone: 1 609 448 6433
Business Description: Modelmakers. Sculptors, excelling in relief (forced perspective, texture, form), for holographic models. We will design and/or work from, photographs, drawings or other material.
****

Baldar
PO Box 4340

Berkeley, CA
Postal Code: 94704
Country: United States of America
Contact: Franz Ross
Voice Phone: 1 510841 2474
Voice Phone: 1 800 367 0930
FAX Phone: 1 5108412695.
e-ma il add ress:franz@baldar.com
Business Desc ripti on: Publi sher of Holography Marketplace and Holography Handbook
****

Barr & Stroud, Ltd.
I Linthouse Rd
Glasgow, Scotland
Postal Code: G51 4BZ
Country: United Kingdom
Contact: George Brown
Voice Phone: 44 141 440 4000
FAX Phone: 44 141 4404001
Business Description: Artistic holography DCG
****

Batelle Paci fic Northwest
PO Box 999
(K7-02 mailbox)
Richland, WA
Postal Code: 99352
Country: Uni ted States of America
Contact: Michael Lind
Voice Phone: I 509 375 4405
FAX Phone: I 509 375 6499
Business Description: Research laboratory de-veloping holography technology, among others.
****

Bbt ln strumenter Aps.
Dronning Olgasvej 6
Postal Code: DK-2000
Country: Denmark
Vo ice Phone: 45 0 I 198 208
FAX Phone: 45 0 I 198 747
Business Description: Branch office of Newport Corporation.
****

Beddis Ken ley (Machinery) Ltd.
Unit 3
Eiland Terrace
Holbeck, Leeds, England
Postal Code: LS I I 9NW
Country: United Kingdom
Contact: S.D. Smith
Voice Phone: 44 11 32 465 979
FAX Phone: 44 11 32 425 400
Business Description: Large sheet hot foil stamping machines for grap hic enhancement and hologram application. Maximum sheet size 29 x 41 (73cm x 104cm)
****

Beijing Institute Of Posts
And Telecommunications
Applied Phys ics (Ho lograp hy)
Posta l Code: 10080
Country: China
Contact: Hsu Da-Hsiung
Voice Phone: 86 1 668 1255.
Business Description: College courses in holog-raphy.
****

Beijing Normal Univers ity.
Analysis And Testing Centre
Postal Code: 100875
Country: China
Contact: Huang Wanyun
Business Description: Industrial and Scientific research , Non-Destructive testing

```
*********************************       London, England                              London, England
Berkhout, Rudie                        Postal Code: SE 19 I JT                       Postal Code: SEI 6BU
223 West 21 St St                      Country: United Kingdom                       Country: United Kingdom
New York, NY                           Contact: Patrick Boyd                         Contact: Dr. Eugene Sweeney
Postal Code: 101 II                    Voice Phone: 44 18 1 670 4160                 Voice Phone: 44 171 403 6666
Country: United States of America       FAX Phone: 44 1952250001                      FAX Phone: 44 171 403 7586
Contact: Rudie Berkhout                e-mail address:patrick@geijubu.tjokuba.ac.jp  Business Description: Technology transfer orga-
Voice Phone: 1 212 255 7569            Business Description: Artist using holography nization that offers holographic laminating film
Business Description: Holographic Fine Artist  and other 3D media.                   among its products.
who has had work exhibited at Whitney Mu-  *********************************         *********************************
seum of American Art (New York) among other  Brainet Corporation                    Burleigh Instruments, Inc.
places.                                4F Asset Bldg                                 Burleigh Park
*********************************       3-31-5 Honkomagome Gunkyo-Ku                  Fishers, NY
BIAS                                   Postal Code: 1 13                             Postal Code: 14453
(Bremer Institute Applied Beam)        Country: Japan                                Country: United States of America
Klagenfurter Str 2                     Contact: Yutaka Inoue                         Contact: Patty Payne
Posta l Code: 28359                    Voice Phone: 81 3 5395 7030                   Voice Phone: 1 716 924 9355
Country: Germany                       FAX Phone: 8 1 3 5395 7029                    FAX Phone: 1 716 924 9072
Contact: Prof. Werner J uptner         Business Description: Distributor and producer  Business Description: Burleigh Instruments, Inc.
Voice Phone: 49 421 218 5002           of all types of film and glass holograms and oth-  is a leading manufacturer of Wavemeters and
FAX Phone: 49 0421 218 5059            er holographic gifts and stationery goods.    Fabry Perot Interferometers and Etalon Systems
Business Description: Industrial research; holo-  *********************************   for la ser diagnostics; Piezo electric based mi-
graphic non-destructive testing.       Brandtjen & Kluge, Inc,                       cropositioning equipment; Color Center Lasers.
*********************************       539 Blanding Woods Road                       *********************************
Blue Bond Sales Promoting In c.        St Croix Fall s, WI                           Burns Holographics Ltd
No 20, Alley 9, Lane 133               Postal Code: 54024                            P.O. Box 377
Nanking East Road, Sec 4               Country: United States of America             Locust Valley, NY
Postal Code: 10569                     Contact: Hank A. Brandtjen                    Postal Code: 11560
Country: Taiwan                        Voice Phone: I 715 483 3265                   Country: Un ited States of America
Voice Phone: 886 2 7122580             FAX Phone: I 715483 1640                      Contact: Joseph Burns
Voice Phone: 886 2 718 881 I           Business Description: Manufacturer of Hot     Vo ice Phone: I 516 674 3 130
FAX Phone: 886 2 7153866               Stamp Machinery. Establ ished in 1919. 80 Em-  FAX Phone: I 5166743 130
Business Description: Hologram phone cards,  ployees at this address.                e-mail address: hologram@lihti.org
stationary items, etc.                 *********************************             Business Description: Since 1972, Holograms/
*********************************       Bridgestone Graphic Technologies, Inc.        Stereograms/Edition s; Si lver halide, Photore-
Bob Mader Photography                  375 Howard Ave.                               sist, Nickel , Embossed with Agam, Dali , Cos-
500 Crescent Court # 160               Bridgeport, CT                                sette, Nunez, Dieter Jung, Sam Moree, Others;
Dallas, TX                             Posta I Code: 06605                           1979 - Injection-Molded Holograms; 1987 - De-
Postal Code: 75201                     Country: United States of America             sign/Development NY % Hologram CreditCard.
Country: United States of America      Contact: Richard Zucker                       *********************************
Voice Phone: 1 21487 1 5511            Voice Phone: I 203 366 1595                   Cambridge Laser Labs
Business Description: Artistic holography;  FAX Phone: I 203 366 1667                853 Brown Road
pulsed laser portraits; marketing consultant.  Bus iness Description: Fully integrated inhouse  Fremont, CA
*********************************       manufacturer of holographic materials and prod-  Postal Code: 94539
Bobst Group                            ucts. Design, mastering, electroforming, em-  Country: United States of America
146-T Harrison Avenue                  bossing and all converting. Security and product  Voice Phone: I 510 65 1 0 II 0
Rose land, NJ                          authentication programs.                      FAX Phone: I 510651 1690
Posta l Code: 07068                    *********************************             Business Description: We do a very large busi-
Country: United States of America      Brighton Imagecraft                           ness in laser repair. Also sell used lasers Price
Contact: Bill Seymour                  7 Bath Street                                 list on request. CW lasers
Voice Phone: 1 20 I 226 8000           Brighton, England                            *********************************
FAX Phone: I 20 I 226 8625             Postal Code: BNl 3TB                          Canon Inc. R&D Headquarters
Business Description: One of the world's largest  Country: United Kingdom            890, Kawasaki-Shi
manufacturers of hologram hot stamp machin-  Contact: Jeff Blyth                     Saiwai-Ku Kawasaki
ery. See ad in this issue.             Voice Phone: 44 1273 202 069                  Postal Code: 2 11
*********************************       Business Desc ription: Specialist in producing  Country: Japan
Booth, Roberta                         holographic recording material: red-sensitive  Contact: Tetsuro Kuwayama
Holographic Arti st                    DCG and photopolymer; photochromic plate for  Voice Phone: 81 44 549 5424
5326 Sunset Bl vd                      HeNe lasers.                                  Business Desc ription: Research. Courses in ho-
Los Angeles, CA                        *********************************             lography ..
Postal Code: 90027                     British Aerospace Pic.                        *********************************
Country: United States of America      Sowerby Research Centre                       Carl M. Rodia And Associates
Contact: Roberta Booth                 Fpc: 267 PO Box 5                             13 Locust St
Voice Phone: 1 213 466 5767            Filton, England                              Trumbull , CT
FAX Phone: 1 213 465 5767              Postal Code: BS 12 7QW                        Postal Code: 0661 I
Business Description: I am a holographic art-  Country: United Kingdom               Country: United States of America
ist working in transmission and reflection  Contact: Dr. Steve Parker                Contact: Carl M. Rodia
holograms. Also work as a consultant for  Voice Phone: 44 1179 366 842              Voice Phone: I 203 261 1365
holographic projects and curate holography  FAX Phone: 44 1179 363 733              FAX Phone: I 203 268 161 9
shows.                                 Business Description: R&D in HOE and NDT      Business Description: Comprehens ive engi
*********************************       holography.                                   neering consultation services in preci sion holo-
                                       *********************************             gram manufacturing. Plant design and engineer-
Boyd, Patrick                          British Technology Group (BTG)                ing, process engineering. troubleshooting.
18 Whiteley Rd                         101 Newington Causeway
```

ing and seminar training of manufacturing personnel.
****

Casdin-Silver Holography
99 Pond Avenue Suite D403
Brookline, MA
Posta l Code: 02146
Country: United States of America
Contact: Harriet Casdin-Silver
Voice Phone: 1 617 739 6869
FAX Phone: 1 6 17 739 6869
Business Description: I have been creating holographic art and interactive holographic installations since 1968. Our company specializes in original holograms for advertising, architectural and theater settings, expositions. We are consultants, exhibition organizers/designers.
****

Catalyst Strategic Design, Inc.
14535 Nw 60 Ave
Miami Lakes, FL
Postal Code: 33014
Country: United States of America
Contact: Guillermo Martinez
Voice Phone: I 305 557 5622
FAX Phone: 1 305 557 1109
Business Description: Design studio specializing in packaging holography into new products. Have done Disney, Mattei, Wamer Bros., etc.
****

Cavomit
22 Pipinou Steet
Postal Code: 11257
Country: Greece
Contact: Alkis Lembessus
Voice Phone: 30 I 823 2355
FAX Phone: 30 I 231 4499
Business Description: Hot-stamping equipment (cylinders - platen), hologram registration systems, foils and consumables. Local distributor for Astor Universal, Kluge, Light Impress ions, Applied Holographies, Revere Graphic Products.
****

Central Glass Co. , Ltd.
Kowa-Hitosubashi Bldg
7-IKanda-Nishikicho 3-Chom
Postal Code: 101
Country: Japan
Contact: Chikara Hashimoto
Voice Phone: 81 3 3259 7354
Business Description: Research--Heads Up Display
****

Centre D'Art
Holographique & Photoniq ue
5294 Avenue De L'Esplanade
Montreal, Quebec
Postal Code: H2T 2Z5
Country: Canada
Contact: Philippe Boissonnet
Voice Phone: I 514270 1840
Business Description: Artistic services. Non-profit organization. Exhibitions, catalogues, work shops, education .
****

Cfc/Applied Holographics
500 State St
Chicago Heights, IL
Postal Code: 60411
Country: United States of America
Contact: Dave Beeching

Voice Phone: 1 800 438 4656
Voice Phone: 1 708 89 1 3456
FAX Phone: 1 708 758 5989
Business Description: For security, decoration & packaging: Full range of tamper evident & authentication labels & hot stamping foils. Release and size coat combinations. Print treatments, & custom metallizing available.
****

Checkpoint
Chatham Street
Reading, England
Postal Code: RGI 7JX
Country: United Kingdom
Contact: Keith Hatton
Voice Phone: 44 1734 258 251
FAX Phone: 44 1734 569 988
Business Description: Manufacturers "Holoseal Security Hologram Applicator" used to apply hologrm seals to checks for security.
****

Cherry Opti cal Holograp hy
2047 Blucher Va lley Road
Sebastopol, CA
Postal Code: 95472
Country: United States of America
Contact: Greg Cherry
Voice Phone: 1 707 823 7171
FAX Phone: 1 707 823 8073
Business Description: Highest quality display holography available. Stock and custom reflection/transmission holograms on glass plates or film up to 42" x 72" in size. Limited edition fine art holograms.
****

Chiba Unive rsity
Faculty Of Engineering
1-33 Yayoi-Cho
Postal Code: 260
Country: Japan
Contact: Jumpei Tsujiuchi .
Voice Phone: 81 472 511 111
Business Description: Scientific research; holographic
****

CHIRON Technolas GmbH
Max-Planck Strasse 6
Postal Code: 85609
Country: Germany
Contact: Mr. Junger
Voice Phone: 49 89 9455 14 0
FAX Phone: 49 89 945514 70
Business Description: Ophthalmologic Systems.
****

Chromagem Inc.
573 South Schenley
Youngstown, OH
Postal Code: 44509
Country: United States of America
Contact: Thomas J. Cvetko vich
Vo ice Phone: I 2 16 793 35 15
FAX Phone: 1 2 1679335 15
Business Description: Specializing in photoresist masters for mass-production : one-of-a-kind display pieces; consulting and photoresist lab set-ups. Established in 1981. 4 Employees at this address. Dot matrix stereo grams
****

Cise Spa Technologie Innovat ive
P.O. Box 12081
Postal Code: 1-20134
Country: Italy
Contact: M. Luciana Ri zzi

Voice Phone: 39 2 2167 2634
FAX Phone: 39 2 2 167 2620
Business Description: Various R&D on HOEs and NDT holography.
****

City Chemical
132 West 22nd Street, Dept. H
New York, NY
Postal Code: 100 II
Country: United States of America
Voice Phone: 1 201 653 6900
FAX Phone: 1 212 463 9679
Business Description: Photochemicals and emulsions
****

Class ic City Holography Studio
188 Kevin Way
Hull , GA
Postal Code: 30646
Country: United States of America
Voice Phone: I 706 543 1423
FAX Phone: I 706 543 1423
Business Description: Georgia's only private studio. Holographic art, limited editions, business logo's and education. Studio rental available.
****

Coburn Corporation
1650 Corporate Road West
Lakewood, NJ
Postal Code: 0870 I
Country: United States of America
Contact: John White
Voice Phone: I 908 367 55 11
FAX Phone: I 908 367 2908
Business Description: Embossing & shim making; training programs.
****

Coherent, In c.
Laser Group
5100 Patrick Henry Drive
Santa Clara, CA
Postal Code: 95054
Country: Uni ted States of America
Contact: Sandra Todd
Voice Phone: I 408 764 4983
Voice Phone: I 800 527 3786
FAX Phone: I 408 764 4800
Business Description: Coherent provides lasers to the industry. Call for catalogue .
****

Continental Optica l
15 Power Drive
Hauppauge, NY
Postal Code: 11 788
Country: United States of America
Voice Phone: I 5 16 582 3388
Business Description: Optics and custom orders.
****

Control Module Inc
380 Enfi eld Street
Enfi eld, CT
Postal Code: 06082
Country: United States of America
Contact: Ralph Billeri
Voice Phone: 203 745 2433
FAX Phone: 203 741 6064
Bus iness Description: Embossed holograms for security purposes.
****

Control Optics
13 111 Brooks Dri ve, Unit J

Baldwin Park, CA
Postal Code: 9 1706
Country: United States of America
Voice Phone: 1 818 8 13 1990
FAX Phone: 1 8 188 13 1993
Business Description: Maker of laser optics and accessories.
****

Corion Corp.
73 Jeffrey Ave
Holli ston, MA
Postal Code: 01746
Country: United States of America
Contact: Don McLeod
Voice Phone: 1 508 429 5065
FAX Phone: I 508 429 8983
Business Description: Corion Corp. manufactures volume and one-of-a-kind, custom and stock, optical components including coatings, filters, optics and optical assemblies for use in the UV-Visible-IR spectrum.
****

Corry Laser Technology Inc
414 W Main Street
Po Box 18
Corry, PA
Postal Code: 16407- 1728
Country: United States of America
Voice Phone: 1 8 14 664 72 12
FAX Phone: 1 8 14 664 3689
****

Cossette, Marie Andree
1145 Avenue Des Laurentides, Apt. 2
Quebec City, Quebec
Postal Code: GIS 3C2
Country: Canada
Voice Phone: I 41 8 687 2985
Business Description: Holographic fine artist, one-offs & limited editions. Holography exhibitions. Private gallery. Holography education, tutoring, studio rental, consultant
****

Coulter Optical Company
PO Box K
54 140 Pinecrest Road
Idyllwild, CA
Pos tal Code: 92349
Country: United States of America
Contact: Mary Braginton
Voice Phone: 1 909 659 4621
Business Description: Make telescope mirrors, parabolic mirrors and more. Send for free list of products.
****

Creative Holography Index , The
Intern at iona l Catalog IHolograms
Postfach 200210
Postal Code: 51432
Country: Germany
Contact: Andrew Pepper
FAX Phone: 49 2202 30497
Business Description: The Creative Holography Index is an internatio na l catalogue, in colour, published four times per year. Available by subscription. It fea tures art ists working with holography as a creative medium and in cludes critical essays. Cost US$55.001 25 pounds Sterling/65 German OM.
****

Creative Label
2450 Estes Drive
Elk Grove Vi llage, IL
Posta l Code: 60007
Country: United States of America

Contact: Jerry Kori I
Voice Phone: I 708 956 6960
FAX Phone: I 708 956 8755
Business Description: Bindery application of holograms on Kluge (2 stream) and Bobst (4 stream) machines. Call for more in fo rmation.
****

Crown Roll Leaf, Inc.,
91 Illinois Ave
Paterson, NJ
Postal Code: 07503
Country: United States of America
Contact: George Waitts
Voice Phone: 1 20 I 742 4000
FAX Phone: 1 201 7420219
Busi ness Descript ion: Crown Roll Leaf has been supplying embossing material internationally for 5 years. We have been manufacturing shim s and embossed products for 4 years. Please ca ll Jim Waitts with any questions.
****

CSI
7 Meadowfield Park South
Stocksfield, England
Postal Code: NE43 7QA
Country: United Kingdom
Voice Phone: 44 1661 842 741
fAX Phone: 44 1661 842 288
Business Description: Manufacture mirrors; optics.
****

CVI Laser Corporation
200 Dorado Place
Albuquerque, NM
Postal Code: 87192
Country: United States of America
Contact: Bob Soales
Voice Phone: 1 505 296 9541
fAX Phone: 1 505 298 9908
Business Description: Manufactures ho lographic quality single and multi element lenses, mirrors, wi ndows, and beamsplitters for all standard holographic laser sources. Free 104-page catalog available.
****

Czechoslovak Academy Of Science
In stitute Of Physics
Na Slovance 2
Postal Code: 180 40
Country: Czech
Contact: Josef Horvath
Voice Phone: 42 84 22 419
Business Description: Holography research
****

Dai Nippon Printing Co., Ltd.
Central Research Institute
250-1 Aza-Kahasawa
Kashi wa-City, Chiba
Postal Code: 277
Country: Japan
Contact: Takashi Wada
Voice Phone: 8 1 4 7134 05 12
FAX Phone: 8 1 4 7133 2540
Business Description: Central reserach center. Embossed holography research.
****

Daimler Benz Aerospace
Dornier Medizintechnik GmbH
Industriestrasse 15
Postal Code: 82 110
Country: Germany
Contact: Ms. Thiemon
Voice Phone: 49 89 84108 0
FAX Phone: 49 89 84108 575

Business Description : Industrial Research: holographic non-destructive testing. HOE research.
****

Dan Han Optics
188-261 An Nyeong-Ri Tean-Eup
Country: South Korea
Contact: Chung Song
Voice Phone: 82 033 1 35 1 030
fAX Phone: 82 0331 351 031
Business Description: General optical supplies.
****

Datacard Corporation
I 1111 Bren Road West
Minneapolis, M
Postal Code: 55440
Country: United States of America
Contact: Mark Iverson
Voice Phone: I 6 12 933 1223
FAX Phone: I 612933 797 1
Business Description: Did PA Driver's License Holographic Overlay.
****

Datasights Ltd.
Alma Road
Ponders End
Enfield, England
Postal Code: EN3 7BB
Country: United Kingdom
Contact: frank Sharpe
Voice Phone: 44 181 805 41 5 1
fAX Phone: 44 181 805 8084
Business Description: Manufacture mirrors for use in holography.
****

Dazzle Equipment Company
10305 Hull Street Road Unit C
Midlothian, VA
Postal Code: 23112
Country: United States of America
Voice Phone: 1 804 674 9740
fAX Phone: 1 804 674 9717
Business Description: Manufactures high-precision holographic embossing machinery; silvering, electroforming, hot-stamping, laminating, die-cutting equipment. Embossing, laminating and plating work for customers.
****

Db Electronic In struments S.R.L.
Via Teano 2
Postal Code: [-201 61
Country: Italy
Voice Phone: 39 02 646 934
FAX Phone: 39 02 645 6632.
Business Description: Newport Co. branch office. Laser supplier.
****

De La Rue Holographics Ltd.
Stroudley road
Daneshill Industrial Estate
Basingstoke, Eng land
Postal Code: RG24 8fW
Country: United Kingdom
Contact: P.M.G. Hudson
Voice Phone: 44 1256 463 000
fAX Phone: 44 1256 460 800
Business Descri ption: Integrated manufacturer of holographic security compone nts for protection of brands and value documents. Hot stamp foil, tamper-evident labels & overlaminates from custom-designed origi nation
****

Dec-Art Inc
I 190 Lavallee St.
Prevost, Quebec
Postal Code: JOR 1 TO
Country: Canada
Voice Phone: I 514 224 8505
FAX Phone: I 5142247502
Business Description: We carry all the finest holographic products.
****

Deem, Rebecca
709 112 West Glen Oaks Blvd
Glendale, CA
Postal Code: 9 1202
Country: United States of America
Contact: Rebecca Deem
Voice Phone: 1 818 549 0534
FAX Phone: 1 818 549 0534
Business Description: Holographic artist.
****

Deep Space Holographics
1070 Moss Street # I 05
Victoria, British Colombia
Postal Code: V8V 4P3
Country: Canada
Contact: Karan Wells
Voice Phone: 1 604 384 3927
e-mail address: uj613@free net.victo ria
Business Description: Exotic fine art/commercial sculpture/animation, conceptual/industrial design, display merchandising, exhibits and special effects. Since 1980 secured worldwide distribution of our DCG designs via Holocrafts. Star Trek holograms design.
****

Dell Optics Company, Inc
25 Bergen Blvd
Fairview, NJ
Postal Code: 07022
Country: United States of America
Contact: Belle Steinfeld
Voice Phone: 1 201 94 1 101O
FAX Phone: 1 201 94 1 9524
Business Description: Custom working of precision optical components. Establi shed 1950. 15 Employees at this address.
****

Denisyuk, Yuri N.
A.F.loffe Physicotechnical In stitute
Politechnicheskaya 26
Postal Code: 194021
Country: Russ ia
Contact: Yuri N. Deni syuk
Voice Phone: 7 812 247 9384
Business Description: Holography teacher. One of the founders of holography.
****

Deutsche Gesellschaft fur Holografie
Geschaftsste lle
Lutterdamm 82
Postal Code: 49565
Country: Germany
Contact: Vito Orazem
Vo ice Phone: 49 5461 9 1124
FAX Phone: 49 5461 9 1122
Business Description: The society was founded to promote awareness of holography, and its members are mai nly holographers and artists. To this end, the group intends to organize exhibitions. Interferenzen is a periodical published by thi s organization.
****

Dialectica Ab
Skanegatan 87
6I r
Postal Code: S-11 6 37
Country: Sweden
Contact: Ambjorn Naeve
Business Description: Artistic holography.
****

Diaures S.A. Holography Division
Via I Maggio 262/A
1-41019 Soliera
Country: Italy
Voice Phone: 39 059 567 274.
Business Descripti on: Artistic holography; embossed holography; equipment & supplies.
****

Diavy srl
Via Viva ldi 108
Pos tal Code: 41019
Country: Italy
Contact: Alesandro Dondi
Voice Phone: 39 59 565758
FAX Phone: 39 59 566074
Business Description: Subsidiary of Diaures; producer of holographic metallic paper as part of a venture with Scharr Industries, USA.
****

Die Dritte Dimension.
Frankfurter Strasse 132-134
Pos tal Code: 63263
Country: Germany
Contact: Elke Hein
Voice Phone: 49 6102 33367
FAX Phone: 49 6102 326709
Business Description: Greatest specialized shop for holography in Germany. Always over 1,000 different ho lograms in stock. Very comprehens ive fine art section. Branch office: Nordwest-Zentrum, Tituscorso, 60439 Frankfurt/M. Germany.
****

Dietmar Ohlmann
Bortfelder Stieg 4
Postal Code: 38 116
Country: Germany
Voice Phone: 49 531 352 8 16
FAX Phone: 49 531352816
Business Description: Holographic fine artist. Unique pieces, limited editions and commissions. Courses and publications.
****

Diffraction Company
38 Loveton Circle
Sparks, MD
Postal Code: 21152
Country: United States of America
Contact: Dean Hell
Voice Phone: 1 410 666 1144
FAX Phone: 1 410472 49 11
Business Descri ption: We offer 58 patterns available in 16 colors with a variety of adhesives; Color explosion graphics/micro-etching an alternative to 3D; Custom embossing of holograms; Dazzlers-Stickers.
****

Diffraction Ltd.
P.O. Box 909
Waitsfie ld, VI
Postal Code: 05673
Country: United States of America
Contact: Heather Bevans
Voice Phone: 1 802 496 6642
FAX Phone: 1 802 496 6644
Business Description: Holographic Fine Artist;

multi-color reflection holograms. Diffractive optical elements and electrically switchable holograms.
****

Dimension 3
3380 Francis-Hughes St.
Lava l, Quebec
Postal Code: H7L 5A 7
Country: Canada
Contact: Andrew Gellert
Voice Phone: 1 5 14 662 0610
FAX Phone: 1 5 14 6620047
Business Description: Designer and supplier of large-run holograms.
****

Dimensional Arts
15730 West Hardy St. , Suite 310
Houston , TX
Postal Code: 77060
Country: United States of America
Contact: Larry Pfiel
Voice Phone: I 71 3 448 0 185
FAX Phone: 1 713 448 3961
Business Description: Exclusive manufacturer of the Light Machine, a patent protected digital origination sys tem. Custom stock Dotz(r) dot matrix patterns available. Capable of 20, 3D and full color stereogram work. Can transfer technology worldwide.
****

Dimensional Cinematography Co.
PO Box 5202
Berkeley, CA
Postal Code: 94704
Country: United States of America
Contact: Glenn Gustafson
Voice Phone: I 510 84 1 3578
Voice Phone: 1 510 720 3646
Business Description: Cinema-Photographer for stereograms. Many years of experience with numerous major accounts such as Walt Disney,
****

Dimensional Foods Co.
8 Faneuil Hall Market Place
Boston, MA
Postal Code: 02109
Country: United States of America
Contact: Eric Begleiter
Voice Phone: 1 6 17 973 6465
FAX Phone: 1 617 973 6406
Business Description : Scientific and artistic research. Licensing the process to food manufacturers for producing chocolate and hard candy holograms.
****

Dimensions
Taj Pura
Country: Pakistan
Contact: Mr. Shahjahan
Vo ice Phone: 92 432 85 197
Vo ice Phone: 92 432 66006
FAX Phone: 92 432 558336
Business Description : International agents and importers of holograms, diffraction foils and other holographic products .
****

Dimuken
33 Stapledon Rd
Orton Southgate
Peterborough, England
Postal Code: PE2 6TD
Country: United Kingdom
Vo ice Phone: 44 1733 230 044
FAX Phone: 44 1733 230 0 12

Business Description: Manufactures holographic encoding machinery.
****

Direct Holographics
PO Box 295
Strasburg, PA
Postal Code: 17579
Country: United States of America
Contact: Jacque Phillips
Voice Phone: I 7 I 7 687 9422
FAX Phone: I 717 687 9423
Business Description: The exclusive distributor of film holograms for Third Dimension Ltd. Over 130 different images in stock. Also embossed sticker, magnets & keychains. Quality line of holographic ear rings.
****

Duston Holographic Services
115 Shannon Street
Ottawa, Ontario
Postal Code: KI Z 6Y6
Country: Canada
Contact: Deborah A. Duston
Voice Phone: I 613 722 9004
Business Description: Duston Holographic Services consu lts corporate and government clients on HOEs, remotely sensed Holographic Stereograms and the educational and curatorial aspects of Holography. Deborah Duston is also a well known artist-holographer.
****

Dutch Holographic Laboratory BV
Kanaaldijk Noord 61
Postal Code: 5642JA
Country: Netherlands
Contact: Walter Spierings
Voice Phone: 31 40 817 250
FAX Phone: 31 40814865
e-mail address:walter@iaehr.nl
Business Desc ription: Manufacturer of holoprinter and holotrack equipment. Production of holograms on silver halide, photoresist and photopolymer. Computer-generated holograms and multiple photo-generated holograms (MPGH). Also traditional recording techniques.
****

E.!. DuPont De Nemours & Co
Holographic Materials Division
P. O. Box 80352
Wilmington, DE
Postal Code: 19880-0352
Country: United States of America
Contact: Paula Bobeck
Voice Phone: 1 302 695 2838
Voice Phone: 1 302 695 4893
FAX Phone: 1 302 695 9631
Business Description: Manufacturer of photopolymer emulsions for sa le to holography businesses.
****

Ealing Electro-Optics (UK)
Greycaine Road
Watford, England
Postal Code: WD2 4PW
Country: Un ited Kingdom
Business Description: Supplies optics for holography labs. Branch office for Ealing Co. USA.
****

Ealing Electro-Optics Inc.
89 Doug Brown Way
Holliston, MA
Postal Code: 01746

Country: United States of America
Contact: David Clark
Voice Phone: 1 508 429 8370
FAX Phone 1 508 429 7893
Business Description: Manufacturer of mirrors, mounts & optics.
****

Eastman Kodak Company
Scientific Imaging Dept
343 State St Dept 841-S
Rochester, NY
Postal Code: 14650-081 I
Country: United States of America
Voice Phone: I 800 242 2424
FAX Phone: I 716 781 5986
Business Description: Manufacturer of holographic plates & film.
****

Ed Wesly Holography
5331 N Kenmore Ave
Chicago, IL
Posta l Code: 60648
Country: United States of America
Contact: Ed Wesly
Voice Phone: I 312 539 3672
Business Description: Holographic Fine Artist. I am an artist making candy for the eyes using Holographic Optical Elements and junk (found objects).
****

Edmund Scientific Company
101 East Gloucester Pike
Barrington, NJ
Postal Code: 08007
Country: United States of America
Voice Phone: I 609 547 3488
FAX Phone: I 609 573 6295
Business Description: Mailorder catalogue, wholesale, and retail We offer one of the largest selections of precision optics and optical components and accessories for the optical lab. Holography products for schools, science fairs, etc.
****

Electro Optical Industries, Inc.
859 Ward Drive
Santa Barbara, CA
Postal Code: 93 I II
Country: United States of America
Contact: Joseph Lansing
Voice Phone: I 805 964 6701
FAX Phone: I 805 967 8590
Business Description: Manufacturer of infrared test and calibration instrumentation including: collimators, choppers, blackbody sources, differential temperature sources, FUR lest equipment, radiometers and LLL-TV target simulators.
****

Electro Optics Developments Ltd.
Howards Chase
Pipps Hill Industrial Estate
Basildon , England
Postal Code: SS 14 3BE
Country: United Kingdom
Contact: Chris Varney
Voice Phone: 44 1268 531 344
FAX Phone: 44 1268 531 342
Business Description: Equipment & supplies; optics
****

Elusive Image
603 Munger Street, # 316
Dallas, TX

Postal Code: 75202
Country: United States of America
Contact: Argelia Lopez
Voice Phone: I 214 720 6060
FAX Phone: I 214 754 7009
Business Description: Holography gallery.
****

Embossing Technology Ltd
Steepmarsh, Nr Petersfield
Hants, England
Postal Code: GU32 2BN
Country: United Kingdom
Contact: Graham Ridot
Voice Phone: 44 1730 895 390
FAX Phone: 44 1730 894 383
Business Description: Wide web embossing by contract. Also stock images. Also for sale is complete system for originating embossed holograms including laser.
****

Environmental Research (Erim)
Optical And Infrared Science Lab
POBox 13400 I
Ann Arbor, MI
Postal Code: 48113-400 I
Country: United States of America
Contact: Juris Upatnieks
Voice Phone: 1 313 994 1220
FAX Phone: 1 313 9945704
Business Description: Industrial & academic research. BOD & environmental research.
****

Envision Enterprises
DBA Interactive En tertainment and Education
465 East Las Flores Drive
Altadena, CA
Postal Code: 91001
Country: United States of America
Contact: Jeff Allen
Voice Phone: 1 800 662 0424
Voice Phone: 1 415 332 200 I
FAX Phone: 1 8 18 398 0563
Business Description: Custom designed and priced Holography Shows. One of the largest holographic collections in the world. Combining holography, computer animation , robotics and virtual realities.
****

ETA-Optik Gmbh
Niethausener Strasse 15
Postal Code: D-52525
Country: Germany
Contact: Dr. Wilbert Windeln
Voice Phone: 49 2452 66654
FAX Phone: 49 2452 64433
Business Description: DCG pendants, diffraction gratings and custom HOE
****

Excitek Inc .
277 Coit Street
Irvington, NJ
Postal Code: 07111
Country: United States of America
Contact: Greg Springer
Voice Phone: I 201 372 1669
FAX Phone: I 20 1 372 8551
Business Description: Supplier of re-manufactured argon and krypton ion laser tubes. and used laser systems. Established in 1984. 10 Employees at this address.
****

Expanded Optics Limited
Noon Lane

Barnet, England
Postal Code: ENS SST
Country: United Kingdom
Contact: T.R. Hollinsworth
Voice Phone: 44 181 441 2283
FAX Phone: 44 181 4496143
Business Description: Manufacturer of medical and industrial endoscopes; micro-precision optics.
****

Fantastic Holograms
DlA Terminal - LevelS
8400 Pena Blvd
Denve r, CO
Posta l Code: 80249
Country: United States of America
Contact: RB Osada
Voice Phone: 303 342 3440
FAX Phone: 303 342 3440
Business Description: Retail store in high traffic area specializing in selling holograms.
****

Far East Holographics
12/F Hang Wai Commercial Bldg
231-233 Queen's Road East
Country: Hong Kong
Contact: Adrian 1. Halkes
Voice Phone: 852 2 893 9773
FAX Phone: 852 2 893 0640
Business Description: Finisher and distributor of holograms and holographic products.
****

Fisher Scientific
EMD Divi sion
490 I West Lemoyne Avenue
Chicago, TL
Postal Code: 60651
Country: United States of America
Voice Phone: 1 312 378 7770
FAX Phone: 1 3123787 174
Business Description: Supply science lab equipment, holography kits, lab manuals, lasers and laser related equipment.
****

Flatiron Studio
15 West 24Th Street
7Th Floor
New York, NY
Postal Code: 10010
Country: United States of America
Contact: Frank Bunts
Voice Phone: 1 212 929 7938
e-mail address:traenif@aol.com
Business Description: Painted interference patterns, creating depth and movement effects on painted canvas.
****

FLEXcon Company, Inc .
I FLEXcon Industrial Park
Spencer, MA
Postal Code: 01562
Country: United States of America
Voice Phone: 1 508 885-3973
FAX Phone: 1 508 885 8400
Business Description: Manufacturer of holographic and prismatic materials for packaging, gift wrap and graphic film markets. Wideweb embossing in excess of 60 inch width.
****

Focal Image Ltd.
Number 20 Conduit Place
London, England
Postal Code: W2 I HZ
Country: United Kingdom

Contact: Kaveh Bazargan
Voice Phone: 44 171 706 222 I
FAX Phone: 44 171 706 2223
e-mail address: kaveh@foca l.demon.co.uk
Business Description: Consultancy in holography; display holograms; computer graphics; electronic publishing.
****

FoilMark Holographics
5 Malcolm Hoyt Drive
Newburyport, MA
Postal Code: 01950
Country: United States of America
Contact: David Dion
Voice Phone: 1 508 462 1200
FAX Phone: 1 508 462 0227
Business Description: Produces wide web hot stamp holographic foils and diffraction foils. FHI is joint venture with Embossing Technology Ltd.
****

Fong Teng Technology
No 41 , Lane 63, Hwa Chen Road
Country: Taiwan
Contact: Mark Chiang
Voice Phone: 886 2 2 998 4760
FAX Phone: 886 2 2 992 1240
Business Description: 60 inch hologram and dot-matrix pattern foil manufacturer, service from origination to finished product.
****

Ford Motor Company
Scientific Research Labs (SRL)
20000 Rotunda Drive, #S -1023
Dearborn, MT
Postal Code: 48121
Country: United States of America
Contact: Gordon Brown
Voice Phone: 1 313 323 1539
FAX Phone: 1 3 13 845 0100
e-mail address: gbrown@smail.srl.ford.com
Business Description: Industrial research; Holographic non-destructive testing. Research and development in computer-aided holographic interferometry.
****

Foreign Dimension
Manley Commercial Bldg Rm 190 I
367-375 Queen's Road Central
Country: Hong Kong
Contact: Frederic Schwartzman
Voice Phone: 852 2 542 0282
FAX Phone: 852 2 541 60 11
Business Description: Specialists in manufacturing all kinds of holographic and illusion products (Watches, keyrings), If you are a hologram manufacturer, we can also make top quality products at unbeatable prices using your holograms!
****

Fornari, Arthur David
8 13 Eighth Avenue
Brooklyn, NY
Postal Code: 11 2 15
Country: United States of America
Contact: Arthur David Fornari
Voice Phone: 1 718 965 3956
Business Descrip tion: Art istic holographer; silver halide transmission & reflection holograms.
****

Forth Dimension
36 East Franklin Street
Nashville, IN

Postal Code: 47448
Country: United States of America
Contact: Rob Taylor
Voice Phone: I 812 988 82 12
FAX Phone: I 81298892 11
Business Description: Holographic gal lery and retail shop. Hologram mastering/consulting. Small run si lver hal ide transmission, reflection holograms.
****

Fostec Gmbh Feinmechanik
Opt Systemtechnik
Halberstaedter Str 7
Postal Code: 1000
Country: Germany
Voice Phone: 49 030 891 5077
FAX Phone: 49 030 891 506
Business Description: Laser technology, components , assemblies and systems,Measuring tables
****

Foundat ion Ideecentrum.
PO Box 222
5600 Mk
Country: Netherlands
Business Description : Gallery.
****

Frank DeFreitas Holograp hy Studio
P. O. Box 9035
815 Allen Street
Allentown, PA
Posta l Code: 18105-9035
Country: United States of America
Contact: Frank DeFreitas
Voice Phone: 1 800 458 3525
Voice Phone: 1 610 770 0341
FAX Phone: 1 610 967 8956
Busin ess Descrip tion: Since 1983. Inventors of the photogram, a 3-D hologram portrait from your own camera! Workshops, educational outreach, newsletter, stock images, HeNe Lasers and optics. Free catalogue.
****

Free Uni versity Of Brusse ls.
Faculty Of Applied Sciences
Alna-Tw Plein laan 2
Postal Code : B-1050
Country: Belgium
Contact: Erik Styns
Voice Phone: 32 2 629 3452
FAX Phone: 32 2 629 3450
Business Description: Academic and Scientific research on dim'active elements and HOE's.
****

Fresnel Technologies 1nc
101 West Morningside Drive
Fort Worth, TX
Posta l Code: 761 10
Country: United States of America
Contact: Linda H. Claytor
Voice Phone: 1 8 17 926 7474
FAX Phone: 18179267146
Business Description: Manufactures plastic Fresnel lenses & lens arrays from its POLY IR plastics for use into the infrared; also other optical products for use into the ultraviolet from acrylic & other plastics.
****

Fringe Research Hol ographics
11 79A King Street West
Suite 0 10
Toronto, Ontario
Postal Code: M6K 3C5 ,

Country: Canada
Contact: Michael Sowdon
Voice Phone: 1 416 535 2323
Business Description: Artistic holography; silver halide holograms; pulse portraits; gallery; workshops; traveling exhibit.
****
Fuji Electric Co. Ltd
Mecatronics Division
1-12-1 Yuraku-Cho Chiyoda-Ku
Postal Code: 100
Country: Japan
Voice Phone: 8 1 3 21 I 7 11 I
Business Description: Manufactures CO_2 lasers and related equipment.
****
Fujitsu Laboratories Ltd.
Electronic Systems Division
10-1 Wakamiya Morinosato
Postal Code: 243-0
Country: Japan
Contact: Takehumi Inagaki
Voice Phone: 81 462 48 31 I I
FAX Phone: 81 462 48 3233
Business Description: Embossed Hologram Manufacturer.
****
Full Color Holographies
PO Box 489
Powassan, Ontario
Postal Code: POH I ZO
Country: Canada
Voice Phone: 1 705 724 6164
FAX Phone: I 705 724 6249
Business Description: Full Color Holographies is a Joint Venture of Sil verbridge and Color Holographics. Produces large format DCG holograms.
****
G.M. Vacuum Coating Lab, In c.
882 Production Place
Newport Beach, CA
Postal Code: 92663
Country: United States of America
Contact: Dan Coursen
Voice Phone: I 714 642 5446
FAX Phone: I 7 14 642 7530
Business Description: Will do coatings for front surface mirrors, beamsplitters, etc. for holographic use.
****
Galerie IIlusoria
Schwarztorstrasse 70
Postal Code: CH-3007
Country: Switzerland
Contact: Sandro del-Prete
Voice Phone: 41 3 1 381 773 1
FAX Phone: 41 31 381 773 1
Business Description: Gallery featuring holograms.
****
Gal voptics Ltd .
Harvey Road
Burnt Mills Indu st rial Estate
Basildon, England
Postal Code: SS 13 I ES
Country: United Kingdom
Contact: R. D. Wale
Voice Phone: 44 1268 728 077
FAX Phone: 44 1268 590 445
Business Description: Optics; mirrors, lenses.
****
Gardener Promotion Marketing
4165 Apalogen Road

Philadelphia, PA
Postal Code: 19144
Country: United States of America
Contact: John Gardener
Voice Phone: 1 2 15 849 4049
FAX Phone: I 215 849 4049
Business Description: As the exclusive package goods marketing repre sentative for Bridgestone Graphics, we can show you how holography can be used for problem solving or enhancing opportunit ies compatible with your objectives.
****
General Des ign
2023 - 18th Street
San Francisco, CA
Postal Code: 94107
Country: United States of America
Contact: Brian Kane
Voice Phone: 1 415 5509193
e-mail address:BK94@aol.com
Business Description: Creative Services - Computer Graphics for Print, Video and Holography. 3D Computer Modeling and 20 Computer Composition. General Image Design and Construction.
****
General Holographics, In c.
PO Box 82247
Postal Code: V5C 5P7
Country : Canada
Conta ct: Paula Simson
Voice Phone: 1 604 685 6666
Voice Phone: 1 800 667 9669
FAX Phone: 1 604 685 6678
Business Description: Distributor of dichromate & embossed gift and jewelry items (Holocrafts), silver halide wall and desk decor, and photopolymer for the Canadian market. Custom and stock.
****
Gerald Marks Studio
29 West 26Th Street
New York , NY
Postal Code: 100 I 0 -100
Country: United States of America
Contact: Gerald Marks
Voice Phone: 1 212 889 5994
FAX Phone: 1 212 889 5926
Business Description : General holography laboratory - Working since 1973
****
Glass Mountain Optics
95 17 Old McNeil Road
Austin , TX
Postal Code: 78758 -5225
Country: United States of America
Contact: Don Conklin
Voice Phone: 1 512 339 7442
FAX Phone: I 5123390589
Business Descri ption: Specialize in collimating mirrors.
****
Globa l Images
1 Northumbe rl and Ave
Postal Code: WC2N 5BW
Country: Un ited Kingdom
Contact: Walter Clarke
Voice Phone: 44 171 872 5452
FAX Phone: 44 171 753 2848
Business Description: Specialists in high volume, low cost, quality embossing equipment. ISO 9002 Qualification.
****

Gray Scale Studios Ltd.
63 south 500 west
Richmond, CO
Postal Code: 843 33
Country: United States of America
Contact: George Sivy
Voice Phone: 1 801 258 0709
FAX Phone: 1 801 258 0709
Business Description: Specialists in design and creation of models and sculptures for holographic imaging. Consultant services offered, 10 years experience, samples available upon request.
****
Gress er, E., Kg
An Der Warth 10
Postal Code: 97 199
Country: Germany
Contact: Joachin Muller
Voice Phone: 49 933 1 22 77
FAX Phone: 49 9331 78 41
Businqss Description: Laser measurement techniques, Lasers, medical
****
Hallmark Capita l Corp
230 Park Avenue Suite 5 I 0
New York , NY
Postal Code: 10169
Country: United States of America
Contact: Patricia M. Hall
Voice Phone: I 212 249 9634
FAX Phone: I 2 122499537
Business Description: New York based investment banking firm, specializing in raising debt and equity financing for privately-held companies. Mergers & acqu is itions are a strong secondary activity. Hallmark has raised capital for both public and private holography companies.
****
Harland Check Printers
2939 Miller Rd
Decatur, GA
Postal Code: 30035
Country: United States of America
Contact: Wilili am Borklund
Voice Phone: I 404 593 5132
FAX Phone: I 404 593 5644
Bus iness Description: Offers pre -printed checks with embossed holograms.
****
Harri s, Nick
7 11 East 13Th Street
Houston , TX
Posta l Code: 77008
Country: United States of America
Contact: Nick Harris
Vo ice Phone: 1 71 3 861 2865
Business Description: Artistic holographer; portraits and integrals; consulting.
****
Harvard Apparatus, Canada
60 10 Vanden Abeele
St Laurent, Quebec
Postal Code: H4S I R9
Country: Canada
Voice Phone: I 514 335 0792
FAX Phone: I 5143353482
Business Description: Mirrors, prisms, optical items for holography.
****
Helios Holography
93 15 Santa Fe Dr

Overland Park, KS
Postal Code: 66212
Country: United States of America
Contact: Gene Davis
Voice Phone: 1 913 649 2700
FAX Phone: 1 913 7885106
Business Desc ription: 2000 square foot store/ gallery exclusive ly for holograms. Plans for 1993 include a larger store.
****

Hellenic In stitute Of Holography
28 Dionyssou Street
Postal Code: GR-15 234
Country: Greece
Contact: Alkis l embessis
Voice Phone: 30 1 684 6776
FAX Phone: 30 1 685 0807
Business Description: Established in 1987, the Institute aims at the overall introduction and promotion of holography in Greece. Exhibitions, courses, vocational training and mastering laboratory.
****

High Tech Network
Skeppsbron 2
Postal Code: S-211 20
Country: Sweden
Contact: Chri ster Agehall.
Voice Phone: 46 040 350 75
FAX Phone: 46 040 237 667
Business Description: Art in holography; security applications.
****

HODIC Holographic Display Artists & Engineers Club
Engineering Department Chiba University
1-33, Yayoi-cho
Postal Code: 263
Country: Japan
Contact: Miss Tomoko Sakai
Voice Phone: 81 472 51 1 111
FAX Phone: 81 472 517 337
Business Desc ription: Regular meetings 4 times a year with oral presentations on holography given. Publishes HODIC circular. Membership open to all.
****

HOl 3, Galerie fur Holographie
Europa Center
Postal Code: 10789
Country: Germany
Contact: Frau Konner
Voice Phone: 49 30 261 4490
FAX Phone: 49 30 344 6379
Business Description: Hologram Gallery
****

Hol age
1881 Eighth Avenue
San Francisco, CA
Postal Code: 94122
Country: United States of America
Contact: Brad Cantos
Voice Phone: 1 415 564 1840
Business Description: Fine an holograms; silver halide holograms, microlithography and photoresist consulting.
****

Holart Consultants
18 Bonview Street
San Franc isco, CA
Postal Code: 94 1 10
Country: United States of America
Contact: Gary Zellerbach
Voice Phone: 1 415 282 3646

FAX Phone: 1 415 282 4013
e-mail address:gaz@ix.netcom.com
Bus iness Description: Expert appraisals of holographic art works. Consulting in all aspects of creating, displaying, and marketing custom and stock holograms.
****

Holi con Corporation.
3312 Belle Plain Ave #2
Chicago, IL
Postal Code: 60618
Country: United States of America
Contact: Dr. Hans Bjelkhagen
Voice Phone: 1 312 267 9288
FAX Phone: I 312 267 9288
Business Description: Holograms produced with pulsed lasers. Holographic portraits. Large format reflection and rainbow holograms. Mass production of silver halide holograms.
****

Holo 3
7 Rue du Gal Cassagnou
Postal Code: 68300
Country: France
Voice Phone: 33 89 69 82 08
FAX Phone: 33 89 67 74 06
Business Description: Industrial applications of holography: shock and vibration, non de structive testing, microholography flow visualization, contouring. R&D: study and development of new tools for industrial applications.
****

Holo Gmbh
Lutterdamm 82
Postal Code: 49565
Country: Germany
Contact: Thomas Lucy
Voice Phone: 49 5461 9 1123
FAX Phone: 49 5461 9 11 22
Business Description: Holograms up to 1 x 1m; embossed holography. Holo-design.
****

Holo Images Tech Co. , Ltd.
17, Alley 20, Lane 7, Jong Hwa Road
Country: Taiwan
Contact: Craig Chiou
Voice Phone: 886 66 237 3896
FAX Phone: 886 66 238 4641
Business Description: Embossed holograms and products.
****

Holo Impressions
47-1 Wu Chuan Road
Wu-Ku Industrial Park
Country: Taiwan
Contact: Billy Chiou
Voice Phone: 886 2 299 7576
FAX Phone: 886 2 299 7050
****

Holo Impressions Inc
47-1 Wu Chltan Rd
Wu-Ku Industri al Park
Country: Taiwan
Contact: Jonathan Hsu
Voice Phone: 886 2 299 7576
FAX Phone: 886 2 299 7050
Business Description: Embossed holography.
****

Holo Sciences,LLC
480 East Rudasill Road
Tucson, AZ
Postal Code: 85704

Country: United States of America
Contact: Chuck Hassen
Voice Phone: 1 520 696 0773
FAX Phone: I 520 696 0773
e-mail address:chuckh@dconcepts.com
Business Description: H1 mastering and stereogram creation. See Chapter 2 for additional information .
****

Holo Spectra
7742 Gloria
Van Nyus, CA
Postal Code: 91406
Country: United States of America
Voice Phone: 1 8189949577
FAX Phone: 1 818 9944709
Business Description : Laser repair and resale for all types of lasers.
****

Holo-Laser
6, Rue De La Mission
Ecole
Postal Code: 25480
Country: France
Contact: Dr.Jean Louis Tribillon
Voice Phone: 33 1 45 52 46 52
FAX Phone: 33 I 45 52 46 81
Business Description: Embossed holography and equipment; artistic holography; buying and selling; education .
****

Holo-Or Ltd
PO Box 1051
Kiryat Weizmann
Country: Israel
Contact: Uri Levy
Voice Phone: 972 8 469 687
FAX Phone: 972 8 466 378
Business Description: Manufactures computer-generated dim·active optical elements by VLSI techniques. Catalogue elements and custom designs. Substrates include AnSe, GaAs, various glasses. DOE work station-dedicated work station for element design, mask generation.
****

Holo-Service
Neuensteinerstrasse 19
Postal Code: CH-4153
Country: Switzerland
Contact: Edgar Bar
Voice Phone: 41 502 287
Business Description: Artistic holography.
****

Holo-S ervice.F ries
Therwi ler Strasse 26
Postal Code: CH-4045
Country: Switzerland
Contact: Urs Fries
Voice Phone: 41 61 281 0917
FAX Phone: 41 61 281 09 17
Business Description: Artistic holography, hologram projects.
****

Holo-Spectra
7742-B Gloria Avenue
Van Nuys, CA
Postal Code: 91406
Country: United States of America
Contact: Bill Rankin
Voice Phone: 1 818 994 9577
Voice Phone: 1 800 275 9880
FAX Phone: 1 8189944709

Business Description: Embossed production. Embossed mastering equipment, laser repairs. Optical table and holographic equipment resold.
****

Holo/Source Corporation
11930 Farmington Rd
Livonia, MI
Postal Code: 48 150
Country: Un ited States of America
Contact: Lee Lacey
Voice Phone: 1 313 427 1530
FAX Phone:1 313 525 8520
Business Description: Paperboard sheets of holographic film and paper. Holographic image mastering of all types and finished flezo printed holographic labels.
****

Holocraft International
1155 North Sheridan road
Lake Forest, IL
Postal Code: 60045-0152
Country: United States of America
Contact: William Cri st
Voice Phone: I 708 234 7625
Business Description: Artistic holography, marketing. (not related in any way to Holocrafts in Canada)
****

Holocrafts
Canadian Holograp hic Developme nts Ltd.
Box 1035
Delta, Briti sh Columbia
Postal Code: V4M 3T2
Country: Canada
Contact: Karoline Cullen
Vo ice Phone: I 604 946 1926
FAX Phone: I 604 946 1648
Business Description: Holocrafts manufacturers dichromate holograms. Offering stock and custom production in a variety of formats such as plain discs, watches, keychains, pendants and 3" x 3" plates. Providing a tradition of excellence since 1979.
****

Holocrafts Europe Limited.
Barton Mill House
Balion Mill Road
Canterbury, England
Postal Code: CT I I BY
Country: United Kingdom
Contact: Chris Luton
Voice Phone: 44 0227 463 223
FAX Phone: 44 0227 450 399
Business Description: Specialists in manufacture of dichromate reflection holograms. Also produce holographic gift products as well as selling photopolymer.
****

Holodesign Studies
Rebenstrasse 20
Postal Code: CH-4 125
Country: Switzerland
Business Description: Marketing consulting.
****

Holofar Lab (SrI)
Piazza Acil ia No 3 Int 3
Postal Code: 00]99
Country: Italy
Business Description : Artistic holography
****

Holoflex Company
1413 East Old Church Rd

Urbana, IL
Postal Code: 6 180]
Country: United States of America
Contact: Donald Barnhardt
Voice Phone: 1 217 684 2321
Business Description: Holographic velocimetray particle image (HPIV) testing.
****

Holografia Pol ska
Sw Mikolaja 161I 7
Country: Poland
Contact: Boguslaw Stich
Voice Phone: 48 71 343 46
FAX Phone: 48 71 339 48
Business Description: Practical applications of Holography.
****

Hol ografica
30 I South Light Street
Baltimore, MD
Postal Code: 21202
Country: United States of America
Contact: Renee Fee
Voice Phone: I 410 685 3331
Business Description: Retail store with full range of holographic products
****

Hologram Company RAQD GmbH
Mollner Land Str 15
Postal Code: 22969
Country: Germ any
Contact: Wil fried Schipper
Voice Phone: 49 4104 69386
FAX Phone : 49 4104 69349
Business Description: Specializing in production of embossed holograms and the sale of embossing equipment.
****

Hologram Industries
42/44 Rue De Trucy
Postal Code: 94120
Country: France
Contact: Hughes Souparis
Voice Phone: 33 143 94 19 19
FAX Phone: 33 143 94 00 32
Business Description: Communication holograms.
****

Hologram Land
284 E Broadway
Mall Of America
Bloomington, MN
Postal Code: 55425
Country: United States of America
Contact: George Robinson
Voice Phone: 1 6 12 854 9344
FAX Phone: 1 6128547857
Business Description: Retail store specializing in everything holographic. Product range includes artwork, watches & Jewelry, t-shirts, small gift items and optical novelties. framing provided and lighting accessories.
****

Hologram Research
25 East Loop Road
Stony Brook, NY
Postal Code: 11 790-3350
Country: Un ited States of America
Contact: Jody Burns
Voice Phone: 1 516 444 8839
FAX Phone: 1 5 16 444 8825
e-mail address: hologram@lihti.org
Business Description: Since 1972. Holograms/ Stereograms/Editions; Silver halide, Photore-

sist, Nickel, Embossed with Agam, Dali, Cossette, Nunez, Dieter Jung, Sam Moree, Others; 1979 - Injection-Molded Holograms; 1987 - Design/Development NY % Hologram CreditCard.
***** ** *** **** **** **** *** *** **** ***

Hologram World, Inc.
1860 Berkshire Lane North
Plymouth, MN
Postal Code: 55441
Country: United States of America
Contact: Jim Paletz
Voice Phone: I 612 559 5539
Vo ice Phone:] 800 882 4656
FAX Phone: I 612 559 2286
Business Description : One of the largest wholesale distributors of holographic novelties. We represent over 50 holographic manufacturers. Specialize in helping the new retail store owner in all stages of development from start to finish. Free catalogue.
****

Hologramas, S.A. de C.Y.
PINO 343-3
COL. STA. MA. LA RIBERA
Postal Code: 06400
Country: Mexico
Contact: Dan Li eberman
Voice Phone: 52 525 547 9046
FAX Phone: 52 525 547 4084
Business Description : Holography can be used on: packaging, literature inserts, wrapping paper, security labels, bar codes, security paper, stickers, point-of-sale displays. Anything you can imagine can be done in holography.
****

Hologranun Werkstatt & Galerie
Gallerie Fur Hologramme
Via Princlpale 30, Ch
Postal Code: 7649
Country: Switze rland
Contact: Horst Gutekunst
Voice Phone: 414 11 8241 718
FAX Phone: 41 411 824 1268
Business Description: Creative workshop, developments, looking for new and attractive ways for hologram making.
****

Holograms 3D
5 Queens Court
25 -27 Earl 's Court Square
London, England
Postal Code: SW5 9DA
Country: United Kingdom
Contact: Jonathan Ross
Voice Phone: 44 171 370 2239
FAX Phone: 44 171 3702239
Business Description: Jonathan Ross has a personal holography co llection ava ilable for touring shows. he also deals privately in holographic art and consults on commercial applications.
****

Holograms Fantastic & optical Illusions
PO Box 765
Bayswater, Victoria
Postal Code: 3175
Country: Austral ia
Contact: Trevor McGaw
Voice Phone: 61 18 776 226
FAX Phone: 6 1 39 729 6020
Business Description: Glass, film and foil (opp. PET & PVC) 2D, 2D/3D, 3D & multi images & patterns. Services to printers, hot-stampers, packaging, label, sec urity marketing. sales

promotion and advertising. Specialists in foil holography.
****

Holograms International
83 55 On The Mall
Buena Park , CA
Postal Code: 90620
Country: United States of America
Contact: Dave Krueger
Voice Phone: 1 714 536 0608
FAX Phone: 1 7 14 5360608
Business Description: Distributor of all kinds of holograms to retail stores and whole sale accounts. We are known for our fast delivery, friendly consulting and factory-direct prices. Call or write for quote or catalogue.
****

Ho lographic and Photonic World Center
of the Art, Science and Technology In st.
800 K Street, N.W.
Washington, DC
Postal Code: 2000 I
Country: United States of America
Contact: Odile Meulien-Ohlmann
Voice Phone: I 202 667 6322
Business Descripti on: Research and educational organization for development of the art, science and technology of holography, laser and photonic imaging; educational, information and entertaining program services.
****

Holographic Appli cations
21 Woodland Way
Greenbelt, MD
Postal Code: 20770-1728
Country: United States of America
Contact: Susan St. Cyr
Voice Phone: I 30 I 345 4652
FAX Phone: I 30 I 345 4653
Business Description: Design and product engineering services for consumer products and licensed promotions using 3-D imaging technologies. Product specifications and quality assurance. Vendor se lection and product management. General contractor delivering finished, packaged product.
****

Holographic Desi gn Systems
11 34 West Was hington Blvd .
Chicago , IL
Postal Code: 60607
Country: United States of Ameri ca
Contact: Robert Billings
Voice Phone: I 3 12 829 2292
FAX Phone: I 31 2 829 9636
Business Description: Unrivaled creativity, combining artistic imagination with complete technical mastery of all forms of holography resulting in a worldwide reputation for excellence. The most complete labs in the industry with the most powerful and advanced lasers and computers. Our c lients include the most innovative and sophisticated com panies in the US and abroad.
****

Holographic Dimension
16 115 Sw I 17Th Avenue
Unit A-2 1
Miami, FL
Postal Code: 33 177-1615
Country: United States of Ameri ca
Contact: Kevin Brown
Voice Phone: 1 305 255 4247
FAX Phone: 1 305 255 0334
e- mail address: holodi@shadow.net

Business Description: Origination and mass replication of holographic imagery.
****

Holograp hic Images Inc.
521 Michigan Ave
Miami Beach, FL
Postal Code: 33139
Country: United States of America
Conta ct: Larry Lieberman
Voice Phone: 1 305 531 5465
FAX Phone: 1 305 53 1 3029
Business Description: Limited-edition art holograms.
****

Holographic Indu strie s, Inc
PO Box 11 09
Libertyvi ll e, IL
Posta l Code: 60048
Country: United States of Ameri ca
Contact: Robert Pricone
Voice Phone: I 708 680 1884
FAX Phone: I 708 680 0505
Business Description: Designer and operator of retail galleries/gift shops in major shopping centers. We produce our own pulse holographic images, and can obtain nearly any holographic product worldwide.
****

Holographic Label Converting (H LC)
7669 Washington Avenue South
Edina , MN
Postal Code: 55439
Country: United States of America
Contact: Scott Labe lle
Vo ice Phone : I 6 129447408
FAX Phone: 1612944 72 10
Business Description: Full service capabilities, 2D/ 3D holography, designing, embossing, hot-stamping. precision die-cutting. wide variety of foils. Custom holographic labeling, magnetic holograms, packaging and more. You think of it, and we can put it together.
****

Holographic Marketing. Inco
9250 SW First Street
Pl antati on. FL
Postal Code: 33324
Country: United States of America
Contact: Mark Rapke
Voice Phone: 1 305 474 9965
FAX Phone: I 305 474 9965
Business Description: Consultant to foreign and domestic corporations on applications of embossed holography; broker for fine art holograms: exporter of artistic holograms, jewelry and novelties.
****

Holographic Optics Inc
358 Saw Mil l Ri ver Rd
Millwood. NY
Postal Code: 10546
Country: United States of America
Contact: Dr. Jose R. Magarinos
Voice Phone: I 914 762 1774
FAX Phone: I 9147622557
Business Description : Manufacturer of holographic optical elements. particularly holographic filters, holographic mirror and beamsp litters. Design and manufacture of prototypes.
****

Holog raphic Products
171 1 St. Clai r Ave

St Paul, MN
Postal Code: 55105
Country: United States of America
Contact: Stephen Sugarman
Voice Phone : 1 612 698 6893
FAX Phone: 1 612 698 1619
Business Description: Holographic Products. is actively pursuing new product development in educational toys , intermedia print design, ad specialties, promotions, premiums, hands on elementary school workshops, original fine art productions and in structional presentations.
****

Holographic Service
10 Via Cive rchio
Postal Code: 1-20159
Country: Italy
Business Description: Consultant, holograms on packaging material.
****

Ho lographic Studi os
240 East 26Th Street
New York, NY
Postal Code: 100 I 0
Country: United States of America
Contact: Jason Sapan
Voice Phone: I 2 12 686 9397
FAX Phone: 1 212 481 8645
Business Description: New York's only gallery and commercial holographic lab. Custom and stock holograms. Integral portrait cinematography, mastering. and scan copies from small to large format. Single or massproduced holograms.
****

Holographic Sys tems Munchen
Wiegenfeldring 2a
Postal Code: 85570
Country: Germany
Contact: Gunther Dausmann
Voice Phone: 49 8 121 93000
FAX Phone: 49 8 12 1 930099
Business Description: Holog raphic production, including machines .
****

Holographic Technologies
637 South Vinewood Street
Escondido, CA
Postal Code: 92029
Country: United States of America
Contact : Donald C. Broadbent
Vo ice Phone: I 619 746 0976
FAX Phone: I 6 197466141
Business Description: An independent, privately owned holographic facility producing HOE's and display holograms in vario us recording materials. Donald Broadbent has 24 years experience in holography.
****

Holographics (Uk) Ltd.
12 Whidborne SI
London, England
Postal Code: WC I H 8EU
Country: United Kingdom
Contact: Jon Vogel
Voice Phone: 44 171 833 2236
FAX Phone: 44 171 833 2237
Business Description: Holographic & 3-D multimedia , design originat ion and production specialists (est 1982) providing comprehensive service for the corporate, retail, & leisure sectors.
****

Holograph ics Inc.
44-0 I Eleventh Street
Long Island City, NY
Postal Code: 11101
Country: United States of America
Contact: Anna-Marie Nicholson
Voice Phone: I 708 784 3435
FAX Phone: I 71870808 13
Business Description: Holographic portraiture
for any customer. R&D as well
****
Holographics North In c.
444 South Union Street
Burlington, VT
Postal Code: 0540 I
Country: United States of America
Contact: John Perry
Voice Phone: 1 802 658 2275
FAX Phone: 1 802 658 5471
Business Description: Designers/producers of
large format holography up to 44 x 72 inches (1.
1 m x 1.8 m). Known worldwide for the highest
quality commercial and fine art display work.
Design, model building, production, in stallation
and consu lting services.
****
Holographie Anubis
Oberer Kaul berg 37
Postal Code: 96049
Coun try: Germany
Contact: M.T. Frieb
Voice Phone: 49 951 5795 1
FAX Phone: 49 95 1 59529
Business Description: We are producer and dis-
tributor of all formats of holograms. Import and
export. The holograms are in protective covering
from co mpletion through purchase. Request our
over 100 page full color catalogue.
****
Holographie Fachstudio Bad Rothenfeld
Postfach 1304
Niederesch 28
Postal Code: D-49214
Country: Germany
Contact: Gun ter Deutschmann
Voice Phone: 49 5424 5363
FAX Phone: 49 5424 5359
Business Description: Expert consultancy for
integration of holographic products into fin-
ished advertising media, including application,
overprinting etc. Founder and office of the AHT
(Arbeitskreis Hologramm-Techniken und neue
Medien).
****
Holographie Konzept GmbH
Koerberstr 3
Postal Code: 60433
Country: Germany
Contact: Ulrich Anders
Voice Phone: 49 69 531 071
FAX Phone: 49 69 532 055
Business Description: Advertising holograms,
Holographic projects,
****
Holographie Labor
Bertelsmann AG
Auf dem Eickholt 47
Postal Code: 33334
Coun try: Germany
Contact: Saurda Uwe
Voice Phone: 49 5241 580 192
FAX Phone: 49 5241 580 549
Business Descr iption : Holograms. Holographic
projects, .

****
Holography Center of Austria
Kahlenbergstrasse 6
Postal Code: A-3042
Country: Austria
Conta ct: Irmfr ied Wober
Vo ice Phone: 43 2275 82 10
FAX Phone: 43 2275 82 105
Business Description: Our Holography Labora-
tory, founded in 1985, is the first in Austria. We
are the bigge st hologram producers in town. We
organize exhibitions in Austria and Germany and
sell embossed holograms. Transportable pulse-
laser system for portaits.
****
Holography Development Group
23 Nevi lle Park Blvd
Toronto, Ontario
Postal Code: M4E 3P5
Country : Canada
Contact: Andrew Laczynsk i
Voice Phone: I 416 925 5569
Business Description: Designers of numerous
holography projects including coca-cola, du-
pont, Royal Oera, London Zoo, MacDonalds,
etc.
****
Holography Group
Oxford Universi ty
Parks Road
Oxford, England
Postal Code: OX I 3 PJ
Country: United Kingdom
Voice Phone: 44 1865 270 000
Business Description: University research
and courses on a wide range of holography
topics .
****
Holography In stitute
PO Box 24-153
San franc isco, CA
Postal Code: 94124
Country: Un ited States of America
Contact: Jeffrey Murray
Voice Phone: I 4 15 822 7123
e-mail address: hologram@well.com
Business Description: Limited editions; holo-
graphic art; consulting; training.
****
Holography Israe l
21 Hakomemiut Str.
Postal Code: 46683
Country: Israel
Contact: Hameiri Shimon
Voice Phone: 972 09 572 387
Voice Phone: 972 09 559 766
FAX Phone: 972 09 570 569
Business Description: Holography Israel spe-
cializes in exhibitions-lectures and demon-
strations to pupils and students-advertising.
commission, sales and production of art ho-
lograms.
****
Holography News
Runnymede Malthouse
Egham, England
Postal Code: TW20 9BD
Country: Uni ted Kingdom
Contact: Ian M. Lancaster
Voice Phone: 44 1784 497008
FAX Phone: 44 1784 49700 I
e-mail address :
100142.1 164@compuserve.com
Business Description : Holography News is the

international business news letter of this indus-
try. Coverage and dis tribution is worldwide.
Now in its seventh year, it is valued for its obj
ectivity, depth and analysis.
****
Holography Presses On (HPO)
201 North fruitport Road
Box 193
Spring Lake, MI
Postal Code: 49456-0193
Country: United States of America
Contact: Jan Bussard
Voice Phone: 1 616 842 5626
FAX Phone: 1 6 16 842 5653
Business Description: Holographic stock or
custom shapes and sizes applied with heat or
pressure for adhesion to all substrates. Sealed
edges prevent delamination in all weather;
washable/dry cleanable. Worldwide distribu-
tors sought I
****
Holography Studio
40 Springdale Avenue
Broadstone , England
Posta l Code: BH 18 9EU
Country: United Kingdom
Contact: Dr. Margaret Benyon
Voice Phone: 44 1202 698 067
FAX Phone: 44 1202 698 067
Business Description: famous Holographic art-
ist since 1968, with works included in a large
number of pri vate and public collections world
wide. Works are available for exhibition, hire or
sale.
****
Holo land S.c.
Batumi 6 m 43
Postal Code: 02-760
Country: Poland
Contact: Pawel Stepien
Voice Phone: 48 22 427 463
fAX Phone: 48 2 625 5567
e-mail address:stepie n@if.pw.edu. pl
Business Description: Low volume holographic
labels, holographic consultancy, security CGH
research & deve lopment.
****
Hololaser Gallery
PO Box 23386
Country: United Arab Emerates
Contact: Abdul Wahab Baghdadi
Voice Phone: 97 1 4 5 18 989
FAX Phone: 971 45280 15
Business Description: Holography Gallery and
holographic items. We are the first and only gal-
lery in the Gulf Countries and we produce laser
shows.
****
Holomat
74 1 East Gorham Street
Madison, WI
Postal Code: 53703
Country: United States of America
Contact: Matt Hansen
Voice Phone: 1 608 255 3580
Business Description: Aristic holography and
holographic engineering consulting.
****
HoloMedia Ab/Hologram Museum .
PO Box 45012
Postal Code: 10460
Country: Sweden
Contact: Mona Forsberg
Voice Phone: 46 8 411 1108

FAX Phone: 46 8 107 638
Bus iness Description: Broker for embossed and custom made artistic holography; buying & selling holograms; holography education; gallery. Display unit available. Hologram center
****

Holomedia France
16 rue Maurice Fontvielle
Postal Code: 31000
Country: France
Contact: Luigi Castagna
Voice Phone: 33 62 27 17 04
FAX Phone: 33 62 27 17 04
Business Description: Wholesale and distribution of silver halide, jewelry and fine art holograms. Two retail shops in Toulouse and Lyon, France.
****

Holomex Ltd.
4 Borrowdale Avenue
Harrow, England
Postal Code: HA3 7PZ
Country: United Kingdom
Contact: Mike Anderson
Voice Phone: 44 18 1 427 9685
Business Description: Holographic camera design. Holographic viewer design. Supplier of film processing kits and safelights.
****

Holonix
P.O. Box 45577
Seattle, WA
Postal Code: 98145
Country: United States of America
Contact: Joel S. Kollin
Voice Phone: I 206 689 6966
e-mail address :jkolin@wizard.com
Business Description: Optical engineering consultancy in holography, 3-D imaging, projection, signs, scanning and illumination technology. Expertise includes holographic video, aerial image projection ("floating images"), HOEs, retinal scanning, head-mounted and autostereoscopic displays.
****

Holopak Technologies
(Transfer Print Foil s)
P.O. Box 538
East Brunswick, NJ
Postal Code: 08816
Country: United States of America
Contact: Rod Siberine
Voice Phone: 1 908 238 1800
Voice Phone: 1 800 235 FOIL
FAX Phone: 1 908 238 7936
Business Descripti on: Manufacture and development of holographic hot-stamping foils, films and metallized holographic paper and paper board. In-house metallizing, embossing and design center. We offer holographic images as well as TransFraction f.l patterns (gratings).
****

Holophile , In c.
56 Abner Lane
Killingworth , CT
Postal Code: 06419
Country: Unit ed States of America
Contact: Paul D. Barefoot
Voice Phone: 1 203 663 3030
FAX Phone: 1 203 663 3067
e-mail address: barefoot@holo phil e.com
Business Description: Founded in 1975, Holophile provides consulting services in

holography and "Spectral Imagery" (3-D projection of moving im ages) to corporations, museums and display builders. Our company is the premier producer in the world of holography exhibitions for museums, science centers and children's museums.
****

Holopublic Unbehaun
Hirschst rasse 84
Postal Code: 42285
Country: Germany
Contact: Kl aus Unbehaun
Voice Phone: 49 202 84 11 8
Business Description: Consulting, education, newsletters "Holography 3D Software" and "AHT Reflexionen", fine arts (Holofo tografik) book "Holo Show International", founding member "AHT-Association for Holography and New Media" .
****

Holos Art Galerie
4 Place Grenus
Postal Code: 120 I
Country: Switze rland
Contact: Pascal Barre
Voice Phone: 4 1 22 32 5 191
Business Description: Gallery, retail sales.
****

Holosco, Ernest Barnes
Bajada de Viladecols. 2
Barcelon a, Spain
Posta I Code: 08002
Country: Spain
Voice Phone: 34 3 310 71 13
FAX Phone: 34 3 3 19 16 76
Business Description: Holography Lab. Reflection and transmission, transfer to photoresist - embossing facilities. Consulting services.
****

Holostik India Pvt. Ltd.
50, Adhchini
Sri Aurobindo Marg.
Postal Code: 11 0017
Country: India
Contact: Govind Sharma
Voice Phone: 9 1 11 665 690
Voice Phone: 91 11 669 725
FAX Phone: 91 11 686 8828
Business Description: We are among the first to have set up a fully automated plant for manufacture of security and promotional holograms and films in India. Soon setting up master lab and 40 inch wide web machine for holographic packaging.
****

Holotek
55 Sci ence Pkway
Rochester, NY
Posta l Code: 14620
Count ry: United States of America
Contact: Roger O'Brien
Voice Phone: 1 716 244 6000
Voice Phone: 1 800 822 8525
FAX Phone: 1 716 244 6048
Business Description: Engineering and design of laser optic scanning systems.
****

Holovision
Tumblinstr 32 RGB
Postal Code: 80337
Country: Germany
Contact: Julian Fischer
Voice Phone: 49 89 746 9336
FAX Phone: 49 89 746 9382

Business Description: Des igners/Producers of large format holograms (ho lographic stereograms) up to 40 x 40 inches (I sq meter), manufacturer of large format holoprinter and holotrack equipment , produclion of color holograms on silver halide, computer generated holograms, portrait, displays, consulting.
****

Holovision AB
Box 70002
Posta l Code: 10044
Country: Sweden
Contact: Jonny Gustafsson.
Vo ice Phone: 46 8 331 186
FAX Phone: 46 8 331 186
Business Description: Specializing in silver halide holography with pulsed lasers. Denisyuk and transferred-type reflection holograms up to 30 x 40 cm. Rainbow holograms with pulsed laser up to 2 x 1 m.
****

Holovision Systems Inc.
11 9 South Main St
Findlay, OH
Posta l Code: 45840
Country: United States of America
Contact: Roland L. Kirk
Voice Phone: 1 419 422 3604
FAX Phone: 1 419 422 4270
****

Honeywell Technology Center
1070 I Lyndale Avenue South
Bloomington , MN
Postal Code: 55420
Country: Un ited States of America
Contact: Dr. J. Allen Cox
Voice Phone: I 6 12 95 1 7738
e-mail address:Cox@SRC. Honeywell .com
****

HRT (Hologra phie Recording Tec hnologies Gmbh)
Am Steinaubach 19
Postal Code: 36396
Country: Germany
Contact: Ric hard Birenheide
Voice Phone: 49 6663 7668
FAX Phone: 49 6663 7463
Business Description: Silver Halide Emulsions
****

HRT Holographic Reco rding Tec hnologies GMBH
AM Steinaubach 19
Postal Code: 36396
Country: Germany
Contact: Dr. Biren Heide
Voice Phone: 49 6663 7668
FAX Phone: 49 6663 7463
Business Description: Own production of silver halide emulsions with low noise and high diffraction efficiency.
****

Hughes Power Products, in co
1925 East Maple Ave.
Mail Station SC/S 13/G343
El Segundo, CA
Posta l Code: 90245
Country: United States of America
Contact: John E. Gunther
Vo ice Phone: 1 310 414 7086
FAX Phone: 1 3 10 726 0008
Business Descriptio n: Hughes' Power Products department, a pioneer in holography since 1974, offers complete line of holograms.

mastering. development and production services for DCG and photopolymer including HOEs, image master ing, and mass production/replication on photopolymer film.
****

Hyogo Prefectual Museum of Modern Art
Art Curator
Kobe-3- 8-3 Harada-Dori
Country: Japan
Contact: Hitoshi Yamazaki
Voice Phone: 81 78801 159 1
FAX Phone: 81 788614731
Business Description: 20th century Art, History of Art and Holography, Art and Optics, curating a exhibition of holography into Art.
****

I. S. Gill
214 Kailash Hills
East of Kailash
Postal Code: 110065
Country: India
Voice Phone: 91 II 1684 0377
Voice Phone: 91 II 6847 0377
Business Description: Bindry - application of holographic foil and stickers.
****

fBM Almaden Resea rch Center
K24/802
650 Harry Road
San Jose, CA
Posta l Code: 95120
Country: United States of America
Contact: Michael Ross
Voice Phone: 1 408 927 1283
FAX Phone: I 408 927 30 II
e-mail address:mikeross@almaden.ibm.com
Business Description: Scientific holography research; holographic storage.
****

iC Holographics
8 Flitcroft St.
London, England
Postal Code: WC2H 8DJ
Country: United Kingdom
Voice Phone: 44 171 240 6767
FAX Phone: 44 171 240 6768
e-mail address: 100413 .3406@compuserve
Business Description: Holographic Design and Digital Mastering.
****

lei Americas
Concord Plaza
3411 Silverside Road
Wi lmington , DE
Postal Code: 19850
Country: United States of America
Voice Phone: I 302 887 3000
Business Description: Films polyester film.
****

Ill inois Institute Of Technology
Mechanical/Materials & Aerospace Engineering
Engineering Building #1 Rm 252-B
Chi cago, IL
Postal Code: 60616
Country: United States of America
Contact: Cesar Sciammarclla
Voice Phone: I 312 567 3220
FAX Phone: I 312 567 7230
e-mail address: mesciammarella@mimna.iit.ezu
Business Description: Holographic interferometry; industrial holographic research: nondestructive testing.

****

IIluminations
1223 7th Avenue
San Francisco, CA
Postal Code: 94122
Country: United States of America
Contact: Louis Brill
Voice Phone: I 415 664 0694
Business Description: Involved in developing & expanding market & sales efforts for holographic retail/wholesale product lines. Assist in preparation of promotions and collateral sales materials, identify potential sales markets & impl ementation of sales.
****

Imac Internationa l, Inc.
130 I Greenwood
Wilmette, IL
Postal Code: 60091
Country: United States of America
Contact: J. Kauffmann
Voice Phone: 1 708 256 6646
Business Description: Holography marketing consultants.
****

Imagen Holography, Inc
303 Aspen ABC
Suite J
Aspen, CO
Postal Code: 8 1612
Country: United States of America
Contact: Alan P. Morterud
Voice Phone: I 970 925 8044
FAX Phone: I 970 925 9 176
Business Description: Specialized holographic products for main stream marketing app lications, including Advanced Holographic Textiles - a soft, supple, fully washable rendering of reflection holograms in both 2D & 3D, to a variety of fabrics.
****

Imagenes Holograficas De Colum bia
Avenida Quinta Norte No 17-23
AA 3076 PBX 685450
Country: Columbia
FAX Phone: 5790 151084 12695
****

Images Company
P.O. Box 140742
Staten Is land , NY
Postal Code: 10314
Country: United States of America
Conta ct: Barbra Metz
Voice Phone: I 7 18 698 8305
FAX Phone: 1 7189826 115
Business Description: Sells holographic equipment targeted toward schools, students and private holographers. Lase rs, film , development kits, optics, etc.
****

Imagination Plantation
159 Caselli Ave.
San Francisco, CA
Postal Code: 941 14
Country: United States of America
Voice Phone: I 415 487 0841
FAX Phone: I 4 15431 9961
Business Description: 3D imaging and content creation for all media. Experienced in modeling for holographic applicat ions, including direct output to master. For a look at our work, see front cover.
****

Imaging & Design
1 101 Ransom Road
Grand Island, NY
Postal Code: 14072- 1459
Country: United States of America
Voice Phone: 1 716 773 7272
Business Description: Designer using holography.
****

ImEdge Technology
2123 Fountain Court
Yorktown Heights, NY
Postal Code: 10598
Country: United States of America
Contact: Michael Metz
Voice Phone: I 914 962 1774
FAX Phone: I 914 962 1774
e-ma il add ress:
7 1341.3 106@compuserve.com
Business Description: Research, development and manufacturing of edge- lit holograms; creative and innovative holography and optics applications and problem-solving; consulting; hologram brokering.
****

Imperial College Of Science
Optics Section
Blackett Laboratory
London, England
Postal Code: SW7 2BZ
Country: United Kingdom
Contact: J. Dainty
Voice Phone: 44 171 589 5 111
Business Desc ription: Courses in holography; scientific holography research; particle measurement.
****

Industrial Technology Research Inst.
Holography Departm ent
Bldg 44, 195 Chung Hsing Road, Section 4
Postal Code: 3 101 5
Country: Taiwan
Contact: Dr. J.J. Su
Voice Phone: 886 35 9 17 482
FAX Phone: 886 35 9 17479
Business Description: Research in HUD (Head Up Display) and Dot Matrix Hologram systems
****

Infinity Laser Laboratories
2506 Flanders Station
P.O. Box 777
Polk City, FL
Postal Code: 33868
Country: Un ited States of Ameri ca
Contact: Thad Cason
Voice Phone: I 941 984 3108
FAX Phone: I 941 984 4244
e-mail address:infinity@aol.com
Business Description: Full service company from concept and design to finished product. Laboratories include fu ll color photopolymer capability with pulsed, continuous wave and solid state lasers including Argon, Krypton, Nd:YAG:KTP and HeCd
****

Infox Corporation
3rd floor No 283
Sec2 Fu-hsing South Road
Country: Taiwan
Contact: Alex c.T. Chen
Voice Phone: 886 2 7056699
FAX Phone: 886 2 7551800
Business Description: Maker of injected molded holograms.

****

Infrared Optical Products, Inc.
PO Box 3033
South Farmingdale, NY
Posta l Code: 11735-0664
Country: United States of America
Contact: Barry Bassin
Voice Phone: I 5 166946035
FAX Phone: I 516 694 6049
Business Description: Manufacturer of infra-
red lenses, windows, reflectors, beam splitters,
computer-designed IR lens systems. Front sur-
face and non-linear optical coatings for mir-
rors.

****

Inrad, Inc
18 1 Legrand Avenue
Northvale, NJ
Postal Code: 07647
Country: United States of America
Contact: Maria Murra
Voice Phone: I 20 I 767 1910
FAX Phone: I 201 767 9644
Business Description: Manufacturer of nonlin-
ear materials, harmonic generation systems,
electro-optic and acousto-optic devices and
drivers. Also provides optical components, as-
semb lies and optical coatings for the uv. vis-
ible and IR.

****

Institute Of Optical Science
Centra l University
Postal Code: 32054
Country: Taiwan
Contact: Tang Yaw Tzong
Voice Phone: 886 3 425 7681
FAX Phone: 886 3 425 88 16
Business Description: HOEs, academic research.

****

Institute Of Plasma Physics
And Laser M icrofus ion
PO Box 49
Postal Code: 00-908
Country: Poland
Contact: Zbigniew Sikorsky
Business Description: Academic research

****

Integraf
745 North Waukegan Road
Lake forest, IL
Postal Code: 60045
Country: United States of America
Contact: T.H. Jeong
Voice Phone: 1 708 234 3756
FAX Phone: 1 708 615 0835
e-mail address: jeong@lfmial.lfc.edu
Business Description: Distribute holographic
films and plates. We also carry pre-packed pro-
cessing chemicals, and a variety of stock holo-
grams .

****

Interactive Industries
696 Plank Road
Waterbury, CT
Postal Code: 06705
Country: United States of America
Contact: Ronald Phillips
Voice Phone: 1 203 755 2111
FAX Phone: 1 203 755 3999
Business Description: Stock holographic
items such as decals, labels, desk accesso-
ries, etc.

****

Interferens Holografi D.A.
Museum, Gallery, Stud io
Halvor Hoels Gt 6
Postal Code: N-2300
Country: Norway
Contact: Olav Skipnes
Voice Phone: 47 62 25050
Voice Phone: 47 62 30659
FAX Phone: 47 62 30659
Business Description: Ongoing exhibition
of Norway:s largest collection of holo-
grams. Makes glass (mainly reflection) ho-
lograms of museum exhibit s. Contin uous
wave laser. Norway's largest collection of
holograms. Our specialty: museum exhib-
its.

****

International Hologram
Manufacturers Association
Runnymede Malthouse
Runnymede Road, England
Postal Code: TW20 9BD
Country: United Kingdom
Contact: Ian M. Lancaster
Voice Phone: 44 1784 430 447
FAX Phone: 44 1784 431 923
Business Description: Promotes the interests of
its members-full and associate-and the hologra-
phy industry worldwide. founder members are
among the leading manufacturers in each con-
tinent. Current president: Philip Hudson, De La
Rue Holographics.

****

In trepid World Communications
555 South Woodward, Suite 1109
Birmingham, MI
Postal Code: 48009
Country: United States of America
Contact: Ann Marie Harrison
Voice Phone: 1 8 10 642 9885
fAX Phone: 1 810 642 9886
Business Description: All types of holograms. in
particular true-co lor holograms. Subsidiary of
American Propylaea Corp.

****

Ion Laser Technology In c.
3828 South Main
Salt Lake City, UT
Postal Code: 84115
Country: United States of America
Contact: Jeff Smith
Voice Phone: 1 80 I 262 5555
FAX Phone: 1 80 I 262 5770
Business Description: Manufacturer of aircooled
argon lasers.

****

Ishii, Ms. Setsuko
#404,
1-23 26 Kohinata ,Bunkyo-Ku
Postal Code: 102
Country: Japan
Contact: Setsuko Ishii
Voice Phone: 81 03 3945 90 17
FAX Phone: 81 03 3945 9068
Business Description: Holographic fine Artist.

****

James Feroe
1420 45th Street #359
Emeryville, CA
Postal Code: 94608
Country: United States of America
Voice Phone: 1 510 658 9787
Business Description: Consultant with 15 years
hands-on experience in holography.

****

James River Products
800 Research Road
Richmond, VA
Postal Code: 23236
Country: United States of America
Contact: Mike florence
Voice Phone: 1 804 378 1800
fAX Phone: I 804 378 5400
Business Description: World leader in embossed
hologram machinery. Products include: origination lab equ ipment, photoresist plate spinning, electroform facilities, embossing machines, die cutting equipment, supporting
technology and training -- plus custom embossed holograms.

****

Japan Communication Arts Co.
Yonezawa Bldg2f
2-37 Suehirocho, Kita-Ku
Postal Code: 530
Country: Japan
Conta ct: Mineko Fukuma
Voice Phone: 81 06314 19 19
FAX Phone: 8106315 1900
Business Description: Sales of cards with holo-
gram.

****

Jayco Holographics
29-43 Sydney Road
Watford, England
Postal Code: WD I 7PY
Country: United Kingdom
Contact: Rohit Mistry
Voice Phone: 44 1923 246 760
fAX Phone: 44 1923247769
Business Description: Complete production service for embossed holograms. Embossing masters through to fu lly finished product. Sixteen
years of experience enables Jayco to offer out-
standing quality of product and service at competitive prices.

****

Jodon Inc.
62 Enterprise Drive
Ann Arbor, MI
Postal Code: 48103
Country: United States of America
Contact: John Gillespie
Voice Phone: I 313 761 4044
fAX Phone: I 313 761 3322
Business Description: Manufacturer of HeNe
lasers, laser systems, specialty laser tubes, opti-
cal and electro-optical instruments and systems.
Holographic films , plates and chemicals. Engi-
neering services.

****

Jr Holographics
Suite 2065
100 Wilshire Blvd
Santa Monica, CA
Postal Code: 90401 -11 35
Country: United States of America
Contact: Judy Roberts
Voice Phone: 1 3 10 393 2388
fAX Phone: 1 3103938611
Business Description: Marketer and consul-
tant for clients around the world who wish
to display an image, logo, or product in three
dimensions with expertise in entertainment li-
censing.

****

K.C. Brown Holographics
17 Salisbury Road
New Malden

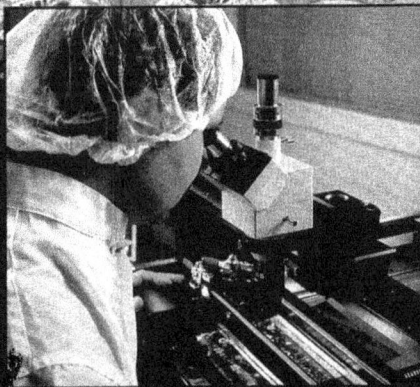

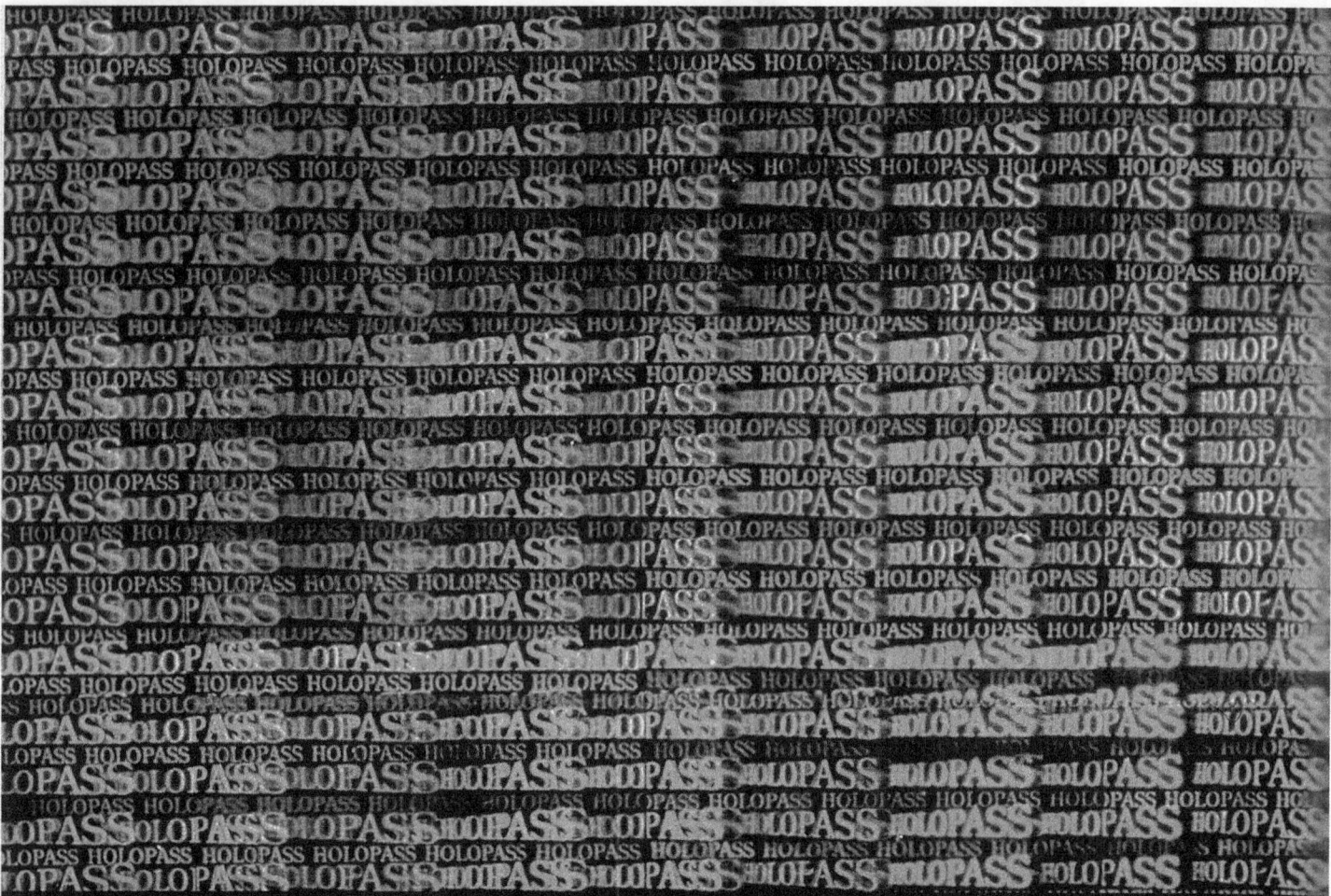

Surrey, England
Postal Code: KT3 3HZ
Country: United Kingdom
Contact: Kevin C. Brown
Voice Phone: 44 181 942 8294
FAX Phone: 44 181 942 8294
Business Description: Pulse portraits; artistic holography.
****

Kaiser Optical Systems, Inc.
POBox 983
371 Parkland Plaza
Alm Arbor, MI
Postal Code: 48106
Country: United States of America
Contact: James McNaughton
Voice Phone: 1 313 665 8083
FAX Phone: 1 313 665 8199
Business Description: We specialize in HUD (Holographic Head-Up display) projects and Holographic Optical Elements. Many years of experience with military projects as well.
****

Karas Studios S.L.
Hospital, 12
Postal Code: 28012
Country: Spain
Contact: Ramon Benito
Voice Phone: 34 1 530 89 88
FAX Phone: 34 1 530 89 88
Business Description: Established in 1988; art exhibitions, art gallery, private collection.
****

Karolinska Institutet
School Of Dentistry
Box 4064
Postal Code: S-14104
Country: Sweden
Contact: Hans Rvden
Voice Phone: 46 -8 774 0080
Business Description: Holography research applied to dentis try.
****

Kauffman , John
Box 477
Point Reyes Station, CA
Postal Code: 94956
Country: United States of America
Contact: John Kauffman
Voice Phone: 1 415 663 1216
FAX Phone: 1 415 663 1216
Business Description: Holographic Fine Artist. Specializes in multi color reflection holograms.
****

Keio University
Dept Of Electrical Engineering
3-14-1 Hiyoshi Kohoku-Ku
Postal Code: 223
Country: Japan
Contact: Dr. Masato Nakajima
Voice Phone: 81 045 563 1141
FAX Phone: 81 045 563 3421
Business Description: Research using HNDT
****

Kendall Hyde Ltd.
Kingsland Industrial Park
Stroud ley Road
Basingstoke, England
Postal Code: RG24 8UG
Country: United Kingdom
Contact: M. Kendall
Voice Phone: 44 1256 840 830
FAX Phone: 44 1256 840 443

Business Description: Optical coating specialists. front surface mirrors, etc.
****

Keystone Scientific Co.
PO Box 22
Thorndale, PA
Postal Code: 19372-0022
Country: United States of America
Contact: Ed Kelly
Voice Phone: 1 610 269 9065
fAX Phone: 1 6102694855
Business Description: Distributor for Agfa and Kodak holographic films plates and chemicals. Manufacturer of holography kits and automatic processors.
****

Kimmon Electric Co. , Ltd.
TM21 Building
1-53-2 Itabashi , Itabashi-Ku
Postal Code: 173
Country: Japan
Voice Phone: 81 3 5248 4822
fAX Phone: 81 3 52480018
Business Description: Manufacturer of Helium Cadium lasers with 10cm and 30cm coherence for use in holography. Also manufacturer HeNe and air cooled argon lasers.
****

Kinetic Systems, Tnc.
20 Arboretum Road
Boston, MA
Postal Code: 02131
Country: United States of America
Contact: Moss Blosvern
Voice Phone: 1 617 522 8700
Voice Phone: 1 800 992 2884
fAX Phone: 1 617 522 6323
Business Description: Manufacturers of Vibraplane standard and special Honeycomb optical tables in four grades up to 5' x 12' x 24". Larger sizes available by butt splicing. Also vibration isolation support systems.
****

Kolbe-Druck mit Tochtergesell sc haften
In Industriegelande 50
Postal Code: 33775
Country: Germany
Contact: Roland Pahnue
Voice Phone: 49 5423 9670
fAX Phone: 49 5423 41 230
Business Description: Bindry: Hot stamping holographic foil
****

Kreischer Optics, Ltd.
906 North Draper Rd
Mchenry, IL
Postal Code: 60050
Country: United States of America
Contact: Cody Kreischer
Voice Phone: 1 815 3444220
FAX Phone: 1 815 344 4221
Business Description: Custom Manufacturer of master and production test glasses, optical flats, lenses, condensers, cylinders, windows, filters, prisms, mirrors, beamsplitters, substrates, magnesium fluoride coatings. Consulting services in optical design.
****

Krystal Holographics International
555 West 57th Street
New York, NY
Postal Code: 10019
Country: United States of Ameri ca
Voice Phone: 1 212 261 0400

FAX Phone: 1 2122620414
Business Description: International distributor of a wide variety of holograms .
****

Krystal Holographics Vertriebs GmbH
Birnenweg 15
Postal Code: 72766
Country: Germany
Contact: Richard Stooss
Voice Phone: 49 7121 9461 85
fAX Phone: 49 7121 9461 10
Business Description: Di stribution of innovative photopolymer holograms based on DuPont-Omnidex film. High quality combines with low cost mass-production. Main product lines include photopolymer film with custom and stock hologram s, premium/gift items, wall-art and security applications.
****

L.A.S.E.R. Co.
1900 Gore Drive
Haymarket, VA
Postal Code: 22069
Country: United States of America
Contact: Jim Bowman
Voice Phone: 1 703 754 2526
Business Description: Fine art holograms and point of sa le holograms & display.
****

L.A.S.E.R. News
Laser Arts Society for Education and Research
PO Box 24-153
San Francisco, CA
Postal Code: 94124
Country: United States of America
Contact: Jeffrey Murray
Voice Phone: 1 415 822 7123
e-mail address:hologram@well.com
Business Description: Volunteer staffed nonprofit organization dedicated to holography and laser education and research. Members receive the quarterly L.A.S.E.R. News. One year membership USA $30, outside USA $40.
****

Label Systems/OSI
56 Cherry Street
Bridgeport, CT
Postal Code: 06605
Country: United States of America
Contact: John Halotek
Voice Phone: 1 203 333 5503
fAX Phone: 1 203 336 8570
Business Description: Holography labels from start to finish.
****

Labor Dr. Steinbichler
Am Bauhof 4
Postal Code: 83115
Country: Germany
Contact: Dr. H. Steinbichler
Vo ice Phone: 49 8035 87040
fAX Phone: 49 8035 1010
Business Description: Holographic investigations,
developments on contract basis; application laboratory for: vibration analysis, nondestructive testing, deformation measurements, contour measurements, image analysis; pulsed and CW-lasers, motor test bench, computer based evaluation.
****

Laboratories of Image Information
Science and Technology
LC Bldg. 1Of 1-4-2 Shin-senri Higashi-mach i
Postal Code: 565

Country: Japan
Voice Phone: 81 06 873 2053
FAX Phone: 81 06 873 2056
Business Description: R&D on Holographic Display, Dynamic Holography, Holographic Optical Elements
****

Laborato ry Vinck iner
Holography Workshop Univ Gent
41 St Pietersnieuwstraat
Postal Code: B-9000
Country: Belgium
Contact: Pierre Boone
Voice Phone: 32 9 264 3242
FAX Phone: 32 9 223 7326
Business Description: Consultancy, education, problem-solving for display holography. Museum applications and (mainlyl) non-destructive testing.
****

Lake Forest Col lege
Center for Photonics Studies
Sheridan & College Road
Lake Forest, IL
Postal Code: 60045
Country: United States of America
Contact: Dr. Tung H. Jeong.
Voice Phone: I 708 735 5160
Voice Phone: I 708 735 5 163
FAX Phone: I 708 735 6291
e-mail address: jeong@lfmail.lfc.edu
Business Description: Each summer during July, Lake Forest Coll ege offers as-day hands-on workshop for participants who have no prior experience in holography. An advanced 6-day workshop follows. Write for information.
****

Laminex/High Tech Uk Ltd.
Bromfield Industrial Estates
Mold , England
Postal Code: CH7 I JR
Coun try: United Kingdom
Contact: Keith Green
Voice Phone: 44 1352 584 44
Business Description: Artistic holography; security holographic applications
****

Lasart Ltd.
PO. Box 703
Norwood, CO
Postal Code: 8 1423
Country: United States of America
Contact: Steven Siegel
Voice Phone: 1 970 3274701
FAX Phone: 1 970 327 4045
Business Desc ription: Lasart, Ltd. specializes in custom DCG work, from modelmaking, mastering and quality finishing. This includes production and limited ed ition jewelry, watches, and medium format composite sculpture.
****

Laser Affiliates
2047 Blucher Valley Road
Sebastopol, CA
Postal Code: 95472
Country: United States of America
Conta ct: Nancy Gorg lione
Voice Phone: 1 707 823 7 171
FAX Phone: 1 707 823 8073
Business Description: Laser Affiliates is an award-winning non-profit organization that designs innovative holographic and laser theatrical productions, in sta llations and exhibitions.

Services include curatorial guidance, videotapes and media lectures.
****

Laser Arts
1712 Cathedral Street
Plano, TX
Postal Code: 75023
Country: United States of America
Contact: M. Talbott
Voice Phone: I 2144230158
Business Description: Holographic consultants and implementers. Commercial utilization of holography, trade shows, unique promotions and museum exhibits (design, build , rent or sell). Venture capitalists consu ltants. Professionals in business, art, technology and applications.
****

Laser Focus World
10 Tara Boulevard 5th floor
Nashua, NH
Postal Code: 03062
Country: United States of America
Voice Phone: I 603 89 1 0 123
FAX Phone: I 603 891 0574
Business Description: Laser trade magazine; annual catalogue.
****

Laser Holography Works hop
320 South Willard Street
New Buffalo, MICHIGAN
Postal Code: 491 17
Country: United States of America
Contact: Joseph A. Farina
Voice Phone: 1 6 16 469 4658
FAX Phone: 1 616 469 4658
e- mail address:
102017,1330@compuse rve.com
Business Description: The laser holography workshop specializes in the creation of models for holography. We provide the international holography industry with quality work, timely service and reasonable rates.
****

Laser Innovations
668 Flinn Ave
Moorpark, CA
Postal Code: 93021
Country: United States of America
Voice Phone: I 805 529 5864
FAX Phone: I 805 529 6621
Business Description: Laser repair.
****

Laser Inspeck
360 rue Franquet, Bur 20
St Foy, Quebec
Postal Code: DIP 4N3
Country: Canada
Contact: Roger Lessard
Voice Phone: 1 4186502112
FAX Phone: 1 4186502141
Business Description: Marketer of the RoTech holographic interferometry camera and thermoplastic films. Also photo resist plates. Digital interferomet ry. 3-D digit izer (scans into computer the coordinates of object and color).
****

Laser Institute Of America
Educat ion Divi sion
12424 Research Parkway # 125
Orlando, FL
Postal Code: 32826
Country: United States of America
Contact: Jackie Thomas

Voice Phone: 1 407 380 1553
FAX Phone: 1 407 380 5588
Business Description: Lase r sa fety courses. Publishes Joulllal of Laser Applications. Hosts annual International Congress on Applications of Lasers and Electro-Optics (ICALEO), including holographic applications.
****

Laser International
19 ormanton Rise
Holbeck Hill
Scarborough, England
Postal Code: YOII 2XE
Country: United Ki ngdom
Contact: Keith Dutton
Voice Phone: 44 1723 364 452
Business Description: Holography Gallery. Specializing in associated laser display systems for hologram gallery exhibitions. Auto & manual control available . Any colorlaser. Inexpensive units from stock. Full display system to order.
****

Laser Ionies In c.
(Main Office)
70 I South Kirkman Road
Orlando, FL
Postal Code: 3281 I
Country: United States of America
Contact: Drew Nelson
Voice Phone: 1 407 298 1561
FAX Phone: 1 407 297 4167
Business Description: Manufacturer of gas ion lasers in cluding Argon, Krypton and mixed gases. Specializing in high power requiremen ts needing stable power in a compact package.
****

Laser Las Vegas
5725 N Fort Apaphe
Las Vegas, NV
Posta l Code: 89 129
Country: United States of America
Voice Phone: I 702 645 0477
FAX Phone: I 702 367 1209
Business Description: Laser sa les, repairs, rentals
****

Laser Light Designs
2412 Kennedy Way
Antioch, CA
Postal Code: 94509
Country: United States of America
Contact: Michael Malott
Vo ice Phone: I 510 754 3 144
Business Description: New product designs using embossed foil , tinsel and holographic films. I specialize in jeweled novelty designs. Designer of the original Rainbow Flasher, inventor of original Rainbow Sparkler.
****

Laser Light Ltd.
28 Old Fulton st
Brooklyn Heights , NY
Postal Code: 1120 I
Country: United States of America
Contact: Abe Rezny
Vo ice Phone: 1 212 226 7747
FAX Phone: J 718 858 2062
e-mail address:ddt-Iaser@aol.com
Business Description: Al l formats of holography. Designers and producers of 3-D imagIng.
****

Laser Movement
ZA 57. Ie Trou Grillon

Postal Code: 91280
Country: France
Contact: Patrice Pecheux
Voice Phone: 33 1 60 75 67 27
FAX Phone: 33 1 69 89 08 43
Business Description: Laser show.
****
Laser Optics, Inc.
III Wooster St
Bethel, CT
Postal Code: 0680 I
Country: United States of America
Contact: Jim Larim
Voice Phone: I 203 7444160
FAX Phone: I 203 798 7941
Business Description: A complete line of laser and optical components for ultraviolet, visible and infrared applications from 250 nm to 16 microns, including focusing lenses, windows, cavity components, prisms, beamsp litters, mirrors and coatings.
****
Laser Photonics, Inc.
12351 Research Parkway
Orlando, FL
Posta l Code: 32826
Country: United States of America
Contact: Steve Qualls
Voice Phone: I 407 281 4103
FAX Phone: I 407 380 3479
Business Description: Manufacturer of sc ientific and medical lasers. CW lasers. carbon dioxide & ruby lasers; equipment & supplies for holography.
****
Laser Reflections
25 North Second Street
San Jose, CA
Postal Code: 95113
Country: United States of America
Contact: Ron Olson
Vo ice Phone: I 408 292 7484
FAX Phone: I 408 292 81 IS
Business Description: Holographic portraiture, limited editions on glass up to 14" x 24", selfstanding displays and custom installations. Advanced Nd:YAG laser recording technology producing visibly superior holography of living subjects.
****
Laser Resa le Inc.
54 Balcom Road
Sudbury, MA
Postal Code: 01776
Country: United States of America
Contact: Jack Kilpatrick
Voice Phone: I 508 443 8484
FAX Phone: I 508 443 7620
Business Description: Laser Resale provides a marketplace for buying and selling pre-owned lasers, laser systems, optical tables and associated equipment for holographers.
****
Laser Technical Services
13 96 Ri ver Road, Box 248
Upper Black Eddy. PA
Postal Code: 18972
Country: United States of America
Contact: Dan Morrison
Voice Phone: 1 610 982 0226
FAX Phone: 1 6109820226
Busi ness Description: Teclulical consultant and field repair of lasers. Full customer service of laser equipment - Specifically Pulsed Ruby Holographic lenses.

****
Laser Technology, Inc.
IOSS West Germantown Pike
Norristown, PA
Postal Code: 19403
Country: United States of America
Contact: TOm Gleason
Voice Phone: I 610 631 5043
FAX Phone: I 6 10 631 0934
Business Desc ription: Manufacture equipment for laser-based NDT; Ho lography and Shearography equipment and inspection services. Portable and production units available.
****
Laserfilm Eckhard Knuth
Multi -Plex Holographie
Milchstrasse 12
Postal Code: 8 1667
Country: Germany
Contact: Eckhardt Knuth
Voice Phone: 49 89 480 77 14
FAX Phone: 49 89 485 666
Business Desc ription: Multiplex holograms (ste reograms).
****
Lasermedia
6833 Arizona Circle
Los Angeles, CA
Postal Code: 90045
Country: United States of America
Voice Phone: I 310 338 9200
FAX Phone: I 3103389221
Business Description: Install laser light show exhibiti ons.
****
Lasermetrics, Inc.
220 Midland ave
Saddl ebrook, NJ
Postal Code: 07662
Country: United States of America
Vo ice Phone: I 201 894 0550
FAX Phone: I 201 478 6115
Business Description: Laser components and electronic drivers for laser components.
****
Lasersmith , Inc.(The)
1000 West Monroe Street
Chicago, IL
Postal Code: 60607
Country: United States of America
Contact: Steven L. Smith
Voice Phone: I 312 733 5462
FAX Phone: I 312 733 5926
Business Description : Specialists in holographic imaging. In-house service: ali work origination, photo shoots of art/object, 2DI 3D full color 2D & 3D masteri ng . Stereogram filming; separations & mastering; computer modeling & rendering for full color.
****
Laserworks
PO Box 2408
Orange. CA
Postal Code: 92669
Country: United States of America
Contact: Selwyn Lissack
Voice Phone: I 714 832 2686
FAX Phone: I 714832 1451
Business Description: Responsible for major applications involving innovative holographic systems. Produced numerous holographic exhibitions in addition to all the Salvador Dali holograms.
****

Lasing S.A. ,
Marques De Pico Velasco 64
E-28027
Country: Spain
Voice Phone: 34 0 1 268 3643
Business Descri ption: Branch office of Newport Corporation
****
Lasiris Inc.
Main Office
3549 Ashby
Ville St Laurent, Quebec
Postal Code: H4R 2K3
Country: Canada
Contact: Alain Beauregard
Voice Phone: I 514 335 1005
FAX Phone: I 5 14 3354576
Business Description: Embossed holography; HOE optics in stock Diffusers, gratings. HOE special projects.
****
Lauk Kommunikation
A ugustinysstr 96
Postal Code: 50226
Country: Germany
Contact: Laug Mathias
Voice Phone: 49 02234 963 120
FAX Phone: 49 02234 963 1255
Busi ness Desc ription: Holograms, Holographic projects, Hologram Museum
****
Lawrence Berkeley Laboratory
Un ivers ity Of California
I Cyclotron Road
Berkeley, CA
Postal Code: 94720
Country: United States of America
Voice Phone: I 510 486 4000
Business Description: Industrial & academic holography research. Will do commercial research projects.
****
Laza Holograms Ltd.
6 Marlborough Road
Richmond Surrey, England
Postal Code: TWIO 6JR
Country: Un ited Kingdom
Contact: Jonathan Cope
Voice Phone: 44 18 1 332 1080
FAX Phone: 44 181 332 2990
Business Descript ion: manufact urer of high quality mass produced reflection holograms. 120 stock images in various sizes. Production of custom holograms on silver halide, photopolymer and embossed.
****
Lazart Holographics.
22 Erina Valley Road
Erina, NSW
Postal Code: 2250
Country: Australia
Contact: Brett Wilson
Voice Phone: 61 043 676 245
FAX Phone: 61 043 652 306
Business Description: Artistic holography; buying & selling holograms. Wholesale distribution and retail sa les of artist editions and stock images. Production of jewelry items and novelty products from embossed images. Gallery exhibition open 7 days.
****

Lazer Wizardry
5805 West 6th Ave
Suite B
Lakewood, CO
Postal Code: 80214
Country: United States of America
Conta ct: Richard M. Osada
Voice Phone: 1 303 274 0706
Voice Phone: 1 800 793 0506
FAX Phone: 1 303 274 0733
Business Description: Wholesale holography.
Servicing specialty holographic and other retail
stores. One of the largest se lections available
anywhere, including some industry exclusives.
Custom broker, all types of holography all mar-
kets.
****

Leitfoil In c.
Box 489
Postal Code: POH IZO
Country: Canada
Contact: Bruce Loucks
Voice Phone: 1 705 724 6 164
Voice Phone: 1 800 4656 8735
FAX Phone:1 705 724 6249
Business Description: Holographic foil applica-
tion onto textiles, letifoil (TM) can be combined
with screen printing, manufacturing plants in
Canada and Asia.
****

Lenox Laser
12530 Manor Road
Glen Arm, MD
Postal Code: 21057
Country: United States of America
Contact: Joseph P. D'Entremont
Voice Phone: 1 410 592 3106
FAX Phone: 1 410 592 3362
Business Description: Laser- systems laboratory
specializing in laser drilling, electron beam
welding, EDM machining, and water jet. Offers
pre-fabricated aperture kits.
****

Leonhard Kurz GmbH
Schwabacher Strasse 482
Postal Code: 90763
Country: Germany
Contact: Werner Reinhart
Voice Phone: 49 911 7 141 0
FAX Phon e: 49 911 7141507
Business Description: Manufacturer of em-
bossing equipment; broker for hologram em-
bossing.
****

Les Productions Hol olab!
3970 Boulevarde St Laurent
Montreal, Quebec
Postal Code: H2W I Y3
Country: Canada
Contact: Marie-Christiane Mathieu
Voice Phone: I 514 849 4325
Business Description: Artistic holography
****

Leseberg, Dr. Detlef
Kamener Str 172
Postal Code: W4670
Country: Germany
Contact: Dr. Detlef Leseberg
Voice Phone: 49 2306 1794
FAX Phone: 49 2306 1793
Business Description: Scientific holography
research , HOE, computer-generated hologra-
phy.
****

Letterhead Press, Inc .
W226 N880 Eastmound Drive
Waukesha, WI
Postal Code: 53186-1689
Country: United States of America
Contact: Mark Mul vaney
Voice Phone: I 414 574 17 17
FAX Phone: I 414574 1719
Business Description: Full service trade finisher
with 24-hour, 7-days/week manufacturing. Fea-
turing 19 x 25 inch and 40 inch formats for holo-
graphic stamping. Complete projects from print
to final bindery assuring single-source respon-
sibility.
****

Levine, Chris
6B Wescott Rd
Kennington
London, England
Postal Code: SE 17 3QY
Country: United Kingdom
Contact: Chris Levine
Voice Phone: 44 171 582 8 14
FAX Phone: 44 171 490 2693
Business Description: Fine art holograms.
****

Lexel Laser, Inc.
48503 Milmont Drive
Fremont, CA
Postal Code: 94538
Country: United States of America
Voice Phone: I 510 770 0800
FAX Phone: I 510 65 1 6598
Business Description: Lexel produces the highest quality Argon , Krypton and mixed gaas laser
systems. In particular, Lexel specializes in production of single frequency systems which are
very stable over a variety of environmental
situations.
****

Liconix
3281 Scott Boulevard
Santa Clara, CA
Posta l Code: 95054
Country: United States of America
Contact: Michael Fisk
Voice Phone: I 408 496 0300
FAX Phone: I 408 492 1303
Business Description: LiCONiX, long the recog-
nized leader in Helium Cadmium laser technolo-
gy, also supp lies semiconductor diode laser sys-
tems and a recently introduced line of ion lasers.
****

Light Dimension, Inc.
Sunfamily Hongo #403
5-10 , Hongo 4-Chome, Bunkyo-ku
Postal Code: I 13
Country: Japan
Contact: Mariko Oishi
Voice Phone: 81 3 38 12 920 I
FAX Phone: 81 338129422
Business Description: Handling the whole range
of holograms/holographic products from em-
bossed holograms to fine art images; also focus-
ing on exhibitions on holography.
****

Light Fantastic Pic
4E/F Gelders Hall Road
Shepshed, England
Postal Code: LEI2 9NH
Country: United Kingdom
Contact: Peter H.L. Woodd
Voice Phone: 44 509 600 220
FAX Phone: 44 509 508 795

Business Descriptio n: A fully -integrated
holographic business providing the creative and technical services that produce
innovative standard and custom-designed
holograms of the highest quality. Total service covers embossing and finished product.
****

Light Impressions Europe
333 East Ontario Street, Suite [902 B
Chicago, IL
Postal Code: 6061 I
Country: United States of America
Contact: Pame la Jamison
Voice Phone: I 312 255 1320
FAX Phone: I 312 155 1321
Business Description: Distributor of a large va-
riety of Holographic Products. Please call for
catalogue.
****

Light Impress ions Europe Ltd.
5 Mole Business Park 3
Leatherhead , England
Postal Code: KT22 7BA
Country: United Kingdom
Contact: John Brown
Voice Phone: 44 372 386 677
FAX Phone: 44 372 386 548
Business Description: Distributor of a large va-
riety of Holographic Products. Please call for
catalogue.
****

Light Wave Gallery
2801 Leavenworth St.
San Francisco, CA
Posta l Code: 94133
Country: United States of America
Voice Phone: I 41 54740133
Business Description: Complete line of holo-
grams. Watches and 3-D images as well.
****

Light Wave Gall ery
North Pier
435 East Illinois Street
Chi cago, IL
Postal Code: 6061 1
Country: United States of America
Contact: Jim Harden
Voice Phone: I 3 12 32 1 1123
Business Description: Gallery, retail shop. Com-
plete line of holograms .. 2 locations: Chicago
and San Francisco
****

Lightrix
377 Oyster Point Blvd #1 [
South San Francisco, CA
Postal Code: 94080
Country: United States of America
Contact: Deborah Robinson
Voice Phone 1 415 244 9791
FAX Phone: 1 415 244 9795

Business Description: Lightrix manufactures
and sells high quality holographic puzzles,
stickers, and other various holographic toys and
wall decor. We offer a full custom program for
embossed or photopolymer.
****

Lone Star Illusions
290 I Capital Of Texas Highway, # 191
Austin , TX
Postal Code: 78746
Country: United States of America
Contact: Alan L i fshen
Voice Phone: 1 512 328 3599

Business Description: Gal lery and retail shop.
****

Lopez's Gallery Inte rnational
500 North Michigan Ave.
Chicago, [L
Postal Code: 6061 1-3704
Country: United States of America
Contact: Mark Holtzblatt
Voice Phone: 1 3 12 975 2052
Voice Phone:1 800 4D-F INE ART
FAX Phone: 1 312 248 9527
Business Description: Creating the world's most collectible holographic Fine Art. Specialists in 4 dimensional female figure studies and holographic sculptures. Models, masters, and transfers created for artistic , scientific and commercial applications.
****

Los Angeles School Of Holography
PO Box 85 1
Woodland Hill s, CA
Postal Code: 91365
Country: United States of America
Contact: Mandy Fox
Voice Phone: 1 8 18 703 1111
FAX Phone: 1 818 703 1182
Business Description: The Los Angeles School of Holography offers a 3 day class. Students learn all phases of holography, and produce both laser viewable transmission and white light viewable holograms in silver halide format
****

Loughborough Univ. Of Tech.
Dept Of Phys ics
Loughborough, England
Postal Code: LEll 3TU
Country: United Kingdom
Contact: Prof. Nick Phillips
Voice Phone: 44 509 263 171
FAX Phone: 44 509 219 702
Business Description: Embossing masters/ shims; Scientific, industrial research. The Univers ity and Markem Systems (UK) are participating in a joint ve nture, Advanced Holographic Laboratories.
****

Lulea Un ive rsity Of Technology
Dept Of Mechanical Engineering
Postal Code: S-951 87
Country: Sweden
Contact : Nils-Erik Molin
Business Description: Industrial research; holographic non-destructive testing.
****

Lumenx technologies
PO box 2 19
New Durham, NH
Postal Code: 03855
Country: United States of America
Voice Phone: 1 603 859 3800
FAX Phone: 1 603 859 3800
Business Description: We manufacture laser equipment. Laser repairs.
****

Lumonics Inc.
105 Schneider Road
Kanata, Ontario
Postal Code: K2K I Y3
Country: Canada
Voice Phone: 1 613 592 1460
FAX Phone: 1 613 592 5706
Business Description: Lumonics is a manufacturer of high power pulsed ruby lasers for portrait holography and engineering holocameras for NOT. Other products include laser marking and materials processing systems.
****

Lumonics Ltd
Cos ford Lane
Swift Valley
Rugby, England
Postal Code: CV2 1 I QN
Country: United Kingdom
Contact: George Synowiec
Voice Phone: 44 1788 570 321
FAX Phone: 44 1788 579 824
Business Description: Lumonics manufactures pulsed lasers for a range of industrial and scientific applications including pulsed ruby lasers for Holography. Single pulse and multiple pulse units available for commercial, research and NDT applications.
****

Lund In stitute Of Tech.
Department Of Phys ics
Box 118
Postal Code: S-221
Country: Sweden
Contact: Sven-Goran Patte rsson
Voice Phone: 46 046 222 7656
FAX Phone: 46 046 222 4017
e-mail address: sevengoran. pattersson@fysik.lth.se
Business Description: Color H1 ; holography education; academic research.
****

M.LT.
Media Laboratory/ Spatia l Imaging
20 Ames Street # E 15-416
Cambridge, MA
Postal Code: 02139
Country: United States of America
Contact: Menssa Yoon
Voice Phone: I 617 253 0632
FAX Phone: 1 6 17 253 8823
e-mail address: sab@media.mit.edu
Business Description: College holography courses; Computer Generated Holography research.
****

M.LT. Museum
265 Massachuesetts Ave
Cambridge, MA
Postal Code: 02139-4307
Country: United States of America
Contact: Warren Seamans
Voice Phone: 1 617 253 4444
FAX Phone: I 617 253 8994
Business Desc ription: Museum has 1,500 holograms. Some of the most historically-significant holograms made are on display here.
****

M.O .M. [nco
2436 Forest Green Rd.
Baltimore, tvID
Postal Code: 2 1209
Country: United States of America
Contact: Alan Evan
Vo ice Phone: 410 484 0496
Business Description: Maryland Optical Man-ufacturing. Highest quality. 37 years experi-ence.
****

MacShane Holography
CIO Laser Arts Programs
512 West Braeside Dri ve

Arlington Heights, IL
Postal Code: 60004
Country: United States of America
Contact: Jim Macshane
Voice Phone: 1 708 398 4983
Business Description: Design and manufactur-ing of Sunbows; sculptural, arc hitectural , and gift embossed holographic products; educational programs and artistic holography.
****

Magic Laser
Quarrier De L'Horloge
4 Rue Brantome
Postal Code: 75003
Country: France
Contact: Anne-Marie Chri stak is
Voice Phone: 33 I 42 74 35 78
FAX Phone: 33 I 42 74 33 57
Business Description: Importer and wholesaler of all holographic products--travelling exhibi-tion.
****

Magick Signs Holograp hie
Sudliche Ringstrasse 234
Postal Code: 63225
Country: Germany
Contact: Andreas Wollenwerber
Voice Phone: 49 61 03 25248
FAX Phone: 49 61 03 26623
Business Description : Design and production of embossed hologram s.
****

Man Environment, Inc.
2251 Federal Avenue
Los Angeles, CA
Postal Code: 90064
Country: United States of America
Contact: Gary Fisher
Voice Phone: I 3 10477 7922
e-mail address:gfisher@netcom.com
Business Description: Silver Halide R&D proj-ects, create optical printers, and any holographic systems. Multiplex machines.
****

Markem Systems Ltd.
Astor Road
(Eccles New Road)
Salford, England
Postal Code: M5 2DA
Country: United Kingdom
Contact: Jane Oliver
Voice Phone: 44 61 789 8131
FAX Phone: 44 617075315
Business Description: One-stop-Shop for em-bossed hot stamping foil and laminating film, including everything from origination to foil manufacture
****

Martinsson Elektronik Ab.
Instrum entvagen 16
Box 9060
Postal Code: S-126 09
Country: Sweden
Contact: Mikael Hell
Voice Phone: 46 08 744 0300
FAX Phone: 46 08 744 3403
Business Description: Equipment & supplies.
****

Marubun Corporation
8-1 Nihombashi Odemmacho
Chuo-Ku
Postal Code: 103
Country: Japan
Voice Phone: 81 03 648 8115

FAX Phone: 81 03 648 9398
Busi ness Description: Branch office of Newport
Corporation, Fountain Valley, CA USA
****

Maste rPrint Holography, Inc.
25P Exec uti ve Drive
Edgewood, NY
Postal Code: 11717
Country: United States of America
Contact: Michael Liu
Voice Phone: I 516 243 0170
FAX Phone: I 516243 01 80
Business Description: Specialists in photoresist
mastering for mass production.
****

Mazda Motor Corp.
Technical Research Center
POBox 18
Postal Code: 730 91
Country: Japan
Contact: lchiro Masamori
Voice Phone: 81 082 282 1111
FAX Phone: 81 082 252 5343
Business Description: Holographic Interferom-
etry
****

McCain Marketing & Graphic Design
10962 North Wauwatosa Road
76W
Mequon , WI
Postal Code: 53097
Country: United States of Ameri ca
Contact: Richard Mccain
Voice Phone: 1 414 242 4023
Business Description: Graphic designer and
litho printing consultant, 20 years experience.
Extensive understanding of holographic effects
to educate or sell product using holography
with print. Set up proj ect, customer buys
direct from manufacturer.
****

Mcmahan Electro-Optic
2 160 Park Avenue
(Orlando Division)
Winter Park, FL
Postal Code: 32789
Country: United States of America
Contact: Robert McMahan
Voice Phone: I 407 645 1000
FAX Phone: 1 407 644 9000
Business Description: McMahan Electro-Optics
manufactures a laser-based NDT system for test-
ing composite aerospace components and assem-
blies.
****

Melissa Crenshaw
2525 York Avenue
Vancouver, British Colombia
Postal Code: V6K lE4
Country : Canada
Contact: Melissa Crenshaw
Voice Phone: I 604 734 1614
FAX Phone: I 604 734 1614
Business Description: Color reflection holo-
grams mastering and mass production. 8x 10,
4x5 film. Lighting design, Architectural ele-
ments.
****

Melles Griot
1770 Kettering Street
Irv in e, CA
Postal Code: 92714
Country: United States of America
Voice Phone : 1 714 26 1 5600

FAX Phone: I 714261 7589
Business Description: Melles Griot is a major
manufacturer of off-the - shelf and custom tables
and isolation equipment, laser, lenses, mounting
hardware, pos itioners, polarizers, coated optics,
detectors, co llimators and spatial filters.
****

Melles Griot Gmbh
Kleyerstrasse 14
Postal Code: 64295
Country: Germany
Contact: Burckhardt Baier
Voice Phone: 49 06151 81 340
FAX Phone : 49 06151 899 352
Business Description : Condensers (optics), fi-
ber optics construction components, laser diodes
laser optics, optical filters ,optical lenses, opti-
cal mirrors, optical parts, bulk, optoelectronic
components, planar optics , planar parallel op-
tics, prisms
****

Meredith Instruments
5035 North 55th Avenue
Suite 5
Glendale, AZ
Posta l Code: 8530 I
Country: United States of America
Contact: Chad Andersen
Voice Phone: 1 602 934 9387
FAX Phone: I 602 934 9482
Business Desc ription : Specializing in surplus
inventories of He-Ne la se rs as well as argon
and diode la sers, Meredith Instruments is the
USA's largest laser discount dealer. Free Cata-
logue. Laser repair
****

Mesmerized Hol ograp hic Marketing
P.O. Box 984
White Plains, Y
Postal Code: 10602 -0984
Country: United States of America
Contact:
Jeffrey Levi ne
Voice Phone: 9149486138
FAX Phone: 9 14 948 9509
Business Desc ription: Des ign and manufactur-
ing of finished hologram products, including
desk premiums, award s, direct mailers, and
displays. Has developed proprietary design,
production and application techniques for produc-
ing the hi g he st quality fini shed products at the
lowest possible price. Makes over six different
types of holograms.
****

MetroLaser
18006 Skypark Ci rc le # I 08
Irvine, CA
Postal Code: 927 14
Country: United States of Ameri ca
Contact: James D. Trolinger
Voice Phone: I 7 14553 0688
FAX Phone: I 714 553 0495
Business Desc ription : Holographic non de-
structive testing services. Measurements and in
struments based on ho lography and holographic
interferometry. Holographic particle and flow
diagnostics.
****

Metrologic Gmbh
Dorni erstrasse 2
Postal Code: 82 178
Country: Germany
Voice Phone: 49 89 890 19 0
FAX Phone: 49 89 890 19 200

Business Description: see Metrologic, USA.
****

Metrologic In struments, Inc .
PO Box 1458
Coles Road @ Route 42
Blackwood, NJ
Postal Code: 08012
Country: United States of America
Contact: Betty Williams
Voice Phone: I 609 228 8886
FAX Phone: I 609 228 6673
Business Description: Retail/industrial: laser
bar-code scanners including portables, compact
mini-slots and high-speed projection scanners.
Education: HeNe la sers; optics lab, sandbox ho-
lography lab; optics bench system.
****

MGD Productions
(George Dyens)
5982 Ave Durocher
Outremont, Quebec
Postal Code: H2V 3Y4
Country: Canada
Contact: George Dyens
Voice Phone: I 5 14 278 4593
FAX Phone: I 514 987 4651
Business Description: Indoor and outdoor holog-
raphy integrated to architectural sites.
****

MGM Converters Inc
10600 Pioneer Blvd
Santa Fe Springs, CA
Postal Code: 90670
Country: United States of America
Contact: Steve Meyer
Voice Phone: I 3 10 946 3441
FAX Phone: I 3 10 944 0524
Business Desc ription: Quality graphics serving
the holography market. Foil stamping applica-
tion.
****

Microelectronics and Computer Technology
Corp.
3500 W Ba lcones Center Dr.
Austin , TX
Postal Code: 78759
Country: United States of America
Contact: William Stotesbery
Voice Phone: I 5 12 338 3400
FAX Phone: I 512 338 3897
Business Description: Research - Parent com-
pany of Tanarack (tel. 512 338 3100) - maker
of Hol ographic Storage devices for comput-
ers.
****

Midwest Laser Products
PO Box 2 187
Bridgeview. IL
Posta l Code: 60455
Country: United States of America
Contact: Steve Garrett
Voice Phone: I 708 460 9595
FAX Phone: I 708 430 9280
e-ma il address:m lp@mcimail.com
Business Description: New and used laser equip-
ment inc luding: HeNe, Argon , HeCd, NdYAG,
visible diode lasers. Manufactures low-cost
HeNe lasers for lab use. Distributes ho lography
kit. and related materials.
****

Mind's Eye
Holographic Consultants
17329 Zola Street
Granada Hill s, CA

bu...
tive te...
*****

Mitutoyo Measuring Instrument.
18 Essex Road
Paramus, NJ
Postal Code: 07652
Country: United States of America
Contact: Joe Scriff
Voice Phone: I 20 I 368 0525
Business Description: Manufacturers of precision measuring instruments including holographic linear tracking systems.
****

Moeller Wedel Optische Werk
Rosengarten 10
Postal Code: 22880
Country: Germany
Voice Phone: 49 4103 7090 I
FAX Phone: 494103 709370
Business Description: Optics, medical equipment.
****

Moonbeamers
1/5 Gibbons Street
Telopea, NSW
Postal Code: 2117
Country: Australia
Contact: John Tobin
Voice Phone: 61 612 890 1233
FAX Phone: 61 612 890 1243
Business Description: Since 1984, we have been producing commercial holograms and diffractions for security and display applications. We offer a complete service from artwork creation through appl ication & printing within Australia and Asia.
****

Morning Light Holograms
106 Xi Huan Middle Street
Cang Zhou
Postal Code: 061001
Country: China
Contact: Chen Guo Tong
Voice Phone: 86 317 226 164
FAX Phone: 86 317 226 167
Business Description: "Largest Hologram producer in China" (HN 2.95)
****

Mu's Laser Works
1328 Dunsterville Avenue

the Multiple... holograms for trad...
360 degree and flat format white light viewable holograms made to your specifications. Stock images also available. Ask for our catalogue! In business since 1973 I
****

Musee De L'HolographLe
15 a 21 Grand Balcon
Forum Des Hailes Bp 180,
Postal Code: 75001
Country: France
Contact: Anne-Marie Christakis
Voice Phone: 33 I 40 33 68 76
FAX Phone: 33 1 40 33 68 76
Business Description: Permanent & traveling exhibition; educational courses.
****

Museu D' Holografia
Jaume I, I
Postal Code: 08002
Country: Spain
Voice Phone: 343 3 102 172
FAX Phone: 343 3 319 1676
Business Description: Holographic Gallery, Itinerant exhibitions , sale of holograms. Teaching. Holographic courses. Holographic laboratory.
****

Museum Of Holography/Chicago
1134 West Washington Blvd
Chicago, IL
Postal Code: 60607
Country: United States of America
Contact: Loren Billings
Voice Phone 1 312226 1007
FAX Phone 1 3 12 829 9636
Business Description: Founded in 1978, the MOHC is now the world's oldest institution devoted to the display, acquisition and maintenance of holography as well as education and research in the field. Permanent collection is now the largest in the world. At least two major exhibitions a year featuring artists from around the world.
****

...America
...2 1
...1
...nology R&D in em-
...n stallation.
****

...America
...5
...New York Hall of Sci-
...ands-on science and
...s and optics demonst
...n depicting quantum
****
...ldg
...graphic ne...

Nippon Polaroid K.K.
Business Development Division - Mori Bldg
No. 30
3-2-2 Toranomon, Minato-ku
Postal Code: 105
Country: Japan
Contact: Makoto ide
Voice Phone: 81 03 3438 8883
FAX Phone: 81 03 5473 8637
Business Desc ription: Subsidiary of Polaroid Corp., Cambridge, MA USA
****

lippondenso Co. , Ltd.
System Deve lop Engineering Dept
I-I Showa-Cho Kariya-Slu
Postal Code: 448
Country: Japan
Contact: Toru Mizuno
Voice Phone: 81 0566 256 924
Business Description: Hologram manufacturer.
****

Nippondenso Co. , Ltd.
System Deve lopment Engineering
I-I Showa-Cho Kariya-Shi
Postal Code: 448
Country: Japan
Contact: Hiroshi Ando
Voice Phone: 81 0566 256 924
Business Description: Man...
...splay. Al...
****

Neovision Productions
PO Box 74277
Los Angeles, CA
Postal Code: 90004
Country: United States of America
Contact: Bill Hillard
Voice Phone: 1 2133870461
Business Description: Fine art originals, producing holograms for home and industry, consulting.
****

New Clear Imports Ltd.
27 Burrard Street
St Helier
Jersey, England
Country: United Kingdom
Contact: Anthony Hopkins
Voice Phone: 44 534 30614
Business Description: Gallery; retail shop.
****

New Dimension Holographics.
27 Nurses Walk, The Rocks
Sydney, NSW
Postal Code: 2000
Country: Australia
Contact: Tony Butteriss
Voice Phone: 61 2 743 3767
FAX Phone: 61 2 743 3241
Business Description: Retail shop. Wholesale distribution. Origination consultant. Educational consultant.
****

New Hori zons (Thai land), Ltd.
460/33 Chiangmai Land Changklan Road
Postal Code: 50000
Country : Thailand
Contact: Silvio Aprile
Vo ice Phone: 66 53 252 007
FAX Phone: 66 53 252 007
Business Description: Mass produce dichromates for jewelry, etc.
****

2148 North 86Th Street
Seattle, WA
Postal Code: 98 103
Country: United States of America
Contact: Edward Aites
Voice Phone I 206 526 5752
Business Desc ri ption: Holographic fine imagery in transmission and reflection forma... Unique and limited-edition artwork.
****

Northwestern University
Dept of Biomedical Engineering
Evansto n, IL
Postal Code: 60208
Country: United States of America
Contact: Dr. Hans Bjelkhagen
Voice Phone: I 708 491 2946
FAX Phone: I 708 491 4133
Busi ness Description: Artistic, commercial, interferometry, scientific, medical, and industria... research. Portraits, fiberoptics, pulsed laser, holographic non-destructive testing, educational holography.
****

Numazu College O...
Dept Of M...
3600 O...
...

al Code: 91344
ntry: United States of America
tact: Stephen Roth
ce Phone 1 818 360 6023
iness Description: Marketing consultant
****

istry Of I nternational Trade
ctrotechnical Laboratory
ical Information Section
tal Code: 305
untry: Japan
ntact: Dr. Satoshi Ishihara
ice Phone: 81 0298 58 5625
X Phone: 81 0298 58 5627
siness Description: Research using HOEs
****

tsubishi Heavy Industries Ltd.
agasaki Technical In stitute
l Akunoura-Machi
stal Code: 850-91
untry: Japan
ntact: M. Murata
siness Description: Holographic non-destruc-
sting; industrial research.
****

Victoria, British Colombia
Postal Code: V8Z 2Xl
Country: Canada
Contact: Ron Meuse
Voice Phone: I 604 479 43
Business Description: Holc
tographic services, Can p
technical assistance. Lase
tion and rental.
****

Multifacet
Paul Gilsonlaan 450
Postal Code: 1620
Country: Belgium
Contact: Desmet Guy
Voice Phone: 32 2 331 24
FAX Phone: 32 2 331 300
Business Description: A
which concentrates on hol
****

Multiplex Moving Hologr
746 Treat Street
San Francisco, CA
Postal Code: 94110
Country: United States of
Contact: Peter Claudius
Voice Phone: 415 285 90
Business Description: W
Hologram
de sho

New Light Industries
West 97 13 Sunset Hwy
Spokane, WA
Postal Code: 99204
Country: United States of
Contact: Steve McGrew
Voice Phone: I 509 456 83
FAX Phone: I 509 456 835
Business Description: Tec
bossing. Complete system
****

New York Hall Of Science
48th and I I I Th Street
Corona, NY
Postal Code: 11368
Country: United States of
Contact: John Driscoll
Voice Phone: I 718 699 00
Business Description: The
ence is New York's only
technology museum. Laser
rated daily. Color hologra
atom is on display.
****

Newport (Asian Office)
Kyokuto Boeki Kaisha
7Th Floor New Otemachi B
Postal Code: Tokyo 100-
Country: Japan
Business Description: Branch office of Newport
Corp. , Fountain Valley CA, USA.
****

Newport Corporation
1791 Deere Ave
Irvine, CA
Postal Code: 927 14
Country: United States of America
Voice Phone: I 800 222 9980
Vo ice Phone: 1 714 863-3144
FAX Phone: I 714 252 1680
Business Description: Designer and manufacturer of laser/holographic systems, E/O
components, optics, spatial filters, optical &
beamsteering instruments, magnetic bases,
fi ber optic components, vibration isolation
systems, and holographic recording materials.
****

Newport Gmbh
European Headquarters
Holzhofallee 19
Postal Code: 64295
Country: Germany
Voice Phone: 49 6151 362 10
FAX Phone: 49 6151 362 152
Business Description: Designer and manufac-
turer of laser/holographic systems, E/O compo-
nents, optics, spatial filters , optical & beam-
steering instruments, magnetic bases, fiber optic
components, vibration isolation systems, and
holographic recording materia ls.
****

Nihon University
Dept Electronic Engineering
7-24-1 Narashinodai
Funabashi-Shi , Chiba
Postal Code: 274
Country: Japan
Contact: Dr. Hiroshi Yoshikawa
Voice Phone: 81 047469 5391
FAX Phon e: 81 0474 67 9683
Business Desc ription: Research NDT
****

fracture HeadsUp
Disp so Mr. Tor Mizuno, Mr. Tatsuya Fu-
jita, or Mr. Shinji Nanba.
****

Nissan Motor.
Central Research Lab
Natsush ima Machi
Postal Code: 237
Country: Japan
Voice Phone: 81 0468 625 182
FAX Phone: 81 046 654 183
Business Description: Hologram manufacturer
head-up display
****

Norges Tckni ske Hogskole.
In stitute For Almen Fysikk
Sem Saelandsv 7 -7034
Country: Norway
Voice Phone: 47 07 5
FAX Phone: 47 07 592 886
Business Description: Holographic Non-destruc-
tive testing
****

Norland Produ cts, Inc.
PO Box 7145
North Brunswick , NJ
Postal Code: 08902
Country: United States of America
Contact: Jean Spalding
Voice Phone: 1 908 545 7828
FAX Phone I 908 545 9542
Business Description: Optical adhesives (which
cure with UV light). Light sources and fiber op-
tics.
****

Northern IIIinois University
Department Of Physics
Dekalb, IL
Postal Code: 60115
Country: United States of America
Contact: Thomas Rossing
Voice Phone: I 815 753 1772
Business Descri ption: Scientific holography re-
search. Proj ects vary in nature.
****

Northern Lightworks

Of Technology
chanical Engineering
oka
Numazu-City, Shizuoka
Postal Code: 410
Country: Japan
Contact: Dr. Koji Lkegami
Voice Phone: 81 0559 212 700
Business Description: Holographic research.
****

OIE Research
7227 Eastwood Street
Philade lphia , PA
Postal Code: 19149
Country: United States of America
Contact: Len Stockier
Voice Phone: I 2 15 331 5067
Business Description: Technical and educational
consulting. Holography work shops, resource
center. Conceptual design and production of display holograms. H-1 and H-2 mastering. Custom
optical table components. Touring programs, exhibits and workshops.
****

Odhner Holographics
PO Box 56-8574
Orlando, FL
Posta I Code: 32856-8574
Country: United States of America
Contact: Jeff Odhner
Voice Phone: I 407 856 7665
FAX Phone: I 407 856 9003
Business Description: Exclusive distributor of
the Stabilock 11 inch fringe stabilizer (used
to make brighter holograms) , manufacture of
custom holograms (translre fl.). Specializing in
HOE arrays (to 8" x 10") on silver halide.
****

Ojasmit Holographics
409 Vardhman Market Sector 17, VASHI
Postal Code: 400703
Country: India
Contact: Kail esh Shah
Voice Phone 91 22 768 3526
Voice Phone: 91 22 763 0373
FAX Phone: 91 22 768 1936
Business Description : Manufacturing and mar-
keting of embossed holograms, photopolymers
and dichromates for va ried appli cations.
****

Omnichrome
Quality Lasers And E/O Systems
13580 Fifth Street
Chino, CA
Postal Code: 91710
Country: United States of America
Contact: Kevin Rankin
Voice Phone: 1 909 627 1594
FAX Phone: 1 909 59 1 8340
Business Desc ription: Manufacturer of Ar-
gon, Krypton and HeCd lasers ranging in
wavelength from 325nm to 752nm, hands-off
operation lasers find appl ications in semicon-
ductors, optical disk mastering, medicine and
holography.
****

Ontario College Of Art
100 McCaul Street
Toronto, Ontario
Postal Code: M5T I WI
Country: Canada
Contact: Michael Page
Voice Phone: 1 416 977 531 1
FAX Phone: 1 416977 0235
Business Description: General Holography
courses
****

Ontario Science Centre
770 Don Mills Road
Don Mills, Ontario
Postal Code: M3C I T3
Country: Canada
Contact: Alena Kottova
Voice Phone: 1 416 429 4 100 ex. 2820
FAX Phone: 14166963197
e-mail address:
alena_kottova@fcgatel.osc.on.ea
Business Description: We have gallery of 15
holograms on permanent display and laser
demonstration area. Holography workshops
cover theory and practical uses of holography.
Participants make their own reflection holo-
gram.
****

Op-Graphics (Ho lography) Ltd.
Unit 4 Technorth
7 Harrogate Road
Leeds, England
Postal Code: LS7 3NB
Country: United Kingdom
Contact: Valerie Love
Voice Phone: 44 11 3 2 628 687
FAX Phone: 44 11 32374 182
Business Desc ription: Manufacturer of display
holograms. Large selection of stock images in
variety of formats and sizes. Commissioned
work undertaken. Copying work for hologra-
phers undertaken.
****

OpSec
4500 Cherry Creek Drive South, Suite 900
Denver, CO
Postal Code: 80222
Country: United States of America
Contact: Yoram Curiel
Voice Phone: 1 303 759 5756
FAX Phone: 1 303 759 1046
Business Descripti on: Produces holographic la-
bels as optical security devices.
****

Optical Coating Laboratory GmbH
MMG Division
Alte Heerstrasse 14
Postal Code: 38644

Country: Germany
Contact: Mr. Koch
Voice Phone: 49 5321 359 0
FAX Phone: 49 532 1 359 103
Business Description: Flat glass, refined front
surface mirrors, glass components, mirrors,
optical mirrors, surface-coated mirrors
****

Optical Corporation Of America
3-A Lyberty Way
Westford, MA
Postal Code: 0 1886
Country: United States of America
Contact: John Ward
Voice Phone: 1 508 692 3220
FAX Phone: 1 508 692 9416
Business Description: Products: Precision, large
aperture (to 36 inch diam.) aspheric mirrors for
holographic production systems.
****

Optical Research Services
3280 East Foothill Blvd, Suite 300
Pasadena, CA
Postal Code: 9 1107-3 103
Coun try: United States of America
Voice Phone: 1 8 18 7959101
FAX Phone: 1 818 705 9102
Business Descrip tion : We create Holographic
Optical Elements to your specifications using
our software. Software can be leased.
****

Optical Security Industry
4D E- I F Gelders Hall Road
Shepshed, England
Postal Code: LEI2 9NH
Country: United Kingdom
Contact: Peter H.L. Wood
Voice Phone 44 1509 600 220
FAX Phone: 44 1509 508 795
Business Description: A total secure service
from concept design artwork to finished prod-
uct-specializing in customer service and deliv-
ering quality embossed security and nonsecurity
work on time.
****

Optical Society of America (OSA)
2010 Mass Avenue NW
Washington, DC
Postal Code: 20036-1023
Country: United States of America
Voice Phone: 1 202 223 8 130
FAX Phone: 1 2024166100
Business Descripti on: Publications for the Opti-
cal Industry.
****

Optical Surfaces Ltd .
GodslOne Road
Ken ley, England
Postal Code: CR2 5AA
Country: Uni ted Kingdom
Contact: John Mathers
Voice Phone: 44 81 668 612
FAX Phone: 44 81 660 7743
Business Description: Manufacturer of precision
optics
****

Optical Works Ltd.
Eal ing Science Centre
Treloggan Lane
Newquay, England
Postal Code: TR7 I HX
Country: United Kingdom
Contact: E.O. Frisk
Voice Phone: 44 0637 877222

FAX Phone: 44 0637 8772 11
Business Description: Make optical components
and lenses.
****

Optics Plus Inc
1369 East Edinger Avenue
Santa Ana, CA
Postal Code: 92705
Country: United States of America
Contact: Allison Valdivia
Vo ice Phone: 1 714 972 1948
FAX Phone: 1 7 14 83565 10
Business Description: Manufacture optics; pre-
cision tool mounts (including lens and mechani-
cal mounts).
****

Optilas B. V.
PO Box 222
2400 Ae Alphen
Country: Netherl ands
Contact: A. Kooi
Vo ice Phone: 3 1 1720 3 1234
FAX Phone: 31 172 43414
Business Description: Sales/service/engineering
of electro- optical and vacuum related products.
****

Optimation
123 Nashua Road # 172
Londonderry, NH
Postal Code: 03060
Country: United States of America
Contact: Dean Jorgensen
Voice Phone: 1 603 623 2800
Voice Phone: 1 603 429 2800
FAX Phone: 1 603 429 1923
Business Description: Specializes in excellent
quality, burr free, pinholes.
****

Optimation Hol ographi cs
3200 South Haskell , Suite 160
Lawrence, KS
Postal Code: 66046
Country: United States of America
Contact: John Trackeberry
Voice Phone: 9 13 841 1642
FAX Phone: 9 13 84 1 0439
Business Description: Large format holographic
embossing. Complete in-house system including
res ist master and shim making.
****

Opti sche F enomenen
Nederlandse Stichting Voor Waarn
Warenarburg 44
Postal Code: NL 2907 CL
Country: Netherlands
Contact: Jan M. Broeders
Business Description: Monthly Newsletter sub-
sidiary of Dutch Foundation of Perception &
Holography
****

Optitek
100 Ferguson Dri ve
Mailstop 5G6 1
Mountain View, CA
Postal Code: 94039
Country: United States of America
Contact: Mitch Henrion
Voice Phone: 1 415 966 3 194
FAX Phone: 1 4 15 966 3200
Business Description: Holographic data SlOrage
research and development.
**********************************.

Oriel Corporation
250 Long Beach Boulevard
Stratford , CT
Posta l Code: 06497
Country: United States of America
Contact: Scott Heidemann
Voice Phone: I 203 377 8282
Business Description: Holography artist.
****

Oriel Scientific Ltd.
(Division Of Oriel Corporation)
PO Box 3 I
Leatherhead, England
Postal Code: KT22 7 AU
Country: United Kingdom
Voice Phone: 44 0372 378822
Business Description: Artistic holography.
****

OWlS Gmbh/ FOSTEC
1m Gaisgraben 7
Postal Code: 79219
Country: Germany
Contact: Hubert Munzer
Voice Phone: 49 7633 9504 0
FAX Phone: 49 7633 9504 44
Business Description : Development, Manufac-
turing and Marketing of precision mechanical
and optical systems.
****

Oxford Holographics
71 High Street
Oxfo rd, England
Postal Code: OX I 4BA
Country: United Kingdom
Contact: Nick Cooper
Voice Phone: 44 1865 250 505
FAX Phone: 44 0865 250 505
Bus iness Desc ription: Oxford Holographics
has both a very well established retail and an
expanding distr ibution operation, focusing on
silver halide & innovati ve embossed products.
****

Oxford University
Dept Of Engineering Science
Parks Road
Ox ford , England
Postal Code: OXI 3PJ
Country : Un ited Kingdom
Contact: D. 1. Cooke
Voice Phone: 44 0865 273 805
Bus iness Description: Holography education;
Industrial research.
****

P.S.A Peugeot Citroen
ISMEI CEI EVM
LaSam - Chemin de la Mahnaison
Postal Code: F-9 1578
Country: France
Contact: M. Fe ingold
Vo ice Phone: 33 1 69 35 8 I 78
FAX Phone: 33 1 69 35 81 94
Business Description: Industri al research; holo-
graphic non-destructive testing.
****

Panat ron Inc.
1736 Terrace Lane
P.O. Box 2687
Pomona, CA
Postal Code: 91769-2687
Country : United States of America
Voice Phone: 1 909 629 0748
FAX Phone: 1 805 620 0378
port, P311S and service on al I lasers. Also man-
ufactures

mirrors, lenses, rods and other parts for lasers.
Laser Repair and used lasers.
****

Pennsylvania Pulp & Paper Co.
Prismatic Sq uare
2874 Lime Kiln Pike
Glensid e, PA
Postal Code: 19038
Country: United States of America
Contact: Brian Monaghan
Voice Phone: I 2 15 572 8600
FAX Phone: I 215 572 8154
Business Description: Manufactures "Prismatic
Illuions" holographic paper line. Stocking holo-
graphic paper and board for printing and packag-
ing. 5 stocked patterns and custom patterns avai
lable.
****

Phantastica
Suchtener Strasse 4a
Postal Code: 59757
Country: Germany
Contact: Gerd M. Albrecht
Voice Phone: 49 2932 8 I 9 I 7
FAX Phone: 49 2932 2944 I
Business Desc ription: Makers and distributors
of articles re lated to embossed holograms and
diffraction foil , including earrings and other
jewelry. badges, pens, mobiles. Main focus is
street, crafts and Christmas markets.
****

Photon Cantina Ltd.
PO Box 1098
La Canada, CA
Postal Code: 9 10 I 2-1 098
Country: United States of America
Contact: Roy Chiarot
Voice Phone: I 8 I 8 790 6735
FAX Phone: I 8 18 790 708 1
e-mail address:CPAX@IX.Netcom.COM
Business Description: Full service producers of
high quality artistic and commercial silver ha-
lide reflection holograms. Mastering and custom
work available.
****

Photon League Of Holographers
110 Sudbury Street, Unit B
Toronto, Ontario
Postal Code: M6J I A 7
Country: Canada
Contact: Nicola Woods
Voice Phone: I 4 16 53 I 7087
Business Desc ription: Arti st run non -profit ho-
lography studio. Technical work shops through-
out the year.
****

Photoni cs Directory
Laurin Publi shing Co Inc
PO Box 4949 Berkshire Common
Pittsfield , MA
Postal Code: 0 1202
Country: Un ited States of America
Contact: Wendy La uren
Voice Phone: 1 413 499 05 14
FAX Phone: 1 4 13 4423 180
Business Description: Publishers of information
on optical components; Optics, ElectroOptics,
Lasers, and Imaging tech nology.
****

Photonics Systems Laboratory
7 Rue De L'Uni versite
Postal Code: 67000
Business Description: Supplies complete sup-
Country : France

Contact: P. Meyrueis
Voice Phone: 33 88 65 50 00
FAX Phone: 33 88 65 52 49
Business Description: Education. NDT
****

Physical Optics Corporation.
2545 W 237Th Street, Suite B
Torrance, CA
Postal Code: 90505
Country: United States of America
Contact: Kevin Rankin
Voice Phone: 1 3 10530 1416
FAX Phone: I 3105304577
Business Description: Photonics -based high
tech company involved in research, develop-
ment and manufacture of holographic optical
elements, including reflect ion and transmission
HOE's, diffraction gratings and display holo-
grams.
****

Physics In stitute. Latvian
SST Academy Of Sciences
Posta l Code : 229021
Country: Latvia
Contact: Dr. Kurt Shvarts
Voice Phone: 37 I 007 947 642
Busi ness Descri pt ion: Scientific research on
recording materials
****

Physik Instrumente (Pi) Gmbh & C
Polytecplatz 5-7
Posta l Code: 76337
Country: Germany
Contact: Dr. Karl Span ner
Vo ice Phone: 49 7243 6040
FAX Phone: 49 7243 604145
Business Description: Holography, laser, optical
instruments, miscellaneous, oscillation insula-
tors, vibrating dampers
****

Pilkington Optronics
Glascoed Road
St Asaph
Clwyd, England
Postal Code: LL 17 011
Country: United Kingdom
Contact: Andrew Hurst
Voice Phone: 44 745 588 344
FAX Phone: 44 745 584 258
Business Description: Manufacturer of DCGI
photopolymer
****

Pink, Patty
PO Box 24-153
San Francisco, CA
Postal Code : 94 I 24
Country: United States of Ameri ca
Contact: Patty Pink
Voice Phone: I 4 I 5 822 7123
Business Description: Fine arts holography.
High quality glass plate transmission and reflec-
tion holograms. Classes for children and adults.
Tec hni ca l holography writing and editing.
L.A.S.E.R. News ed itor.
****

Planet 3-D
20 I Si lver Fox Lane
Downingtown, PA
Posta l Code: 19335
Country: United States of Ameri ca
Contac t: Rich Cossa
Voice Phone: I 6 I 0 873 6 I 92
FAX Phone: I 6 108736194

Business Description: Marketing of holographic products.
****

PMI Data Ltd
Units 5 & 6
Sta ion Industri al Estate, Oxford Rd
Workingham, England
Postal Code: RG II 2YQ
Country: United Kingdom
Voice Phone: 44 734 772255
FAX Phone: 44 734 772296
Business Description: Supplier of Halogram (copyrighted) diffraction foil.
****

Point Of View Dimension s, Ltd.
45-2903 Ri ver Drive South
Jersey City, NJ
Postal Code: 07310
Country: United States of America
Contact: Neal Lubetsky
Voice Phone: I 20 I 626 8844
Business Description: Point of View Dimensions specializes in the conceptualization, design, and execution of holography (all formats and all sizes) for exhibitions and shows, premiums, points of sale, brochures, and annual reports.
****

Point Source Product ions
14670 Highway 9
P.O. Box 55
Boulder Creek, CA
Postal Code: 95006
Cou ntry: United States of America
Contact: Bob Hess
Voice Phone: 1 408 338 1304
FAX Phone: 1 408 338 3438
Business Description: We are an independent recording studio offering product design and technical imaging consultations, Master and ganged holograms for photopolymer production, and "short-run" or limited edition transfer services. See Chapter 2 for add iti onal information.
****

Polaroid Corporation
2 Osborn Street - 2nd Floor
Cambridge, MA
Postal Code: 02139
Country: United States of America
Contact: Doug Marks
Voice Phone: 1 617 386 8676
FAX Phone: I 617 386 867 1
Business Description: fully integrated supplier of highest quality, mass produced photopolymer holograms. Services include design, modeling, origination, manufacturing and converting. Our industrial di vision provides the highest efficiency, mass produced holographic optical elements available, including reflective and transmissive diffusers, projection screens and depixellators.
****

Polymer Image
2 1411 North 11th Ave
Suite 4
Phoenix , AZ
Postal Code: 85027
Country: Uni ted States of America
Conta ct: Dan Norton
Voice Phone: 1 602 780 4882
FAX Phone: 1 602 780 0360
Business Description: full -service holographic lab, including mastering and production of custom images. Products include "Whole-FX" hologram mugs, made with pol ymer film which are

heat-, UV,- water-resistant, dishwasher/microwave-safe.
****

Portson , Inc . (Laser Images)
9201 Qui vira
Overland Park, KS
Postal Code: 662 15
Country: United States of America
Contact: Steve Larson
Voice Phone: 1 9 13 492 70 I 0
FAX Phone: 1 913 492 7099
Business Description: Manufacturers of stock and custom holograms and holographic products; total in-house production capabilities in dichromate, silver halide, and photo-resist. Stock products include jewelry, watches, calculators, and framed art.
****

Print-M-Boss
5/24 Kirti Nagar Indl. Area
Postal Code: 110017
Country: India
Contact: Ravinder Singh
Voice Phone: 91 11 530586
fAX Phone: 9 1 11 544 1144
Business Description: Manufacturers of embossed holograms with in-house facility for shim making. Would be interested in buying copyrights for vario us images and patterns. Like to make contacts for mastering.
****

PTi
Brett Drive
Bexhill on Sea, England
Postal Code: TN40 2JP
Country: United Kingdom
Voice Phone: 44 424 733 128
fAX Phone: 44 424 733129
****

Quantel
17, av de l'Atlantiq ue
ZA de Courtaboeuf, BP 23
Postal Code: 91941
Country: France
Contact: Ala in Orszag
Voice Phone: 33 I 69 29 17 00
fAX Phone: 33 I 69 29 17 29
Business Description: Manufacturer of lasers.
****

Rainbow Symphony Inc.
6860 Canby Ave. # 120
Reseda, CA
Postal Code: 91335
Country: United States of America
Contact: Mark Margolis
Voice Phone: 1 8 18 708 8400
FAX Phone: 1 8 187088470
Business Desc ription: Manufacturers of uniquely designed ho lographic and diffraction products for the gift, novelty, advertising, specialty, premium incentive, souvenir and museum markets.
****

Ralcon
Box 142
8501 South 400 West
Paradise, UT
Pos tal Code: 84328
Country: United States of America
Contact: Richard Ralli son
Voice Phone: 1 80 I 245 4623
fAX Phone: 1 80 1 245 6672
Business Description: Des ign, de ve lopment

and fabrication of volume holographic optical elements, (HOEs) including gratings, sca nners, multi focus devices, heads up and down displays and notch filters formed in dichromated gelatin or photopolymer
****

Ralph Cullen Holographics
C/O Uk Optical Supply
84 Wimborne Road West
Wimborne, England
Postal Code: BH21 2DP
Country: Un ited Kingdom
Contact: Ralph Cull en
Voice Phone: 44 202 886 831
fAX Phone: 44 202 742 236
Business Descrip ti on: A Consultancy-Design Service which in association with UK Optical Supplies (Manufactur ing) provide customized holographic optical components. Advice on component selection and laboratory/ studios designed to any budget is available.
****

Randy James Holography
503 Caledonia Street
Santa Cruz, CA
Postal Code: 95062
Country: United States of America
Vo ice Phone: I 408 458 4213
Business Description: Commercial and fine art holography since 1974. Extensive background in all forms of disp lay holography: design , mastering, and production. Custom quotes, stock price list available.
****

Reconnaissance, Ltd.
Runnymede Mal thouse
Runnymede Road
Egham, England
Postal Code: TW20 98D
Country: United Kingdom
Contact: Ian Lancaster
Voice Phone: 44 1784 497008
FAX Phone: 44 1784 49 7001
Business Description: We are the leading international consultancy for market, industry information and analysis Publisher of Holography News and secretariat to the International Hologram Manufacturers Association. All clients studies are fully confidential.
****

Red Beam, In c.
90 II Skyline Bl vd
Oakland, CA
Postal Code: 946 11
Country: United States of America
Contact: Lon Moore
Voice Phone: 1 510 482 3309
FAX Phone: 1 5 10482 1214
Busi ness Desc ription: Specializes in the design and production of master (H1) holograms for mass production. Produces his own line of holograms. Clients include Activision, AT&T, NFL(Superbowl) and Polaroid. Also uses the name Lightrix - see listing in this book.
****

Reel Image
PO Box 566
Pacifica, CA
Postal Code: 94044
Country: United States of America
Contact: Roy Bradshaw
Voice Phone: I 415 355 8897
FAX Phone: I 4 15 355 5427
Business Description: Fine art holograms. in -

corporation of patented holographic designs into fishing tackle and fishing lures.
****

Regal Press Inc
Holographics Division
129 Guild Street
Norwood, MA
Postal Code: 02062
Country: United States of America
Contact: William Duffey
Voice Phone: I 617 769 3900
FAX Phone: I 617 551 0466
Business Description: Holographic embossing, application; Artistic holography.
****

Reva's Holographic Illusions
446 South Main Street
Frankenmouth, MI
Postal Code: 48734
Country: United States of America
Contact: Reva Krick
Voice Phone: I 517 652 3922
FAX Phone: I 517 652 6503
Business Description: Gallery/Retail store, with over 250 holograms on di splay. We feature a full line of holographic jewelry, gifts apparel, toys, etc. Established in 1992.
****

Reynolds Metal Co
Flexible Packaging Division
6603 West Broad St
Richmond, VA
Postal Code: 23230
Country: United States of America
Contact: Rich Patterson
Voice Phone: 804 281 3969
Voice Phone: 804 743 6608
FAX Phone: 804 281 2238
Business Description: Wide Web embossing of holographic images. Laminating, printing gravure .
****

Rice Systems
1820 East Garry Ave
Santa Ana, CA
Postal Code: 92705
Country: United States of America
Contact: Dr. Colleen Fitzpatrick
Voice Phone: I 714553 8768
FAX Phone: I 714 553 0307
Business Description: Laser metrology and diagnostic measurements, HNDT fluid measurements. Combustion diagnostics. Integrated optics and non linear optical material (R&D and product development). Very successful SBIR company.
****

Richard Bruck Holography
33 12 West Belle Plaine #2
Chi cago, IL
Postal Code: 60618-2316
Country: United States of America
Contact: Richard Bruck
Voice Phone: I 312 267 9288
FAX Phone: I 312 267 9288
Business Description: Specialists in large format holography. Extensive experience with live models and commercial work. We are accustomed to the advertising world, and know the importance of quality and service.
****

Richmond Holographic Studios
Augustine's Hall
6 Yorkton Street

London, England
Postal Code: E2 8NH
Country: United Kingdom
Contact: Edwina Orr
Voice Phone: 44 171 739 9700
FAX Phone: 44 171 739 9707
Business Desc ription: Holograms up to 1 sq. meter. Custom, stoc k, R&D, technology transfer. Autostereoscopic Di splay Technology.
****

Robert Sherwood Holo Design
400 West Erie Street, Suite #405
Chi cago, IL
Postal Code: 60610
Country: United States of America
Contact: Robert Sherwood
Voice Phone: I 312 944 3200
Business Description: Our designers, holographers, and account service personnel provide you with the high est quality standards this new and exciting technology can offer. Now offering holography for apparel.
****

Rochester Inst. Of Technology
CIS
Rochester, NY
Postal Code: 14623-5604
Country: United States of America
Contact: P. Mouroulis
Voice Phone: I 7164756678
FAX Phone: I 716 475 5988
e-mail address: pzmpph@rit.edu
Business Description: Research on HOEs, Holographic materials, CGHs. Instruction in holography and related topics in the department of imaging and photographic technology and the center for imaging science.
****

Rochester Photonics Co.
330 Clay Road
Rochester, Y
Postal Code: 14623
Country: United States of America
Contact: Dean Faklis
Voice Phone: I 7162723010
FAX Phone: I 716272 9374
e-mail address:
fakl is@joker.optics .rochester.edu
Business Description: Designer and manufacturer of diffractive optical elements for use in precision electro-optical systems. Supplier of optical system design services for military and commercial uses.
****

Rofin-Sinar Laser Gmbh
Berzeliusstrasse 85
Postal Code: 22093
Country: Germany
Voice Phone: 49 40 733 630
FAX Phone: 49 40 7336 3 I 00
Business Description: NdYag lasers for materials processing, Laser components, Laser processing devices and machines.
****

Rolls-Royce Pic
Advanced Research Laboratory
POBox 31
Derby, England
Postal Code: DE2 8Bl
Country: United Kingdom
Contact: Ric Parker
Voice Phone: 44 332 242 424
FAX Phone: 44 332 249 936
Business Desc ription: NDT for aircraft engines.

****
Rolyn Optics
706 Arrow Grand Circle
Covina, CA
Posta l Code: 9 1722-2 1 99
Country: United States of America
Voice Phone: 1 818 9 1 5 5707
FAX Phone: 1 8 18915 1379
Business Description: General selection of optical items. Catalogue available.
****

Ross Books
P.O . Box 4340
Berkeley, CA
Posta l Code: 94704
Country: United States of America
Contact: Alan Rhody
Voice Phone: 1 800 367 0930
Voice Phone: 1 5 1084 12474
FAX Phone: 1 510 841 2695
Business Desc ription: Publisher of the HOLOGRAPHY MARKETPLACE Editions 1-5 (a worldwide database and sourcebook), the HOLOGRAPHY HANDBOOK (world's best selling lab manual), and other related titles. Educational and information services.
****

Rowland In stitute For Science
100 Edwin H. Land Blvd.
Cambridge, MA
Postal Code: 02 142
Country: United States of America
Contact: l ea n-Marc Fournier
Voice Phone: I 6 17 497 4657
Business Description: Scientific holography research. NDT
****

Royal College Of Art
Holography Unit
Darwin Building, Kensi ngton Gore
London, England
Postal Code: SW7 2EU
Country: United Kingdom
Voice Phone: 44 17 1 584 5020
FAX Phone: 44 171 584 8217
Business Description: MA (RCA) Holography; two-year post-graduate course. Explores creative holography in both fine art and design-related field s. Facilities include Pulse, Krypton, HeNe, Stereogram and Computer Graphics.
****

Royal Holographic Art Gallery
I Market Square
560 Johnson Street
Victoria, British Colombia
Postal Code: V8W 3C6
Country: Canada
Contact: Derek Galon
Voice Phone: I 604 384 0123
FAX Phone: I 604 384 0123
Business Description: Gallery offers holographic art, gifts , limited editions, custom work, se lection from best studios, and fashon accessories. We ship worldwide. Also some film , plates, deve loper, lasers and holographic kits.
****

Royal In stitute Of Technology
Dept Of Industrial Metrology
Posta l Code: 10044
Country: Sweden
Contact: Lennart Svennson
Voice Phone: 46 08 790 7823

FAX Phone: 46 08 790 8219
Business Description: Holography education. Holographic Non-Destructive testing.
****
Royal Photographic Society
Salisbury College Of Art
Southampton Road
Salisbury, England
Postal Code: SPI 2PP
Country: United Kingdom
Voice Phone: 44 0722 237 11
Business Description: Artistic holography ; holography education.
****
Ruey-Tung, Miss. Hung
A 202
Chigasati-Coat Nango 6-7-12
Chigasaki-Shi, Kanagawa
Postal Code: 253
Country: Japan
Contact: Hung Ruey-Tung
Voice Phone: 81 0467 857 750
Business Description: Holographic Fine Artist.
****
Rutherford & Appleton Labs
Chilton
Didcot, England
Postal Code: OX II OQX
Country: United Kingdom
Contact: Robert Sekulin
Voice Phone: 44 0235 821 900
Business Description: Particle measurements; Holographic non-destructive testing.
****
Saab-Scania
S-581
Postal Code: 88
Country: Sweden
Contact: Sven Malmqvist
Voice Phone: 46 013 129 020
Business Description: Holographic non-destructive testing; scientific holography research.
****
Saginaw Valley State University
2250 Pierce Road
University Center, MI
Postal Code: 487 10-000 I
Country: United States of America
Contact: Dr. Hsuan Chen
Voice Phone: 1 517 790 4000
FAX Phone: 1 51779027 17
Business Description: Course instruction on holography; research includes HOEs, multiplex and rainbow holography.
****
SAM Museum
3-27-3 Isoji, Minato-ku
Postal Code: 552
Country: Japan
Contact: Akinobu Fukuda
Voice Phone: 81 6 572 0036
FAX Phone: 81 65748 136
Business Description: Museum that exhibits holograms.
****
Sandia National Laboratories
Organization 1000
P.O. Box 5800
Albuquerque, NM
Postal Code: 87185
Country: United States of America

Voice Phone: I 505 845 8822
FAX Phone: I 505 844 5716
Business Description: Sandia National Laboratories is able to do research in all phases of holography.
****
Scharr Industries
40 East Newberry Road
Bloomfield, CT
Postal Code: 06002
Country: United States of America
Contact: Peg Home
Voice Phone: 1 203 243 0343
Voice Phone: I 800 284 7286
FAX Phone: 1 203 242 7499
Business Description: Scharr holographic embossing, wide web, on polyester, polypropylene, polyethylene, PVC, nylon and paper. We coat, laminate, metal ize standard and custom patterns. Products include film to paper, board, transfer film, PSA and static cling.
****
School Of Holography
Museum Of Holography/Chicago
1134 W Washington Blvd
Chicago, IL
Postal Code: 60607
Country: United States of America
Contact: Loren Billings
Voice Phone: I 312 226 1007
FAX Phone: 1 312 829 9636
Business Description: Founded in 1978, the oldest continuous school of holographic instruction in the world. Basic courses in holography have been taught to thousands of students. In addition there are special workshops and tutorials for advanced study.
****
Science Kit & Boreal Labs
777 East Park Drive
Tonawanda, NY
Posta l Code: 14150-6784
Country: United States of America
Voice Phone: 1 716 874 6020
FAX Phone: 1 716 874 9572
Business Description: Suppliers and mail-order cataloguers of holography educational materials including holography kits, books and more.
****
Semicon Austria
Morellenfeldgasse 41
Postal Code: A-80 I 0
Country: Austria
Voice Phone: 43 0316 38 25 41
FAX Phone: 43 0316 38 24 03
Business Description: Collection of Russian art holograms for sale.
****
SETEC Oy
PO Box 31
Country: Finland
Voice Phone: 358 0 89 411
FAX Phone: 358 891 887
****
Severum Design
59 Farm View
Yateley, Camberley, England
Postal Code: GUI 77JA
Country: United Kingdom
Contact: Sally-Anne Bennett
Voice Phone: 44 252 890 37 1
FAX Phone: 44 252 890371

Business Description: Photoresist hologram mastering facility.
****
Shandong Academy of Sciences
Keyuan Road
Postal Code: 2500 14
Country: China
Contact: Prof. Zhu De Shun
Voice Phone: 86 6 15 615102 3 16
Business Description: Laser & holography exhibit.
****
Sharon McCormack Holography
PO Box 38
White Salmon, WA
Postal Code: 98672
Country: United States of America
Contact: Sharon McCormack
Voice Phone: 1 509 493 4850
Voice Phone: I 509 493 1334
FAX Phone: 1 509 493 4830
Business Description: Holographic fine artist.
****
Sharp Corp.
Tokyo Research Laboratories
Research Dept2 ; 271 , Kashiwa
Postal Code: 227
Country: Japan
Contact: Shunichi Sato
Voice Phone: 81 0471 346 166
FAX Phone: 81 0471 346 119
Business Description: Research
****
Shipley Chemical Co.
455 Forrest Street
Marlboro , MA
Postal Code: 01752
Country: United States of America
Contact: Stu Price
Voice Phone: 1 800 343 6288
Voice Phone: 1 508 481 7950
FAX Phone: 1 508 485 9113
Bus iness Description: Manufacture holographic photochemistry, film, plates and more.
****
Shires, Mark
5561 North Navajo ave
Glendale, WI
Postal Code: 53217
Country: United States of America
Voice Phone: I 414332 9243
FAX Phone: 1 4143329243
Business Description: Independent holography consultant.
****
Siemens (USA)
186 Wood Avenue South
Iseling, NJ
Postal Code: 08830
Country: United States of America
Voice Phone: 1 212 258 4000
Business Description: Manufacturer of lasers and components.
****
Siemens Ltd.
Siemens House
Windmill Road
Sunbury-On-Thames, England
Postal Code: TW 16 7HS
Country: United Kingdom
Business Description: Manufacturer of lasers and components.

**************MarketPlace************

Silhouette Techno logy Inc .
10 Wilmot Street
Morristown , NJ
Postal Code: 07962-1479
Country: United States of America
Contact: Francine Hannigan
Voice Phone: I 201 53921 10
FAX Phone: I 20 I 539 5797
Business Description: Produces custom HOEs
under contract.
****

Sillcocks Plastics International
3 I 0 Snyder Aven ue
Berkeley Heights, NJ
Postal Code: 07922
Country: United States of America
Voice Phone: I 908 665 0300
FAX Phone: I 908 665 9254
Business Description: Producer of flat plastic
products, printed or unprinted, which can fea-
ture hot-stamped holograms. Products include
credit cards, promotional cards and other custom
specialties and POP products.
****

Silver Dragon Holography
PO Box 2470
2520 Professional Road
Richmond, VA
Postal Code: 23202
Country: United States of America
Contact: Clare Brown
Voice Phone: I 804 272 9284
FAX Phone: I 804 320 2100
Business Description: Exclusive marketer of
Dazzle Equipment Company's embossing equip-
ment to the People's Republic of China.
****

Silverbridge Group Inc.
Box 489
Powassan, Ontario
Postal Code: POH IZO
Country: Canada
Contact: Bruce Loucks
Voice Phone: I 705 724 6164
Voice Phone: I 800 4656 8735
FAX Phone: I 705 724 6249
Business Description: Limited edition DCG
holograms in large fomlat size . Licensee
Star Trek, ext Generation,and more. Joint
venture with Color Holographics (UK) in
Full Color Holographics In c. (see related
listings.)
****

Simian
298 Harvey West
Santa Cruz, CA
Postal Code: 95060
Country: United States of America
Contact: Ken Haynes
Voice Phone: 1 408 457 9052
FAX Phone: I 408 457 9051
Business Desc ription: Mastering for embossed
hologram s. Ken Haynes has over 20 years ex-
perience in the fi eld of embossed holography
and has one of the best optical printers in the
business.
****

Siavich Joint Stock Company
2 Mende leeva Square
Postal Code: 152140
Country: Russia
Business Description: Makers of ultra fine grain
silver halide emulsion for use in holography.

USA distributors include 3Deep Hologram and
Holicon.
****

Smith & McKay Printing Co. Inc.
96 North Almaden Boulevard
San Jose, CA
Posta l Code: 95110-2490
Country: United States of America
Contact: Dave McKay
Voice Phone: 1 408 292 890 I
FAX Phone: I 408 292 0417
Business Description: Parent Company: Holographic Impressions. Hot-stamps foi l holograms
onto paper products. Dimensional printing and
fine lithography. Assists in the coordination of
printing projects featuring embossed holography. Holds holography seminars for graphic designers.
****

Societa Olografica Italia (Soi)
Via Degli Eugenii 23
Postal Code: 00178
Country: Italy
Contact: Luigi Attardi
Voice Phone: 39 6 718 0976
FAX Phone: 39 6 718 5 172
Business Description: Produces artistic/com-
mercial holograms of any type.
****

Sophia Univers ity
Faculty Of Science & Technology
7-1 , Kioi-Cho Chiyoda-Ku
Postal Code: 102
Country: Japan
Contact: Kazue Ishikawa
FAX Phone: 81 03 3238 3341
Business Description: Holography research.
****

Sopra
26 & 28 rue Pierre Joingnequx
Postal Code: 92270
Country: France
Contact: Robert Stehle
Voice Phone: 33 1 47 81 09 49
FAX Phone: 33 1 42 42 29 34
Business Desc ription: Laser equipment, holo-
graphic kit and camera for interferometry.
****

Southern Indiana Hol ograph ics
6841 Newburgh Rd
Evansville, IN
Postal Code: 47715
Country: United States of America
Contact: Larry Johann
Voice Phone: 1 812 474 0604
FAX Phone: 1 8124730981
Business Description: Holographic Fine Artist
****

Space Age Designs Inc.
PO Box 72
Carvers ville, PA
Postal Code: 18913
Country: United States of America
Contact: Val li Rothaus
Voice Phone: I 215 297 8490
Business Description: Conceive, de sign, manu-
facture products for markets which vary--con-
sulting services available.
****

Spatial Holodynamics (India) Pvt. Ltd.
10411 05 Shah & Nahar Estate
Off. Dr. E. Moses Road
Postal Code: 400 018
Country: India

Contact: Yogesh Desai
Voice Phone: 91 22 493 0975
Voice Phone: 91 22 492 1069
FAX Phone: 91 22 495 0585
Business Description: Holographic embossing
using the latest Dl-HO System. Total service
from designing of holograms up to holographic
shims. Alternately, if desired, glass Photo Resist
Masters.
****

Spatial Imaging Ltd.
8A Wheatash Road
Addlestone, England
Postal Code: KTl5 2ER
Country: United Kingdom
Contact: Rob Munday
Vo ice Phone: 44 1932 564 899
FAX Phone: 44 932 564 899
Business Description : Developers of DJ-HO(r),
Computer/Video stereogram system. DI-HO em-
bossed hologram origination and stock images,
3D, 2D/3D, Moviegram , Cyclgram, combination
holograms. Large format, portraits and , through
Laza Holograms Ltd, mass reproduced film ho-
lograms.
****

Specac Ltd
6A River House
Lagoon Road, St Mary Cray
Orpington, England
Postal Code: BR5 3QX
Country: United Kingdom
Voice Phone: 44 689 873 134
Business Description: Artistic holography.
****

Spectra-Physics Lasers, Inc.
1330 Terra Bella Ave.
Mountain View, CA
Postal Code: 94039-7013
Country: United States of America
Contact: Alfred Feitisch
Voice Phone: 1 800 775 5273
Voice Phone: 1 415 966 5596
FAX Phone: 1 415 964 3584
Business Description: World's largest supplier
of CW and pulsed gas and solid state laser sys-
tems, inc ludin g a comprehensive optical ac-
cessories line and a worldwide customer service
network.
****

Spectratek Inc .
1508 Cotner Avenue
Los Angeles, CA
Postal Code: 90025
Country: United States of America
Contact: Mike Wanlass
Voice Phone: I 310 473 4966
FAX Phone: I 310477 67 10
Business Description: Specialists in high-reso-
lution, low-cost holograms. We have delivered
hundreds of millions of holograms.
****

Spectrogon Ab
(Company Headquarters)
Box 2076
Postal Code: S-18302
Country: Sweden
Business Description: Manufactures & designs
interference filters for IR, visible & UV spectral
regions; narrow and bandpass, long/shortwave-
pass, isolation-line, & NO filters; atmospheric
windows; diffraction gratings: AR & metallic
coatings .

****

Spectrolab International Ltd.
PO Box 25
Newbury, England
Postal Code: RG 16 8BQ
Country: United Kingdom
Contact: Bill Vince
Voice Phone: 44 635 248080
FAX Phone: 44 635 248745
Business Descr ipti on: Manufacture holographic systems including laser tables, optics. optical and laser positioning systems.
****

Spectrum Corporation
608 Sangita Complex
Postal Code: 380 006
Country: India
Voice Phone: 91 79 419 080
FAX Phone: 91 79 436 457
Business Description: Distributor of James River Embossing machines.
****

SPIE
International Society for Optical Engineering
PO Box 10
Bellingham, WA
Postal Code: 98227
Country: United States of America
Voice Phone: I 206 676 3290
Voice Phone: I 800 483 9034
FAX Phone: I 206 647 1445
e-mail address: info-optolinkrequest@ mom.spie.org (use "help")
Business Description: Technical publications on holography and other high-tech topics.
****

SPIE Holography Working Group
The Internatinal Society for Optical Engineering
PO Box 10
Bellingham, WA
Posta l Code: 98227-3290
Country: United States of America
Voice Phone: 206 676 3290
FAX Phone: 206 647 1445
Business Description: Monthly publication containing news and projects related to holography.
****

Spindeler & Hoyer Gmbh
Koenigsallee 23
Postal Code: 37081
Country: Germany
Contact: Mr. Keilholz
Voice Phone: 49 551 6935 0
FAX Phone: 49 551 6935 166
Business Description: Manufacturer of preci sion optics. mechanics and laser technology.
****

Springer- Verlag New York
175 Fifth Ave.
NY, NY
Postal Code: 100 I 0
Country: United States of America
Voice Phone: I 212 460 1500
Voice Phone: I 800 774 6437
FAX Phone: I 20123484505
Business Description: Publishers of an "Optical Sciences" series, including "Silver Halide Recording Materials for Holography", by H. Bjelkhagen.
****

Star Magic
745 Broadway
New York, NY
Postal Code: 10003
Country: United States of America
Voice Phone: I 212 228 7770
Business Description: Space Age Gifts. Holograms, novelties, etc.
****

Star Magic
Newport Center Mall
Jersey City, NJ
Postal Code: 07302
Country: United States of America
Voice Phone: I 20 I 626 2222
Business Description: Space Age Gifts. Holograms, novelties, etc.
****

Star Magic
4026 24th Street
San Francisco, CA
Postal Code: 94 11 4
Country: United States of America
Voice Phone: I 415 641 8626
Business Description: Space Age Gifts. Holograms, novelties, etc.
****

Star Magic
275 Amsterdam St.
NY, NY
Postal Code: 10023
Country: United States of America
Voice Phone: I 212 769 2020
Business Description: Space Age Gifts. Holograms, novelties, etc.
****

Star Magic
1256 Lexington (85th St)
New York, NY
Postal Code: 10028
Country: United States of America
Voice Phone: I 2 12 988 0300
Business Description: Space Age Gifts. Holograms, novelties, etc.
****

Star Shop
Media Art Holographie
Arnemannstrasse 5
Postal Code: 22765
Country: Germany
Contact: Claus Cohnen
Voice Phone: 49 40 3990 0039
FAX Phone: 49 40 390 5064
e-mail address: starshop@t42.ppp.de
****

Starcke, Ky.
Ratastie 6
Posta l Code: 32800
Country: Finland
Contact: Ari-Veli Starcke
Voice Phone: 358 39 5460 700
FAX Phone: 358 39 5467 230
Business Description: Starcke K Y is the leading company selling holograms in Scandinavia. Hologram Hot Stamping.
****

Starlight Holocom
73 Stable Way
Kanata, Ontario
Postal Code: K2M I A8
Country: Canada
Contact: Stephen Leafloor
Voice Phone: I 613 592 5617
FAX Phone: I 6 13 592 7647

Business Description: Starlight Holocom Inc. is one of Canada's leading repre se ntati ves of the internati onal holographic community, in both stock and custom holography.
****

Steinbichler Optotechnik Gmbh
Am Bauhof 4
Postal Code: 83 1 15
Country: Germany
Conta ct: Dr. H. Steinbichler
Voice Phone: 49 8035 87040
FAX Phone: 49 8035 1010
Business Description: Holographic analyzers and related equipment.
****

Stephens, Anai t
1685 Fernald Point Lane
Santa Barbara, CA
Postal Code: 93108
Country: United States of America
Contact: Anait Stephens
Vo ice Phone: I 805 969 5666
Business Description: Artist in hands-on holography; reflection holography, portraiture, comm iss ions.
****

Steuer Kg Gmbh & Co.
Ernst Mey Strasse 7
Postal Code: 70771
Country: Germany
Contact: Herr Seitz
Voice Phone: 49 71 I 16068 0
FAX Phone: 49 7 11 16068 50
Business Desc ription: Manufacturer of holographic hot-stamping machines.
****

Stichting Voor Holographie En Laseropiek
Prinsengracht 675
Country: Netherlands
****

Stoltz Ag
Tafernstrasse 15
Postal Code: CH-5405
Country: Switzerland
Contact: Beat Ineichen
Voice Phone: 41 056 840 151
Business Description: Holographic non-destructive testing.
****

Studio Creaius
204 Sheridan Avenue
Palo Alto, CA
Posta l Code: 94306
Country: United States of America
Contact: Stephan Gunning
Voice Phone: I 415 941 7233
Business Desc ription: Custom-designed jewelry pieces and holographic consulting servIces.
****

Studio Fur Holographie
Waldfriedenweg 10
Posta I Code: 82223
Country: Germany
Contact: Dr. Carlo Schmelzer
Vo ice Phone: 49 8141 7083 1
FAX Phone: 49 8 141 82268
Business Description: Products: mastering and copy se rvices (rainbow/reflection) , production of mass-run embossed holograms and shims, open stock :mages, art-pieces.
****

Studio Weil-Alvaron
Ostra Tullgatan 8
S-21 I 28
Country: Sweden
Contact: Lektor H. Herman Weil
Voice Phone: 46 08141 70831
Business Description: Hans Weil's inventions were made in the period 1933-1 937; while Gabor invented holography in 1948, the laser was invented 1962 and the first laser-illuminated hologram was exposed as late as 1964.
****

Supastrip International
PP Payne Ltd
Giltway, England
Postal Code: NG 16 2GT
Country: United Kingdom
Voice Phone: 44 602 384 800
FAX Phone: 44 602 384 851
Business Description: Provider of holographic tear tape for use in tamper-resistant packaging.
****

Superbin Co. Ltd
3F-339
Section 2 Ho Ping E Road
Postal Code: 10662
Country: Taiwan
Contact: Edward Hwang
Voice Phone: 886 02 70 I 3626
FAX Phone: 88602 701 3531
Business Description: Exclusive Chinese representative of Coherent (Argon, Krypton Laser, Dye Laser); Continuum (ruby Laser, Nd:VAG laser); Newport (optical components).Also suppl y embossed hologram- manufacturing equipment/ material and consulting se rvice.
****

Swede Holoprint
Duvhoksgatan 6A
Postal Code: 21460
Country: Sweden
Contact: Bjorn Wahlberg
Voice Phone: 46 040 898 21
Bus iness Description: Artistic holography; Artistic marketing consultant.
****

Swiss Federal Inst Of Technology
Laboratory Of Photoelas ticity
Ramistrasse 101
Postal Code: CH-8092
Country: Switzerland
Contact: Walter Schumann
Voice Phone: 41 0380 246 000
Business Description: Holographic non-destructive testing; Scientific and industrial research.
****

Synchronicity Holograms
Box 4235 , Route I
Linco lnville, ME
Postal Code: 04849
Country: United States of America
Contact: Arlene Jurewicz
Voice Phone: I 207 763 3 J 82
Business Description: Synchronicity Holograms provides outreach educational presentations on all aspects of holography for primary grades through high sc hool. Available for workshops and presentation on education. Research on educational aspects of holography.
****

Tama Art Umversity
Department Of Physics
1723 Yarimizu Hac hiouji-Shi
Country : Japan
Contact: Hidetoshi Katsuma
Voice Phone: 81 0426 768 6 11
FAX Phone: 8 I 0426 762 935
Business Description: Resea rch on Holographic TV, Holography Movie
****

Tamarack Storage Devices
12112 Technology Blvd., Suite 101
Austin, TX
Postal Code: 78727
Country: United States of America
Contact: John Stockton
Voice Ph one: I 5 12 250 3 I I I
Business Description: R&D on optical memory for computers
****

Technical University @ Eindhoven
Faculty Of Architecture
Calibre Institute PO Box 513
Postal Code: NL-5600MB
Country: Netherl ands
Contact: Geert T. A. Smelzer
Business Description: Academic research , computer generated holograms
****

Technical University Of Budapest
In stitute Of Preci sion Mechanics
Applied Biophysics Laboratory
Postal Code: H-1 62 1
Country: Hungary
Business Description: Medical holography research.
****

Technical University Of Wroc law.
In st itute Of Physics
Wybrvzeze Wyspi anskiego 27
Postal Code: PL-50-370
Country: Pol and
Contact: Henryk Kasprzak
Business Description: Academic, scientific research
****

Techni sche Fachhoc hschule Berlin
Fb2 / Labor Fur Laseranwendungen
Seestrasse 64
Postal Code: 13347
Country: Germany
Contac t: Dr. J urgen Eichler
Voice Phone: 49 30 304 504 3917
Voice Phone: 49 30 4504 3918
FAX Phone: 49 0049 304 504 3959
Business Description: Holographic Interferometry, Display Holography, Medical Applications
****

Technoexan Ltd
Polytec hni cheskaya, 26
Posta l Code: 194021
Country: Russ ia
Contact: Igor Lovygin
Voice Phone: 7 8 12 247 9383
Voice Phone: 78122475273
FAX Phone: 78 12 2475333
Business Description: Power semiconductors, Lasers, optoelectronics devices (IR range), many channel1-and p-diodes, equipment for high format art and picture hologram, holographic regi sters, school packages for showing optic effects.

****
Technolas Laser Technik Gmbh
Loc hhamer Schlag 19
Postal Code: 8032
Country: Germany
Voice Phone: 49 089 854 5040
FAX Phone: 49 089 854 561
Business Description : Lasers, medical
****

Textile Graphics, Inc.
Holography Press On (HPO)
201 North Fruitport Road
Spring Lake, MI
Postal Code: 49456
Country: United States of America
Contact: Jan Bussard
Voice Phone: 1 6 16 842 5626
FAX Phone: I 616 842 5653
Business Description: Holographic stock or custom images (all shapes and sizes) applied with heat or pressure for adhesion to all substrates. Sealed edges prevent delamination in all weather; washable/dry cleanable. Worldwide distributors sought!
****

The Foreign Dimension
The Peak Galleria
Le vel 2, Shops 29 & 42, The Peak
Country: Hong Kong
Vo ice Phone: 852 2 849 6361
Business Desc ription: Holography shop offering all varieties of holograms for sale to the public. We also offer holograms for sale wholesale to other businesses - see our wholesale listing and ad.
****

The Hologram Company # I
900 North Point Street
San Francisco, CA
Postal Code: 94109
Country: United States of America
Voice Phone: I 415 775 9356
Bus iness Desc ription : Located in hi storic Ghiradelli Square on the waterfront, selling all types of holographic wall art, jewelry and gift items.
****

The Hologram Company #2
Barefoot Landing Suite 4722A
Highway 17 South
North Myrtle Beach, SC
Postal Code: 29582
Country: United States of America
Voice Phone: I 803 272 3583
Business Description: Located in the golfing capital of the US, selling all types of holographic wall art, jewelry and gift items.
****

The Hologram Company #3
The Florida Mall Shopping Center Room 344
8001 South Orange Blossom Trail
Orlando, FL
Postal Code: 32809
Country: United States of America
Voice Phone: 1 407 856 9072
Business Description: Central Florida vacation land location, selling all types of holographic wall art, jewelry and gift items.
****

The Hologram Company #4
370 HOrlon Plaza
San Diego, CA
Postal Code: 9210 I

Country: United States of America
Voice Phone: I 6195578371
Business Desc ripti on: Sunny Southern Cali fornia location, se lling all types of holographic wall art, jewelry and gift items.
****

The Hologram Company #5
Downtown Plaza Suite B-87
547 L Street
Sacramento, CA
Postal Code: 95815
Country: United States of America
Voice Phone: I 9164980305
Business Description: California Capital loca tion, selling all types of holographic wall art, jewelry and gift items.
****

The Hologram Company #6
New Orleans Center Room
New Orleans, LA
Postal Code: 70 11 2
Country: United States of America
Voice Phone: I 504 529 5700
Business Description: Located in the mall ad joining the Super Bowl, selling all types of ho logrphic wall art, jewelry and gift items.
****

The Hologram Company #7
Broadway at the Beach
1215 Celebrity Circle M 147
Myrtle Beach, SC
Postal Code: 29577
Country: United States of America
Voice Phone: 1 803 444 3583
Business Description: Located in the brand new, exc iting Myrtle Beach Mall, selling all types of holographic wall art, jewelry and gift items.
****

The Hologram Store, Ltd
#2673, 8770 - 170 St
Edmonton, Alberta
Postal Code: T5T 412
Country: Canada
Voice Phone: I 403 3333
FAX Phone: I 403 4455
Business Description : Retail stores located in Canada and U.S. specializing in holographic and science products. Wholesale and mai l order available.
****

The Holographic Image Studio
9 Warple Mews Warple Way
Warple Way
London, England
Postal Code: W3 ORF
Country: United Kingdom
Contact: Dr. Martin Richardson
Voice Phone: 44 181 740 5322
FAX Phone: 44 181 740 1733
Business Description: Commissioned holograms up to 1 x 2 meters, pulse-portraiture, movie ste reograms and mass production of silver halide holograms. Catalogues on request.
****

Third Dimension Arts [n co
1241 Andersen Drive
Suites C
San Rafael, CA
Postal Code: 9490 I
Country: United States of America
Contact: Dara Haskell
Voice Phone: I 415 485 1730
Voice Phone: [800 622 4656

FAX Phone: I 415 485 0435
Bus iness Description: Third Dimension Arts Inc. manufacturers of dichromate jewelry, gifts, and 3-D Arts (TM) hologram watches. Suppliers to: the gift, jewelry, museum, and entertainment industry; (licensing) markets. Custom designs welcome'
****

Three-D Light Gallery
109-A The Commons
Ithaca, NY
Postal Code: 14850
Country: United States of America
Contact: Jonathan Pargh
Voice Phone: I 607 273 1187
Business Description: Artistic holography; ho lography gallery.
****

Titan Spectron
1820 East Gary Ave suite 116
Santa Ana, CA
Postal Code: 92705-5804
Country: United States of America
Voice Phone: I 7 14 622 1950
fAX Phone: I 714622 195 1
Business Description: Able to do most holo graphic research projects. Design optical sys tems. Currently working on NASA shuttle proj ects.
****

Tjing Ling Industrial Research.
130 Kee lung Road
Section Iii
Country: Taiwan
Voice Phone: 886 86 207 041 856
Business Description: Fine art originals.
****

TNO In stitute of Applied Phys ics
Department Of Optics
PO Box 155
Posta l Code: NL-2600
Country: Netherl ands
Contact: Ruud L. Van Renesse
Business Description: Academic and indu strial applications.
****

Tokai Univers ity
Department of Electro Photo Optics
I 11 7 Kitakaname
Postal Code: 259-12
Country: Japan
Contact: Hideshi Yokota
Voice Phone: 81 0463 58 1211
fAX Phone: 81 0463 59 2594
Busi ness Desc ripti on: Holography research - artistic holography.
****

Tokyo In stitute Of Technology
Imaging Science And Engineering
4259 Nagatsuda Midori-Ku
Postal Code: 227
Country: Japan
Contact: Masahiro Yamaguchi
Voice Phone: 81 045 922 IIII
FAX Phone: 81 045921 1492
Busi ness Description: Holographic Display, 3-D Imaging Science. Also Dr. Toshio Honda.
****

Topac GmbH Holography
Auf dem Eickholt 47
Postal Code: 33334
Country: Germany
Contact: Sven Deutschmann

Vo ice Phone: 49 5241 800
FAX Phone: 49 5241 806 0870
Business Description: Embossed holograph full service. Product and marketing concept creation of artwork and model linQ. Production of holo grams and application to spewi products. Pack aging and di stribution sen ice
****

Topcon Inc.
75- 1
Hasunuma-Machi Ttabasi-Ku
Postal Code: 174
Country: Japan
Contact: Reiji Hashimoto
Voice Phone: 81 03 3966 3141
FAX Phone: 81 03 3966 2140
Business Description: Hologram manufacturer.
****

Toppan Printing Co. , Ltd.
[mage Technology Laboratory
1-3-3 Suido Bunkyo-Ku
Country: Japan
Contact: Teiichi Nishioka
Vo ice Phone: 81 03 3817 2873
FAX Phone: 81 03 5684 7600
Business Description: Research in Holography. 3-D TV, Computer Graphics
****

Toppan Printing Co. , Ltd.
4-2-3 Taka nodai-Minami
Sugito-Machi, Saitama-ken, Kita-Katsuskika-Gun
Postal Code: 345
Coun try: Japan
Contact: Fujio Iwata
Voice Phone: 81 0480 33 9079
FAX Phone: 81 0480 33 9022
e-mai l address: fiwata@tri .toppan.co.jp
Business Description: Holographic Display. manufacturer of embossed holograms. Lippman, and Multiplex. Also Mr. Susumu Takahas hi and Mr. Toshiki Toda at this address.
****

Total Register In c.
3 Quarry Road
Brookfield, CT
Postal Code: 06804
Country: United States of America
Contact: John Gallagher
Voice Phone: I 203 740 0199
FAX Phone: I 203 740 0177
Business Description: Manufacturer of registra tion devices for hot-stamping presses. Manufac turer of registered rotary die cutting equipment. Registered hologram sheeting and die cutting services.
****

Touchwood Holographies
50 Sugworth Lane
Radley
Abi ngdon, England
Postal Code: OXI4 2HY
Country: United Kingdom
Contact: George W. Clare
Voice Phone: 44 0865 735 874
Business Desc ription: Small experimental lab oratory investigating holographic application; for advertising and promotions in the normal sports trade. One-off experimental commis sions can be undertaken. Sorry. no long runs or repeat work.
****

Towne Technologies
6-10 Bell Ave
Somerville, NJ
Postal Code : 08876-0460
Country: United States of America
Contact: Sal LoSardo
Voice Phone: I 908 722 9500
FAX Phone: I 908 722 8394
Business Desc ription: Towne Technologies is a
producer of fi ne quality holographic photoresist
plates wi th or without a sub-layer of IRON-OX-
IDE. These pl ates are spin-coated with striation
free photoresist in sizes up to 16" x 16".
****

Toyama National College Of Marit
1-2 Ebi e-Neri ya
Postal Code: 933 02
Country: Japan
Contact: Dr. Kenji Kinoshita
Voice Phone: 81 0766860511
Business Description : Holographic Stereogram
****

Trend
Miramarska 85
Postal Code: 41000
Country: Yugos lavia
Contact: Dalibor Vukicevic
Voice Phone: 381 041 51 1 42
Business Desc ription: Gall ery.
****

Tyler Group
218 Linden Avenue
Moorestown, NJ
Postal Code: 08057
Country: United States of America
Voice Phone: I 609 234 1800
FAX Phone: I 609 866 0351
Business Desc ription: Artfoil hot- stamping ma-
chinery used for holography hot stamping.
****

U.K. Gold Purchasers, Inc.
DBA Holograms Unlimited
140 Heimer Rd Suite 740
San Antonio, TX
Postal Code: 78232
Country: United States of America
Contact: Marvin Uram
Voice Phone: I 210 545 3920
Voice Phone: I 800 722 7590
FAX Phone: I 2 10 545 3950
Business Desc ription: Full line distributor of
hologram and related products for specialty re-
tailers - representing more than 80 firms. One
stop shopping at competitive prices .
****

U.S. Holographics
365 North 600 West
Logan, UT
Postal Code: 84321
Country: United States of America
Contact: Dave Rayfie ld
Vo ice Phone: I 80 I 753 5775
FAX Phone: I 80 I 753 5876
Business Description: Marketing/manufacturing
for mass-produced and custom dichromate, pho-
topolymer and embossed holograms for retail
and ad specialty needs. Stock and custom prod-
ucts available.
****

Uk Optica l Supplies
84 Wimborne Road West
Wimborne, England
Postal Code: BH21 2DP

Country: United Kingdom
Contact: Ralph Cull en
Voice Phone: 44 0202 886 831
FAX Phone: 44 0202 742 236
Business Desc ription: Supplying probably the
world's largest selection of Holographic/Optical
components whi ch are: best quality; best value;
Designed by experienced holographers. Plus
component se lection and laboratory/studio set-
up advice freely available.
****

Ultrafine.
16 Foster Road
Chis wick, England
Postal Code: W4 4NY
Count ry: United Kingdom
FAX Phone: 44 0 I 995 230
Business Description: Holographic non-destruc-
tive testing; scientific and industrial research
****

United Technologies
Research Center
Si lver Lane
Hartford, CT
Postal Code : 06108
Country: United States of America
Contact: Dr. Karl A. Stetson.
Voice Phone: I 203 728 7000
FAX Phone: I 203 727 7852
Business Description: UTRC electronic hologra-
phy systems display rea l-time fringe patterns
on TV comparable to photographic holography.
They also prov ide data output for quantitative
analysis. Complete systems or retrofits are ava
ilable.
****

Universi dade Do Porto
Laboratorio De Fisica
Praca Gomes Tei xe ira
Postal Code: P-4000
Country: Portugal
Contact: Olive rio Soares
Business Descripti on: Ho lographic non-
destructive testing; Academic holography re-
search.
****

Uni vers ita Di Roma
La Sapienza Dipt Di Fisica
Piazzale Aldo Moro 2
Postal Code : 1-001 85
Country: Italy
Contact: Paolo De Santis
Business Description: Scientific research
****

Universi tat Erlange n - umberg
Physikalisches In stitut/B2
Staudtstrasse 7
Postal Code: 91058
Country: Germany
Conta ct: Gerd Leuchs
Voice Phone: 49 9131 850837 1
FAX Phone: 49 913 1 13508
Bus iness Descripti on: Scient ific hol ography
research; HOE; computer generated hologra-
phy.
****

Universite De Neuchatel
In stitut De Microtechnique
2, Rue A L Breguet
Postal Code: CH-2000
Country: Switzerland
Conta ct: Rene Dandliker
Voice Phone: 4 1 038 246 000

Business Description: Industrial research.
****

Univers ite Lava l
Dept Physique-Cop I
Pavilion Vachon Ste-Foy
Quebec city, Quebec
Postal Code: G I K 7P4
Country: Canada
Contact: Roger A. Lessard
Voice Phone: I 4 18 656 2 13 1
Voice Phone: I 41 8 656 3080
Business Description: Holography education;
workshops.
****

University Of Tsukuba
Insti tu te Of Art & Des ign
I-I, Tennodai , Tsukuba
Postal Code: 305
Coun try: Japan
Contact: Shunsuke Mital11 ura
Voice Phone: 81 298 53 2883
FAX Phone: 8 1 298 53 65 08
Business Desc ription: Artistic holography, ho-
lography education.
****

University Of Aberdeen
Dept Of Engineering
Ki ngs College, Scotl and
Postal Code: AB9 2UE
Country: United Kingdom
Business Description: Holographic non-destruc-
tive testing, in dustrial research.
****

University Of Alabama
University At Huntsville
Center For Applied Optics
Huntsville, AL
Posta l Code: 35899
Country: United States of America
Contact: Chandra Vikram
Voice Phone: I 205 895 6030
FAX Phone: I 205 895-66 18
Business Descrip tion: Scientific holography re-
search, NOT.
****

University Of Alicante
Applied Physics/Cent De Holograf
Facultad De Ciencias
Postal Code: 99
Country: Spain
Contact: A Fimia.
Voice Phone: 34 566 1200
Business Description: Artistic holography;
HOEs; workshops.
****

Unive rsity Of Arizona
Optical Sc ience Center
Tucson, AZ
Postal Code: 8572 1
Country: United States of America
Voice Phone: I 602 621 6997
Business Desc ription: Industrial and scientific
Holography resea rch; Holographic interfe rom-
etry: Holographic non-destructive testing.
****

Uni ve rsi ty Of Bologna
Via Fiul11azzo 347
Postal Code: 1-480 I 0
Country: Ita ly
Contact: Pier Luigi Capucci
Business Description: Artistic holography re-
search & education.
****

Uni vers ity Of California
University At San Diego
Elect ric a l/Computer Enginee ring
La Jolla, CA
Postal Code: 92093
Country: United States of America
Voice Phone: I 619 534 2230
Business Description: Holography research
****
Uni versity Of Dayton
Research In st itute
300 College Park
Dayton, OH
Postal Code: 45469-0 I 02
Country: United States of America
Contact: Lloyd Huff
Voice Phone: 1 513 2293515
FAX Phone: I 513 229 3433
e-mai l address: hu ff@udri.udayton.edu
Business Description: Scientific research, In-
dustrial research; courses.
****
Uni ve rsity Of Michigan
College Of Engineering
Chrys ler Center
Ann Arbor, MI
Postal Code: 48109-2092
Country : United States of Ameri ca
Voice Phone: I 3 13 763 5464
Business Description: Holographic interferom-
etry; Particle measu rement; Holographic sci-
entific & industrial research; Holographic non-
destructi ve testing.
****
Uni versity Of Michigan
Dept Elect ri ca l Engineering
Room 1 108 Eecs Building
Ann Arbor, MI
Postal Code: 48109-2 I 22
Country: United States of America
Contact: Emmet Leith
Voice Phone: I 313 764 9545
FAX Phone: I 313 763 1503
Business Description: Scientific holography
research; Design H.O.E.s; Courses on holog-
raphy.
****
University Of Munich
In stitute Of Medical Optics
Barbarastrasse 16
Postal Code: 80797
Country: Germany
Contact: Mr. Zurek
Voice Phone: 49 89 2105 3000
FAX Phone: 49 89 12406301
Business Description: MedicalHolography, Sci-
entific holographic research.
****
Uni versity Of Munster
Ear, Nose And Th roat Cl inic
Kardinal Von Gal en Ring 10
Postal Code: 48129
Country: Germany
Contact: Gert Von Bally
Voice Phone: 49 0251 836 86 I
FAX Phone: 49 0251 836 882
Business Description: holography and Interfer-
ometry in med icine. Environmental research
and cultural heritage protection.
****
Uni versity Of North Carolina
Department Of Cell Biology
Chapel Hill, NC

Postal Code: 27599
Country: United States of America
Vo ice Phone: I 9 199662941
Business Description: Medical holography.
****
University Of Oxford
Holography Group
Department Engineering Science
Parks Road, England
Postal Code: OXI 3PJ
Country: Uni ted Kingdom
Contact: Paul Hubel
Voice Phone: 44 865 273 099
FAX Phone: 44 865 273 305
Business Description: Scientific holography re-
search ; workshop s.
****
Unive rsity Of Rochester
Inst itute Of Optics
Rochester, NY
Posta l Code: 14627
Coun try: United States of America
Contact: Dr. Duncan Moore
Voice Phone: I 7 1 6 275 5248
Business Description: Scientific and industrial
holography research; interferometry; particle
testing & measurement.
****
University Of Southern Calif.
Department Of Physics
Uni vers ity Park
Los Angeles, CA
Postal Code: 90089-0484
Country: United States of America
Contact: Jack Feinberg
Voice Phone: I 213 740 11 34
Business Description: Scientific holography re-
search; Interferometry.
****
University Of Strathclyde
Mechanica l Engineering Group
Glasgow, Scot land
Country: United Kingdom
Contact: P. Waddell
Business Description: Scientific research; in-
dustrial research.
****
University Of Stuttgart
Institute Of Applied Optics
Pfaffenwaldring 9
Postal Code: 70569
Country: Germany
Contact: Hans Tiziani.
Voice Pho ne: 49 71 I 685 6075
FAX Phone: 49 71 1 685 65 86
Business Description: Scientific holography re-
search; interferometry, NDT, HOE's
****
University Of Tokyo
Faculty Of Engineering
Hongo 7-3- 1 Bunkyo-Ku
Country: Japan
Contact: T. Uyemura
Business Description: Scientific and Medical
holography research ; Interferometry.
****
University Of Wisconsin
College Of Engineering
432 North Lake Street
Madison, WI
Postal Code: 53706
Country: United States of America
Contact: Francis P. Drake

Voice Phone: I 608 263 7427
FAX Phone: I 608263 3 160
Business Description: Courses on Laser System
Design. Covers laser operation, techniques for
using & modifying laser output; types of lasers;
with material on scanning, modulation, & detec-
tion of laser radiation; designing practical laser
systems.
****
Unterseher & Associates
709 1/2 West Glen Oaks Blvd
Glendale, CA
Postal Code: 9 1202
Country: United States of America
Contact: Fred Unterseher.
Vo ice Phone: I 8 18 549 0534
Business Description: Artistic holography and
holography educati on.
****
Van Leer Metalli zed Products
24 Forge Park
Franklin, MA
Postal Code: 02038
Country: United States of America
Contact: Randy Jacobs
Voice Phone: I 508 54 1 7700
FAX Phone: I 508 541 7788
Business Description: Manu facturer of Holo-
PRISM holographic metal lized papers. Van Leer
produces diffraction patterns or multi-dimen-
sional holographic images on direct metallized
papers. Existing patterns or customized images
are available.
****
Vincennes University
1002 North First Street
Vincennes, IN
Postal Code: 47591
Country: United States of America
Contact: Richard Duesterberg
Vo ice Phone: I 812 885 5294
Business Description: Offering holography
workshops for high school teachers, & co llege
level courses in ho lography. We have 4 re-
search-grade optical tables, as well as argon and
krypton lasers. Call for more details
****
Vinten Electro Optics Ltd
Unit 28
Ashfield Way
Whetstone , England
Postal Code: LE8 3NU
Country: United Kin gdom
Voice Phone: 44 533 867 11 0
Business Description: Manufacture/Design mir-
rors and optics.
****
Visual Visionaries
20 I I Clement St.
Suite 4
San Francisco, CA
Postal Code: 94121
Country: United States of America
Voice Phone: 415 666 0779
Business Description: Consulting and market-
ing firm. Exhibitions and Educational servic-
es. Over 10 years experience with holographic
production and sales. Professional and reli-
able.
****
Volkswagen Ag
Forschung Und Entwicklung
Messtechnik

Country: Germany
Contact: Dr. M.A. Beeck
Voice Phone: 49 5361 925 824
FAX Phone: 49 5361 972 443
Business Description: Industrial research, Interferometry; Holographic non-destructive testing
****

Volvo-Flyg motor
S-461
Postal Code: 81
Country: Sweden
Contact: Robert Frankmark.
Voice Phone: 46 0520 944 71
Busi ness Description: Holographic non-destructive testing.
****

Waseda University
Dept Of Appli ed Physics
School Of Science & Engineering
Postal Code: 160
Country: Japan
Voice Phone: 81 03 209 321
Business Description: Medical holography research.
****

Wave Mechanics
450 North Leavitt
Chicago, IL
Postal Code: 60612
Country: United States of America
Contact: Deni Drinkwater-Welch
Voice Phone: I 3 12 829 WAVE
FAX Phone: I 312 829 8557
Business Description : Arti st ic holographer; silver halide transmission and reflection; consultant.
****

Wavefront Technology
7343 Adams Street
Paramount, CA
Postal Code : 90723
Country: United States of America
Contact: Joel Petersen
Voice Phone: I 310 634 0434
FAX Phone: I 3106340434
Business Description: Embossed hologram mastering, and diffraction grating patterns. Image combining with minimal scan up to 4 x 8 foot. Prototype, short run embossing. Rigind sheet embossing up to 4 x 8 foot in transmission or reflection.
****

Whiley Foi ls Limited
Firth Road
Houston Industrial Estate
Li vingston, Scotland
Postal Code: EH54 5DJ
Country: United Kingdom
Contact: B J Itch
Voice Phone: 44 0506 386 11
FAX Phone: 44 0506 38262.
Business Description: Whiley Foils Limited is a long-established manufacturer of stamping foils. We have developed special base materials for Holographic emboss ing and market these and other Holographic foils worldwide.
****

White Light Works, In c.
PO Box 85 1
Woodland Hills, CA
Postal Code: 91365
Country: United States of America
Contact: Mandy Fox.

Voice Phone: 1 8 18703 1111
FAX Phone: I 818 703 11 82
Business Description: Full-service holographic production company specializing in low cost embossed holograms and MUltip lex "People Stopper" holographic displays for trade shows and POP applications.
****

Whole Picture
A Gallery Of Holography
634 Parkway
Gatlinburg, T
Postal Code: 37738
Country: United States of America
Contact: Jim Kelly
Voice Phone: I 6 15 436 3650
Voice Phone: I 6 15 436 41 73
FAX Phone: I 615436 8974
Business Descriptio n: Full line of film and plate holograms. Full line of holographic jewelry items.
****

Wolverhampton Polytechnic
54 Strafford Street
Wolverhampton, England
Postal Code: WV I I NJ
Country: United Kingdom
Contact: Graham Saxby
Business Desc ription: Research scientist; author of "Practical Holography"
****

Wonders of Holography Gallery
PO Box 1244
Postal Code: 2 1431
Country: Saudi Arabia
Contact: A.M. Baghdadi
Voice Phone: 966 2 652 0052
Voice Phone: 966 2 653 4004
FAX Phone: 966 2 651 1325
Business Description: Retail holograms of all kinds. We are the first and only holography gallery in the Gulf countries. We resell all types of holograms and we produce laser shows.
****

Worcester Pol ytechnic Institute
Mechanical Engi neering Department
100 Institute Road
Worcester, MA
Postal Code: 01609-2280
Country: United States of America
Contact: Ryszard Pryputniewicz
Voice Phone: I 508 83 1 5536
FAX Phone: I 508 83 1 5483
Business Description: Scientific, Medical & Industrial holography re search; Interferometry; Holographic non-destructive testing.
****

Wyko Corp.
2650 East Elvira Road
Tucson, AZ
Postal Code: 85706-7123
Country: United States of America
Contact: James Wyant
Voice Phone: I 602 741 1044
FAX Phone: I 602 294 1799
Business Description: Scientific holography research; Interferometry and analysis.
****

X-IAL
Les Algorithmes
Parc D'Innovation
Postal Code: F-67400
Country: France

Contact: Dr. Christian D. Liegeois
Voice Phone: 33 88 67 44 90
FAX Phone: 33 88 67 80 06
Business Description: Stereograms for embossing; HOEs des igned and manufactured.
****

Zero Gravity
Pier 39 - KI05
San Francisco, CA
Postal Code: 94133
Country: United States of America
Contact: Scott Sachs
Voice Phone: I 415 989 5277
FAX Phone: I 41 59895277
Business Description: Unique giftware and jewelry featuring Spectore titanium jewelry and a full range of holograms shown in an futuristic setting.
****

Zero Gravity
The Forum Shops at Cesars
3500 Las Vegas Blvd. South Suite F-3
Las Vegas, NV
Postal Code: 89 109
Country: United States of America
Contact: Gene Duvic
Voice Phone: I 702 73 1 3565
FAX Phone: I 702 73 1 3565
Business Description: Spaceage setting for unique jewelry and gifts. Featuring Spectoretitanium jewelry and a mezzanine devoted to a full range of holograms. Zero gravity is a dba (subsidiary) of Galaxies Unlimited. Another Dimension is also owned by this company.
****

Zero Gravity
Aloha Tower # 155
101 ALA Moana Blvd
Honolulu, HW
Postal Code: 96813
Country: United States of America
Contact: Margie Anderson
Vo ice Phone: I 808 545 2355
FAX Phone: I 808 545 2877
Business Description: Seven fantasy rooms of unique giftware and jewelry featuring Spectore titanium jewelry and a full range of holograms shown in an Egyptian tomb setting.
****

Zero Gravity
Mall of America N I 08
Bloomington, MN
Postal Code: 55425
Country: United States of America
Contact: Jinna Lindqui st-Dihn
Voice Phone: 1 612 85 1 9699
Business Description: Unique giftware and jewelry featuring Spectore titanium jewelry and a full range of holograms shown in an Egyptian tomb sett ing.
****

Zone Holografix
5338 B Vineland Ave
North Hollywood, CA
Postal Code: 9160 I
Country: United States of America
Contact: Fred Unterseher
Voice Phone: 1 818 985 8477
FAX Phone: 1 818 549 0534
Business Descripti on: Originates masters for mass production holograms in embossed, DCG and photopolymer materials . Both pulsed and CW lasers available.
****

Businesses By Country

Australia - Australian Holographics Pty Ltd.
Australia - Holograms Fantastic & optical Illusions
Australia - Lazart Holographics.
Australia - Moonbeamers
Australia - New Dimension Holographics.
Austria - Holography Center of Austria
Austria - Semicon Austria
Belgium - Agfa Gevaert
Belgium - Agfa N. V.
Belgium - Free University Of Brussels.
Belgium - Laboratory Vinckiner
Belgium - Multifacet
Canada - Centre D'Art
Canada - Cossette, Marie Andree
Canada - Dec-Art Inc
Canada - Deep Space Holographics
Canada - Dimension 3
Canada - Duston Holographic Services
Canada - Fringe Research Holographics
Canada - Full Color Holographics
Canada - General Holograph ics, Inc.
Canada - Harvard Apparatus, Canada
Canada - Holocrafts: Canadian
Canada - Holography Development Group
Canada - Laser In speck
Canada - Lasiris Inc.
Canada - Leitfoil Inc.
Canada - Les Productions Hololab!
Canada - Lumonics In c.
Canada - Melissa Crenshaw
Canada - Mgd Productions
Canada - Mu's Laser Works
Canada - Ontario College Of Art
Canada - Ontario Science Centre
Canada - Photon League Of Holographers.
Canada - Royal Holographic Art Gallery
Canada - Silverbridge Group Inc.
Canada - Starlight Holocom
Canada - The Hologram Store, Ltd
Canada - Universite Laval
China - Beijing Institute Of Posts
China - Beijing Normal University.
China - Morni ng Light Holograms
China - Shan dong Academy of Sciences
Columbia - Imagenes Holograficas De Columbia
Czech - Czechoslovak Academy Of Science
Denmark - Bbt In strumenter Aps.
Finland - SETEC Oy
Finland - Starcke, Ky.
France - Aerospatiale
lOrance - Atelier Holographique De Paris
France - Holo 3
France - Holo-Laser
France - Hologram Industries
France - Holomedia France
France - Laser Movement
France - Magic Laser
France - Musee De L'Holographie
France - P.S.A Peugeot Citroen
France - Photonics Systems Laboratory
France - Quantel
France - Sopra
France - X-IAL
Germany - 3D Vision
Germany - Academy of Media Arts Cologne
Germany - Adlas G.M.B.H. & Co Kg.
Germany - AKS Holographie-Galerie GmbH
Germany - Arbeitskreis Holografie B. V.
Germany - Armin Klix Holographie
Germany - Artbridge Light Studies
Germany - Assoc iation for Hologram Techniques (AHT)
Germany - BIAS
Germany - CHIRON Technolas GmbH

Germany - Creative Holography Index, The
Germany - Daimler Benz Aerospace
Germany - Deutsche Gesellschaft fur Holografie
Germany - Die Dritte Dimension.
Germany - Di etmar Ohlmann
Germany - ETA-Optik Gmbh
Germany - Fostec Gmbh Feinmechanik
Germany - Gresser, E., Kg
Germany - HOL 3, Galerie fur Holographie
Germany - Holo Gmbh
Germany - Hologram Company RAQD GmbH
Germany - Holographic Systems Munchen
Germany - Holographie Anubis
Germany - Holographie Fachstudio Bad
Rothenfeld
Germany - Holographie Konzept GmbH
Germany - Holographie Labor
Germany - Holopublic Unbehaun
Germany - Holovision
Germany - HRT (Holog raphie Recording
Techno logies Gmbh)
Germany - HRT Holographic Recording Technologies
GMBH
Germany Kolbe-Druck mit Tochtergesellschaften
Germany - Krystal Holographics Vertriebs GmbH
Germany - Labor Dr. Steinbichler
Germany - Laserfilm Eckhard Knuth
Germany - Lauk Kommunik at ion
Germany - Leonhard Kurz GmbH
Germany - Leseberg, Dr. Detlef
Germany - Magick Signs Hol ographie
Germany - Melles Griot Gmbh
Germany - Metrologic Gmbh
Germany - Moeller Wedel Optische Werk
Germany - Newport Gmbh
Gemlany - Optical Coating Laboratory GmbH
Germany - OWLS Gmbh/ FOSTEC
Germany - Phantastica
Germany - Physik Instrumente (Pi) Gmbh & C
Germany - Rofin-Sinar Laser Gmbh
Germany - Spindeler & Hoyer Gmbh
Germany - Star Shop
Germany - Steinbichler Optotechnik Gmbh
Germany - Steuer Kg Gmbh & Co.
Germany - Studio Fur Holographie
Germany - Technische Fachhochschule Berlin
Germany - Technolas Laser Technik Gmbh
Germany - Topac GmbH Hol ography
Germany - Universitat Erlangen - Numberg
Germany - University Of Munich
Germany - University Of Munster
Germany - University Of Stuttgart
Germany - Volkswagen Ag
Greece - Cavomit
Greece - Hellenic Institute Of Holography
Hong Kong - Far East Holographics
Hong Kong - Foreign Dimension
Hong Kong - The Foreign Dimension
Hungary - Ap Holografika Studio
Hungary - Technical Uni versity Of Budapest
India - Advance Photonics
India - Holostik India Pvt. Ltd.
India - IS Gill
India - Ojasmit Hol ographics
India - Print-M-Boss
India - Spatial Holodynamics (India) Pvt. Ltd.
India - Spectrum Corporation
Ireland - 3D Holographics
Israel - Holo-Or Ltd
Israel - Holography Israel
Ita ly - Cise Spa Technologie Innovative
Italy - Db Electronic In struments S.R.L.
Italy - Di aures S.A. Holography Di vision
Italy - Diavy sri
Italy - Holofar Lab (Sri)
Italy - Holographic Service

Italy - Societa Olografica Italia (Soi)
Italy - Uni ve rsita Di Roma
Italy - Uni vers ity Of Bologna
Japan - Asahi Glass Co.
Japan - Brainet Corporation
Japan - Canon Inc. R&D Headquarters
Japan - Central Glass Co., Ltd.
Japan - Chiba University
Japan - Dai Nippon Printing Co., Ltd.
Japan - Fuji Electric Co. Ltd
Japan - Fujitsu Laboratories Ltd.
Japan - HODIC Holographic Di splay Artists
& Engineers Club
Japan - Hyogo Prefectual Museum
Japan - Ishii, Ms. Setsuko
Japan - Japan Communication Arts Co.
Japan - Keio University
Japan - Kimmon Electric Co., Ltd.
Japan - Laboratories of Image Information
Japan - Light Dimens ion, In c.
Japan - Marubun Corporation
Japan - Mazda Motor Corp .
Japan - Ministry Of Internat ional Trade
Japan - Mitsubishi Heavy Industries Ltd.
Japan - Newport (Asian Office)
Japan - Nihon University
Japan - Nippon Polaroid K.K.
Japan - Nippondenso Co., Ltd.
Japan - Nippondenso Co., Ltd.
Japan - Nissan Motor.
Japan - Numazu College Of Technology
Japan - Ruey-Tung, Miss. Hung
Japan - SAM Museum
Japan - Sharp Corp.
Japan - Sophia University
Japan - Tama A11 Umversity
Japan - Tokai University
Japan - Tokyo Institute Of Technology
Japan - Topcon Inc .
Japan - Toppan Printing Co. , Ltd.
Japan - Toppan Printing Co., Ltd.
Japan - Toyama Nation al College Of Marit
Japan - University Of Tsukuba
Japan - Uni versity Of Tokyo
Japan - Waseda University
Latvia - Physics Institute. Latvian
Mexico - Hologramas, S.A. de C.Y.
Netherl ands - Dutch Holographic Laboratory BV
Netherlands - Foundation Ideecentrum.
Netherlands - OptiIas B. Y.
Netherlands - Optisc he Fenomenen
Netherlands - Stichting Voor Holographie En
Laseropiek
Netherlands - Tec hnic a l Uni versi ty @
Ein dhoven
Netherlands - TNO Institute of Applied Physics
Norway - Interferens Holografi D.A.
Norway - Norges Tekniske Hogskole.
Pakistan - Dimensions
Poland - Holografia Polska
Poland - Hololand S.c.
Poland - In stitute Of Plasma Phys ics
Poland - Technical Uni versity Of Wroclaw.
Portugal - Universidade Do Porto
Russia - Denisyuk, Yuri N.
Russia - Siavich Joint Stock Company
Russia - Technoexan Ltd
Saudi Arabia - Wonders of Holography Gallery
South Korea - Dan Han Optics
Spain - Holosco, Emest Barnes
Spain - Karas Studios S.L.
Spain - Lasing S.A.,
Spain - Museu D' Hol ografia
Spain - University Of Alicante
Sweden - Dialectica Ab
Sweden - High Tech Network

Sweden - HoloMedia Ab/Hologram Museum.
Sweden - Holovision AB
Sweden - Karolinska In stitutet
Sweden - Lulea University Of Technology
Sweden - Lund Institute Of Tech.
Swede n - Martinsson Elektronik Ab.
Sweden - Royal In stitute Of Technology
Sweden - Saab-Scania
Sweden - Spectrogon Ab
Sweden - Studio Weil-Alvaron
Sweden - Swede Holoprint
Sweden - Volvo-Flygmotor
Switzerland - Galerie Illusoria
Switzerland - Holo-Service
Switzerland - Holo-Service. Fries
Switzerland - Holodesign Studies
Switzerland - Hologramm Werkstatt & Galerie
Switzerland - Holos Art Galerie
Switzerland - Stoltz Ag
Switzerland - Swiss Federal Inst Of Technology
Switzerland - Uni vers ite De Neuchatel
Taiwan - Blue Bond Sales Promoting In c.
Taiwan - Fong Teng Technology
Taiwan - Holo Images Tech Co. , Ltd.
Taiwan - Holo Impress ions
Taiwan - Holo Impressions Inc
Taiwan - Industrial Technology Research Inst.
Taiwan - Infox Corporation
Taiwan - In stitute Of Optical Science
Taiwan - Superbin Co. Ltd
Taiwan - Tjing Ling Industrial Research.
Thailand - New Horizons (Thailand), Ltd.
United Arab Emerates - Hololaser Gallery
UK - 3D [mages
UK - A.H. Prismatic, Ltd.
UK - Ag Electro-Optics Ltd.
UK - Agfa Ltd
UK - Amazing World Of Holograms.
UK - Applied Holographics, Pic .
UK - Astor Universal - AHL Holographics Di vision
UK - Barr & Stroud, Ltd.
UK - Beddis Kenley (Mac hinery) Ltd.
UK - Boyd, Patrick
UK - Brighton Imagecraft
UK - British Aerospace Pic.
UK - British Technology Group (BTG)
UK - Checkpoint
UK - CSI
UK - Datasights Ltd.
UK - De La Rue Holographics Ltd.
UK - Dimuken
UK - Ealing Electro-Optics (Uk)
UK - Electro Optics Developments Ltd.
UK - Embossing Technology Ltd
UK - Expanded Optics Limited
UK - Focal Image Ltd.
UK - Galvoptics Ltd.
UK - Global Images
UK - Holocrafts Europe Limited.
UK - Holograms 3D
UK - Holographics (Uk) Ltd.
UK - Holography Group
UK - Holography News
UK - Holography Studio
UK - Holomex Ltd.
UK - iC Holographics
UK - Imperial College Of Science
UK - International Hologram
UK - Jayco Holographics
UK - K.C. Brown Holographics
UK - Kendall Hyde Ltd.
UK - Laminex/High Tech Uk Ltd.
UK - Laser International
UK - Laza Holograms Ltd.
UK - Levine, Chris
UK - Light Fantastic Pic
UK - Light [mpressions Europe Ltd.

UK - Loughborough Uni v. Of Tech.
UK - Lumonics Ltd
UK - Markem Systems Ltd.
UK - Nati onal Phys ical Laboratory
UK - New Clear Imports Ltd.
UK - Op-Graphics (Holography) Ltd.
UK - Optical Security Industry
UK - Optical Surfaces Ltd.
UK - Optical Works Ltd.
UK - Oriel Scientific Ltd.
UK - Oxford Holographics
UK - Oxford University
UK - Pilkington Optronics
UK - PMI Data Ltd
UK - PT[
UK - Ralph Cullen Holographics
UK - Reconnaissance, Ltd.
UK - Richmond Holographic Studios
UK - Roll s-Royce Pic
UK - Royal College Of Art.
UK - Roya l Photographic Society
UK - Rutherford & Appleton Labs
UK - Severum Design
UK - Siemens Ltd.
UK - Spatial Imaging Ltd.
UK - Specac Ltd
UK - Spectrolab International Ltd.
UK - Supastrip International
UK - The Holographic Image Studio
UK - Touchwood Holographics
UK - Uk Optical Supplies
UK - Ultrafine.
UK - University Of Aberdeen
UK - University Of Oxford
UK - University Of Strathclyde
UK - Vinten Electro Optics Ltd
UK - Whiley Foils Limited
UK - Wolverhampton Polytechnic
USA - 2 I st Century Finishing [nco
USA - 3 Deep Hologram Company
USA - 3-D Worldwide Holograms, Inc
USA - 3M
USA - 3M Optics Technology Center
USA - A.H. Pri smatic, Inc
USA - ACI Systems
USA - Acme Holography
USA - AD 2000, Inc.
USA - Advanced Optics. Inc.
USA - Advanced Technology Program
USA - Aerotech Inc.
USA - Agfa Div of Bayer Corp.
USA - American Bank Note Holographics
USA - American Holographic In c.
USA - American Laser Corporation
USA - American Propylaea Corporation
USA - Amity Photonics Co.
USA - Ana MacArthur
USA - Another Dimension
USA - APA Optics Inc.
USA - Appl ied Physics Research
USA - Archeozoic Incorporated
USA - Art Institute Of Chicago
USA - Art Lab
USA - Art, Science & Technology Inst
USA - Artigliography Co.
USA - Astor Universal - AHL Holographics Dividion
USA - Astor Universal - AHL Holographics Dividion
USA - Avant-Garde Studio
USA - Baldar
USA - Batelle Pacific Northwest
USA - Berkhout, Rudie
USA - Bob Mader Photography
USA - Bobst Group
USA - Booth, Roberta
USA - Brandtjen & Kluge, In c,
USA - Bridgestone Graphic Technologies, Inc.
USA - Burleigh Instruments, Inc.
USA - Burns Holographics Ltd

USA - Cambridge Laser Labs
USA - Carl M. Rodia And Associates
USA - Casdin-Silver Holography
USA - Catalyst Strategic Design, [nco
USA - Cfc/Applied Holographics
USA - Cherry Optical Holography
USA - Chromagem Inc.
USA - City Chemical
USA - Classic City Holography Studio
USA - Coburn Corporation
USA - Coherent, Inc .
USA - Continental Optical
USA - Control Module Inc
USA - Control Optics
USA - Corion Corp.
USA - Corry Laser Technology Inc
USA - Coulter Optical Company
USA - Creative Label
USA - Crown Roll Leaf, In c.,
USA - CVI Laser Corporation
USA - Datacard Corporation
USA - Dazzle Equipment Company
USA - Deem, Rebecca
USA - Dell Optics Company, Inc
USA - Diffraction Company
USA - Diffraction Ltd.
USA - Dimensional Arts
USA - Dimensional Cinematography Co.
USA - Dimensional Foods Co.
USA - Direct Holographics
USA - E. 1. Dupont De Nemours & Co
USA - Ealing Electro-Optics Inc.
USA - Eastman Kodak Company
USA - Ed Wesly Holography
USA - Edmund Scientific Company
USA - Electro Optica l Industries, Inc.
USA - Elusive Image
USA - Environmental Research (Erim)
USA - Envision Enterprises
USA - Excitek Inc.
USA - Fantastic Holograms
USA - Fisher Scientific
USA - Flatiron Studio
USA - FLEXcon Company, In c.
USA - FoilMark Holographics
USA - Ford Motor Company
USA - Fornari, Arthur David
USA - Forth Dimension
USA - Frank DeFreitas Holography Studio
USA - Fresnel Technologies Inc
USA - G.M. Vacuum Coating Lab, In c.
USA - Gardener Promotion Marketing
USA - General Design
USA - Gerald Marks Studio
USA - Glass Mountain Optics
USA - Gray Scale Studios Ltd.
USA - Hallmark Capital Corp
USA - Harland Check Printers
USA - Han·is, Nick
USA - Helios Holography
USA - Holage
USA - Holart Consultants
USA - Holicon Corporation.
USA - Holo Sciences
USA - Holo Spectra
USA - Holo-Spectra
USA - Holo/Source Corporation
USA - Holocraft International
USA - Holoflex Company
USA - Holografica
USA - Hologram Land
USA - Hologram World, Inc.
USA - Holograms International
USA - Holographic and Photonic World Center
USA - Holographic Applications
USA - Holographic Design Systems
USA - Holographic Dimension
USA - Holographic Images Inc.

USA - Holograph ic Industri es, Inc
USA - Holographic Label Conve rtin g (HLC)
USA - Holographic Marketing, Inc.
USA - Holographic Optics Inc
USA - Holographic Products
USA - Holographic Studios
USA - Holographic Technologies
USA - Holographics Inc.
USA - Holographics North Inc.
USA - Holography In stitute
USA - Ho lography Presses On (HPO)
USA - Ho lomat
USA - Holonix
USA - Holopak Technologies
USA - Holophile, Inc.
USA - Holotek
USA - Holovision Systems Inc.
USA - Honeywell Technology Center
USA - Hughes Holography Unit
USA - IBM Almaden Research Center
USA - Ici Americas
USA - Illinois In stitute Of Technology
USA - Illuminations
USA - Imac International, In c.
USA - Imagen Holography, Inc
USA - Images Company
USA - Imagination Plantat ion
USA - Imaging & Design
USA - ImEdge Techno logy
USA - Infinity Laser Laboratories
USA - Infrared Optical Products, Inc.
USA - Inrad, Inc
USA - Integraf
USA - Interactive Industries
USA - In trepid World Communications
USA - Ion Laser Technology Inc.
USA - James Feroe
USA - James Ri ver Products
USA - Iodon Inc.
USA - Jr Holographics
USA - Kaiser Optical Systems, Inc.
USA - Kauffman, John
USA - Keystone Sc ientific Co.
USA - Kinetic Systems, Inc.
USA - Kreischer Optics, Ltd.
USA - Krystal Holographics Internati ona l
USA - L.A.S.E.R. Co.
USA - L.A.S.E.R. News
USA - Label Systems/OS I
USA - Lake Forest College
USA - Lasart Ltd.
USA - Laser Affiliates
USA - Laser Alts
USA - Laser Focus World
USA - Laser Holography Workshop
USA - Laser In novatio ns
USA - Laser Institute Of America
USA - Laser Ionics In c.
USA - Laser Las Vegas
USA - Laser Light Designs
USA - Laser Light Ltd.
USA - Laser Optics, In c.
USA - Laser Photoni cs, Inc.
USA - Laser Reflections
USA - Laser Resa le Inc.
USA - Laser Technical Services
USA - Laser Tec hnology, Inc.
USA - Lasermed ia
USA - Lasermetri cs, Inc.
USA - Lasersmith, Inc.
USA - Laserworks
USA - Lawrence Berkeley Laboratory
USA - Lazer Wizardry
USA - Lenox Laser
USA - Letterhead Press, Inc.
USA - Lexel Lase r, Inc.

USA - Liconix
USA - Light Im pressions Europe
USA - Light Wave Gallery
USA - Light Wave Gallery
USA - Lightrix
USA - Lone Star Illu sions
USA - Lopez's Gallery International
USA - Los Ange les School Of Holography
USA - Lumenx technologi es
USA - M.LT
USA - M. LT. Museum
USA - MacShane Ho lograp hy
USA - Man Environment, Inc.
USA - MasterPrint Holography, Inc .
USA - McCain Marketing & Graphic Design
USA - Mcmahan Electro-Optic
USA - Melles Griot
USA - Meredi th In struments.
USA - Metrolaser
USA - MetroLaser
USA - Metrologic Instruments, Inc.
USA - Mgm Converters Inc
USA - Microelectronics and Computer Technology Corp .
USA - Midwest Laser Products
USA - Mind 's Eye
USA - Mitutoyo Measuring In struments
USA - Museum Of Holography/Chicago
USA - MWK Industries
USA - Naimark, Michael
USA - Neovision Productions
USA - New Light Industries
USA - New York Hall Of Science
USA - Newport Corporation
USA - Norland Products, Inc.
USA - Northern Illi nois University
USA - Northern Lightworks
USA - Northwestern University
USA - OIE Research
USA - Odhner Holographics
USA - Omnichrome
USA - OpSec
USA - Optical Corporation Of America
USA - Optical Research Services
USA - Optical Society of Ameri ca (OSA)
USA - Optics Plus Inc
USA - Optimation
USA - Optimation Holographics
USA - Optitek
USA - Oriel Corporation
USA - Panatron Inc.
USA - Penn sylvania Pulp & Paper Co.
USA - Photon Cantina Ltd.
USA - Photonics Directory
USA - Physical Optics Corporation.
USA - Pink, Patty
USA - Planet 3-D
USA - Point Of View Dimensions, Ltd.
USA - Point Source Productions
USA - Polaroid Corporation
USA - Polymer Image
USA - Portson, In c. (Laser Images)
USA - Rainbow Symphony In c.
USA - Ralcon
USA - Randy James Holography
USA - Red Beam, Inc.
USA - Reel Image
USA - Regal Press Inc
USA - Reva 's Holographic Illusions
USA - Reynolds Metal Co
USA - Rice Systems
USA - Richard Bruck Holography
USA - Robert Sherwood Holo Design
USA - Rochester In s!. Of Technology
USA - Rochester Photon ics Co.
USA - Rol yn Optics

USA - Ross Books
USA - Rowland Institute For Science
USA - Saginaw Valley State Uni versity
USA - Sandia National Laboratories
USA - SchalT Industries
USA - School Of Ho lography
USA - Science Kit & Boreal Labs
USA - Sharon McCormack Holography
USA - Shipley Chemical Co.
USA - Shires, Mark
USA - Siemens (USA)
USA - Silhouette Technology Inc
USA - Si llcocks Plastics International
USA - Silver Dragon Holograp hy
USA - Simian
USA - Smith & McKay Printing Co. Inc.
USA - Southern Indiana Holographics
USA - Space Age Designs In c.
USA - Spectra-Phys ics Lase rs, Inc.
USA - Spectratek In c.
USA - SPIE
USA - SPIE Holography Work ing Group
USA - Springer-Verlag New York
USA - Star Magic
USA - Star Magic
USA - Star Magic
USA - Star Magic
USA - Star Magic
USA - Stephens, Anait
USA - Studio Creaius
USA - Synchronicity Holograms
USA - Tamarack Storage Devices
USA - Textile Graphics, Inc.
USA - The Hologram Company # I
USA - The Hologram Company #2
USA - The Hologram Company #3
USA - The Hologram Company #4
USA - The Hologram Company #5
USA - The Ho logram Company #6
USA - The Hologram Company #7
USA - Third Dimension Arts Inc .
USA - Three-D Light Gallery
USA - Titan Spectron
USA - Total Register In c.
USA - Towne Tec hnologies
USA - Tyler Group
USA - U.K. Gold Purchasers, Inc.
USA - U.S Holographics
USA - United Technologies
USA - Uni versity Of Alabama
USA - Uni versity Of Arizona
USA - University Of California
USA - Univers ity Of Dayton
USA - Unive rsity Of Mich igan
USA - Uni ve rsity Of Michigan
USA - Uni vers ity Of North Carolina
USA - University Of Rochester
USA - Unive rsity Of Southern Cal if.
USA - Uni versity Of Wisconsin
USA - Unterseher & Associates
USA - Van Leer Metallized Products
USA - Vi ncennes Uni versity
USA - Wave Mechanics
USA - Wavefront Technology
USA - White Light Works, Inc.
USA - Whole Picture
USA - Worcester Polytechnic Institute
USA - Wyko Corp.
USA - Zero Gravity
USA - Zero Gravity
USA - Zero Gravity
USA - Zero Gravity
USA - Zone Ho lografix
Yugos lavia - Trend

USA Addresses by Zip, State and Business Name

01202	MA	Photonics Directory
01420	MA	American Holographic Inc.
01562	MA	FLEXcon Company, Inc.
01609	MA	Worcester Polytechnic Institute
01746	MA	Corion Corp.
01746	MA	Ealing Electro-Optics Inc.
01752	MA	Shipley Chemical Co.
01776	MA	Laser Resale Inc.
01886	MA	Optical Corporation Of America
01950	MA	FoilMark Holographics
02038	MA	Van Leer Metallized Products
02062	MA	Regal Press Inc
02109	MA	Dimensional Foods Co.
02131	MA	Kinetic Systems, Inc.
02139	MA	M.I.T.
02139	MA	Polaroid Corporation
02139	MA	M.I.T. Museum
02141	MA	3 Deep Hologram Company
02142	MA	Rowland Institute For Science
02144	MA	Acme Holography
02146	MA	Casdin-Silver Holography
03060	NH	Optimation
03062	NH	Laser Focus World
03855	NH	Lumenx technologies
04849	ME	Synchronicity Holograms
05401	VT	Holographics North Inc.
05673	VT	Diffraction Ltd.
06002	CT	Scharr Industries
06082	CT	Control Module Inc
06108	CT	United Technologies
06419	CT	Holophile, Inc.
06497	CT	Oriel Corporation
06511	CT	AD 2000, Inc.
06605	CT	Bridgestone Graphic Technologies, Inc.
06605	CT	Label Systems/OSI
06611	CT	Carl M. Rodia And Associates
06705	CT	Interactive Industries
06801	CT	Laser Optics, Inc.
06804	CT	Total Register Inc.
07022	NJ	Dell Optics Company, Inc
07068	NJ	Bobst Group
07111	NJ	Excitek Inc.
07302	NJ	Star Magic
07310	NJ	Point Of View Dimensions, Ltd.
07503	NJ	21st Century Finishing Inc.
07503	NJ	Crown Roll Leaf, Inc.,
07647	NJ	Inrad, Inc
07652	NJ	Mitutoyo Measuring Instruments
07660	NJ	Agfa Div of Bayer Corp.

07662	NJ	Lasermetrics, Inc.
07922	NJ	Sillcocks Plastics International
07962	NJ	Silhouette Technology Inc
08007	NJ	Edmund Scientific Company
08012	NJ	Metrologic Instruments, Inc.
08057	NJ	Tyler Group
08555	NJ	Avant-Garde Studio
08701	NJ	Coburn Corporation
08816	NJ	Holopak Technologies
08830	NJ	Siemens (USA)
08876	NJ	Towne Technologies
08902	NJ	Norland Products, Inc.
10003	NY	Star Magic
10010	NY	Flatiron Studio
10010	NY	Holographic Studios
10010	NY	Springer-Verlag New York
10010	NY	Gerald Marks Studio
10011	NY	City Chemical
10019	NY	American Bank Note Holographics
10019	NY	Krystal Holographics International
10023	NY	Star Magic
10028	NY	Star Magic
10111	NY	Berkhout, Rudie
10169	NY	Hallmark Capital Corp
10301	NY	Art Lab
10314	NY	Images Company
10546	NY	Holographic Optics Inc
10598	NY	ImEdge Technology
11101	NY	Holographics Inc.
11201	NY	Laser Light Ltd.
11215	NY	Fornari, Arthur David
11368	NY	New York Hall Of Science
11560	NY	Burns Holographics Ltd
11701	NY	Amity Photonics Co.
11717	NY	MasterPrint Holography, Inc.
11735	NY	Infrared Optical Products, Inc.
11788	NY	Continental Optical
14072	NY	Imaging & Design
14150	NY	Science Kit & Boreal Labs
14453	NY	Burleigh Instruments, Inc.
14620	NY	Holotek
14623	NY	Rochester Photonics Co.
14623	NY	Rochester Inst. Of Technology
14627	NY	University Of Rochester
14650	NY	Eastman Kodak Company
14850	NY	Three-D Light Gallery
15238	PA	Aerotech Inc.
16407	PA	Corry Laser Technology Inc
17579	PA	Direct Holographics
18105	PA	Frank DeFreitas Holography Stiudio

18913	PA	Space Age Designs Inc.
18966	PA	Archeozoic Incorporated
18972	PA	Laser Technical Services
19038	PA	Pennsylvania Pulp & Paper Co.
19144	PA	Gardener Promotion Marketing
19149	PA	O/E Research
19335	PA	Planet 3-D
19372	PA	Keystone Scientific Co.
19403	PA	Laser Technology, Inc.
19850	DE	Ici Americas
19880	DE	E.I. Dupont De Nemours & Co
20001	DC	Holographic and Photonic World Center
20009	DC	Art, Science & Technology Inst
20036	DC	Optical Society of America (OSA)
20770	MD	Holographic Applications
20899	MD	Advanced Technology Program
21057	MD	Lenox Laser
21152	MD	Diffraction Company
21202	MD	Holografica
22069	VA	L.A.S.E.R. Co.
23112	VA	Dazzle Equipment Company
23202	VA	Silver Dragon Holography
23230	VA	Reynolds Metal Co
23236	VA	James River Products
27599	NC	University Of North Carolina
29577	SC	The Hologram Company #7
29582	SC	The Hologram Company #2
30035	GA	Harland Check Printers
30076	GA	Applied Physics Research
30646	GA	Classic City Holography Studio
32789	FL	Mcmahan Electro-Optic
32809	FL	The Hologram Company #3
32811	FL	Laser Ionics Inc.
32826	FL	Laser Institute Of America
32826	FL	Laser Photonics, Inc.
32856	FL	Odhner Holographics
33014	FL	Catalyst Strategic Design, Inc.
33139	FL	Holographic Images Inc.
33166	FL	3-D Worldwide Holograms, Inc
33177	FL	Holographic Dimension
33324	FL	Holographic Marketing, Inc.
33442	FL	Another Dimension
33868	FL	Infinity Laser Laboratories
35899	AL	University Of Alabama
37738	TN	Whole Picture
44509	OH	Chromagem Inc.
45469	OH	University Of Dayton
45840	OH	Holovision Systems Inc.
46236	IN	Artigliography Co.
47448	IN	Forth Dimension
47591	IN	Vincennes University
47715	IN	Southern Indiana Holographics

48009	MI	Intrepid World Communications
48009	MI	American Propylaea Corporation
48103	MI	Jodon Inc.
48106	MI	Kaiser Optical Systems, Inc.
48109	MI	University Of Michigan
48109	MI	University Of Michigan
48113	MI	Environmental Research (Erim)
48121	MI	Ford Motor Company
48150	MI	Holo/Source Corporation
48710	MI	Saginaw Valley State University
48734	MI	Reva'S Holographic Illusions
49117	MI	Laser Holography Workshop
49456	MI	Textile Graphics, Inc.
49456	MI	Holography Presses On (HPO)
53097	WI	McCain Marketing & Graphic Design
53186	WI	Letterhead Press, Inc.
53217	WI	Shires, Mark
53703	WI	Holomat
53706	WI	University Of Wisconsin
54024	WI	Brandtjen & Kluge, Inc,
55105	MN	Holographic Products
55144	MN	3M
55420	MN	Advanced Optics, Inc.
55420	MN	Honeywell Technology Center
55425	MN	Hologram Land
55425	MN	Zero Gravity
55439	MN	Holographic Label Converting (HLC)
55440	MN	Datacard Corporation
55441	MN	Hologram World, Inc.
55449	MN	APA Optics Inc.
60004	IL	MacShane Holography
60007	IL	Creative Label
60045	IL	Integraf
60045	IL	Lake Forest College
60045	IL	Holocraft International
60048	IL	Holographic Industries, Inc
60050	IL	Kreischer Optics, Ltd.
60091	IL	Imac International, Inc.
60115	IL	Northern Illinois University
60208	IL	Northwestern University
60411	IL	Cfc/Applied Holographics
60455	IL	Midwest Laser Products
60603	IL	Art Institute Of Chicago
60607	IL	Holographic Design Systems
60607	IL	Lasersmith, Inc.
60607	IL	Museum Of Holography/Chicago
60607	IL	School Of Holography
60610	IL	Robert Sherwood Holo Design
60611	IL	Light Impressions Europe
60611	IL	Light Wave Gallery
60611	IL	Lopez's Gallery International
60612	IL	Wave Mechanics
60616	IL	Illinois Institute Of Technology

60618	IL	Holicon Corporation.		90620	CA	Holograms International
60618	IL	Richard Bruck Holography		90670	CA	Astor Universal - AHL Holographics Dividion
60648	IL	Ed Wesly Holography		90670	CA	Mgm Converters Inc
60651	IL	Fisher Scientific		90723	CA	Wavefront Technology
61801	IL	Holoflex Company		91001	CA	Envision Enterprises
66046	KS	Optimation Holographics		91012	CA	Photon Cantina Ltd.
66212	KS	Helios Holography		91107	CA	Optical Research Services
66215	KS	Astor Universal - AHL Holographics Dividion		91202	CA	Deem, Rebecca
66215	KS	Portson, Inc. (Laser Images)		91202	CA	Unterseher & Associates
70112	LA	The Hologram Company #6		91335	CA	Rainbow Symphony Inc.
75023	TX	Laser Arts		91344	CA	Mind'S Eye
75201	TX	Bob Mader Photography		91365	CA	Los Angeles School Of Holography
75202	TX	Elusive Image		91365	CA	White Light Works, Inc.
76110	TX	Fresnel Technologies Inc		91406	CA	Holo Spectra
77008	TX	Harris, Nick		91406	CA	Holo-Spectra
77060	TX	Dimensional Arts		91601	CA	Zone Holografix
78232	TX	U.K. Gold Purchasers, Inc.		91706	CA	Control Optics
78727	TX	Tamarack Storage Devices		91710	CA	Omnichrome
78746	TX	Lone Star Illusions		91720	CA	MWK Industries
78758	TX	Glass Mountain Optics		91722	CA	Rolyn Optics
78759	TX	Microelectronics and Computer Technology Corp.		91769	CA	Panatron Inc.
80112	CO	ACI Systems		92029	CA	Holographic Technologies
80214	CO	Lazer Wizardry		92093	CA	University Of California
80222	CO	OpSec		92101	CA	The Hologram Company #4
80249	CO	Fantastic Holograms		92349	CA	Coulter Optical Company
81423	CO	Lasart Ltd.		92663	CA	G.M. Vacuum Coating Lab, Inc.
81612	CO	Imagen Holography, Inc		92669	CA	Laserworks
84104	UT	American Laser Corporation		92705	CA	Optics Plus Inc
84115	UT	Ion Laser Technology Inc.		92705	CA	Rice Systems
84321	UT	U.S. Holographics		92705	CA	Titan Spectron
84328	UT	Ralcon		92714	CA	Melles Griot
84333	CO	Gray Scale Studios Ltd.		92714	CA	MetroLaser
85027	AZ	Polymer Image		92714	CA	Newport Corporation
85301	AZ	Meredith Instruments.		92714	CA	Metrolaser
85704	AZ	Holo Sciences		93021	CA	Laser Innovations
85706	AZ	Wyko Corp.		93108	CA	Stephens, Anait
85721	AZ	University Of Arizona		93111	CA	Electro Optical Industries, Inc.
87185	NM	Sandia National Laboratories		94039	CA	Optitek
87192	NM	CVI Laser Corporation		94039	CA	Spectra-Physics Lasers, Inc.
87506	NM	Ana MacArthur		94044	CA	Reel Image
89109	NV	Zero Gravity		94080	CA	Lightrix
89129	NV	Laser Las Vegas		94107	CA	General Design
90004	CA	Neovision Productions		94109	CA	The Hologram Company #1
90025	CA	Spectratek Inc.		94110	CA	Holart Consultants
90027	CA	Booth, Roberta		94110	CA	Multiplex Moving Holograms
90045	CA	Lasermedia		94114	CA	Imagination Plantation
90064	CA	Man Environment, Inc.		94114	CA	Star Magic
90089	CA	University Of Southern Calif.		94122	CA	Holage
90245	CA	Hughes Power Products, Inc.		94122	CA	Illuminations
90401	CA	Jr Holographics		94124	CA	Holography Institute
90505	CA	Physical Optics Corporation.		94124	CA	L.A.S.E.R. News

94124	CA	Pink, Patty
94133	CA	Light Wave Gallery
94133	CA	Naimark, Michael
94133	CA	Zero Gravity
94306	CA	Studio Creaius
94404	CA	A.H. Prismatic, Inc
94509	CA	Laser Light Designs
94538	CA	Lexel Laser, Inc.
94539	CA	Cambridge Laser Labs
94608	CA	James Feroe
94611	CA	Red Beam, Inc.
94704	CA	Baldar
94704	CA	Dimensional Cinematography Co.
94704	CA	Ross Books
94720	CA	Lawrence Berkeley Laboratory
94901	CA	Third Dimension Arts Inc.
94954	CA	3M Optics Technology Center
94956	CA	Kauffman, John
95006	CA	Point Source Productions
95054	CA	Coherent, Inc.
95054	CA	Liconix
95060	CA	Simian
95062	CA	Randy James Holography
95110	CA	Smith & McKay Printing Co. Inc.
95113	CA	Laser Reflections
95120	CA	IBM Almaden Research Center
95472	CA	Cherry Optical Holography
95472	CA	Laser Affiliates
95815	CA	The Hologram Company #5
96813	HW	Zero Gravity
98103	WA	Northern Lightworks
98145	WA	Holonix
98227	WA	SPIE
98227	WA	SPIE Holography Working Group
98672	WA	Sharon Mccormack Holography
99204	WA	New Light Industries
99352	WA	Batelle Pacific Northwest

Addendum - Unlisted Cities

Business Name	City
3D Holographics	Dublin 11
3D Vision	Bremen
Academy of Media Arts Cologne	Cologne
Adlas G.M.B.H. & Co Kg.	Lubeck
Advance Photonics	Bombay
Aerospatiale	Bordeaux
Agfa Gevaert	Morsel
Agfa N.V.	Mortsel
AKS Holographie-Galerie GmbH	Essen
Ap Holografika Studio	Budapest
Arbeitskreis Holografie B.V.	Geldern
Armin Klix Holographie	Dusseldorf
Artbridge Light Studies	Braunschweig
Asahi Glass Co.	Tokyo
Association for Hologram Tech.	Bad Rothenfelde
Atelier Holographique De Paris	Paris
Bbt Instrumenter Aps.	Frederiksberg
Beijing Institute Of Posts	Beijing
Beijing Normal University.	Beijing
BIAS	Bremen
Blue Bond Sales Promoting Inc.	Taipei

Brainet Corporation	Tokyo
Canon Inc. R&D Headquarters	Kanagawa
Cavomit	Athens
Central Glass Co., Ltd.	Tokyo
Chiba University	Chiba
CHIRON Technolas GmbH	Dornach
Cise Spa Technologie Innovative	Milano
Creative Holography Index, The	Bergisch Gladbach
Czechoslovak Academy Of Science	Prague
Daimler Benz Aerospace	Germering
Dan Han Optics	Hwasong-Gun
Db Electronic Instruments S.R.L.	Milano
Denisyuk, Yuri N.	St. Petersburg
Deutsche Gesellschaft fur Holografie	Brnsche
Dialectica Ab	Stockholm
Diaures S.A. Holography Division	(Modena)
Diavy srl	soliera (Modena)
Die Dritte Dimension.	Neu Isenburg
Dietmar Ohlmann	Braunschweig
Dimensions	Sialkot
Dutch Holographic Laboratory BV	Eindhoven
ETA-Optik Gmbh	Heinsberg
Far East Holographics	Wanchai
Fong Teng Technology	Taipei
Foreign Dimension	Hong Kong
Fostec Gmbh Feinmechanik	Berlin 31
Foundation Ideecentrum.	Eindhoven
Free University Of Brussels.	Brussels
Fuji Electric Co. Ltd	Tokyo
Fujitsu Laboratories Ltd.	Atsugi
Galerie Illusoria	Bern
General Holographics, Inc.	Burnaby
Global Images	London
Gresser, E., Kg	Ochsenfurt
Hellenic Institute Of Holography	Chalandri
High Tech Network	Malmo
HODIC Holographic Display Artists & Engineers Club	Chiba
HOL 3, Galerie fur Holographie	Berlin
Holo 3	Saint-Louis
Holo Gmbh	Bramsche
Holo Images Tech Co., Ltd.	Yung Kang City
Holo Impressions	Taipei Shien
Holo Impressions Inc	Taipei Shein
Holo-Laser	Miserey
Holo-Or Ltd	Rehovot
Holo-Service	Basel
Holo-Service.Fries	Basel
Holodesign Studies	Riehen
Holofar Lab (Srl)	Rome
Hologram Company RAQD GmbH	Witzhave
Hologram Industries	Fontenay-Sous-Bois
Hologramas, S.A. de C.V.	Mexico D.F.
Hologramm Werkstatt & Galerie	Castesegna
Holographic Service	Milan
Holographic Systems Munchen	Markt Schwaben
Holographie Anubis	Bamberg
Holographie Fachstudio Bad Rothenfeld	Bad Rothenfeld
Holographie Konzept GmbH	Frankfurt
Holographie Labor	Gutersloh
Holography Center of Austria	Wurmla
Holography Israel	Herzlia
Hololand S.C.	Warszawa
Hololaser Gallery	Dubai
HoloMedia Ab/Hologram Mseum.	Stockholm
Holomedia France	Toulouse
Holopublic Unbehaun	Wuppertal
Holos Art Galerie	Geneva
Holostik India Pvt. Ltd.	New Delhi
Holovision	Munchen
Holovision AB	Stockholm
HRT (Holographie Recording Technologies Gmbh)	Steinau
HRT Holographic Recording Technologies GMBH	Steinau
Hyogo Prefectual Museum of Modern Art	Nada-Ku Kobe
I.S. Gill	New Delhi
Imagenes Holograficas De Columbia	Cali

Imagenes Holograficas De Columbia	Cali
Industrial Technology Research	Chutung
Infox Corporation	Taipei
Institute Of Optical Science	Chung-Li
Institute Of Plasma Physics	Wroclaw
Interferens Holografi D.A.	Hamar
Ishii, Ms. Setsuko	Tokyo
Japan Communication Arts Co.	Osaka
Karas Studios S.L.	Madrid
Karolinska Institutet	Huddinge
Keio University	Yokohama
Kimmon Electric Co., Ltd.	Tokyo
Kolbe-Druck mit Tochtergesellschaften	Versmold
Krystal Holographics Vertriebs GmbH	Reutlingen
Labor Dr. Steinbichler	Neubeuern
Laboratories of Image Information	Toyonaka, Osaka
Laboratory Vinckiner	Gent
Laser Movement	Saint Pierre du Perray
Laserfilm Eckhard Knuth	Munich
Lasing S.A.,	Madrid
Lauk Kommunikation	Frechen
Leitfoil Inc.	Powassan
Leonhard Kurz GmbH	Furth
Leseberg, Dr. Detlef	Lunen-Beckinghausen
Light Dimension, Inc.	Tokyo
Lulea University Of Technology	Lulea
Lund Institute Of Tech.	Lund
Magic Laser	Paris
Magick Signs Holographie	Langen
Martinsson Elektronik Ab.	Hagersten
Marubun Corporation	Tokyo
Mazda Motor Corp.	Hiroshima
Melles Griot Gmbh	Darmstadt
Metrologic Gmbh	Puchheim
Ministry Of International Trade	Tsukuba Science City
Mitsubishi Heavy Industries Ltd.	Nagasaki
Moeller Wedel Optische Werk	Wedel
Morning Light Holograms	He Bei
Multifacet	Drogenbos
Musee De L'Holographie	Paris
Museu D' Holografia	Barcelona
New Horizons (Thailand), Ltd.	Chiangmai
Newport (Asian Office)	2-1, 2-Chome.
Newport Gmbh	Darmstadt
Nippon Polaroid K.K.	Tokyo
Nippondenso Co., Ltd.	Aichi-Ken
Nippondenso Co., Ltd.	Aichi-Ken
Nissan Motor.	Yokosuka
Norges Tekniske Hogskole.	Trondheim-Nth
Ojasmit Holographics	New Bombay
Optical Coating Laboratory GmbH	Goslar
Optilas B.V.	A/D Rijn
Optische Fenomenen	Capelle A/D Ijssel
OWIS Gmbh/ FOSTEC	Staufen
P.S.A Peugeot Citroen	Bievres
Phantastica	Arnsberg
Photonics Systems Laboratory	Strasbourg,
Physics Institute. Latvian	Riga-Salaspils
Physik Instrumente (Pi) Gmbh & C	Waldbronn
Print-M-Boss	New Delhi
Quantel	Les Ulis Orsay Cedex
Rofin-Sinar Laser Gmbh	Hamburg
Royal Institute Of Technology	Stockholm
Saab-Scania	Linkoping
SAM Museum	Osaka
Semicon Austria	Graz
SETEC Oy	SF-0671 Vantas
Shandong Academy of Sciences	Jinan Shandong
Sharp Corp.	Kashiwa
Slavich Joint Stock Company	Pereslavl-Zalessky
Societa Olografica Italia (Soi)	Roma
Sophia University	Tokyo
Sopra	Bois-Colombes
Spatial Holodynamics (India) Pvt. Ltd.	Worli. Bombay
Spectrogon Ab	Taby
Spectrum Corporation	Ahmedabad
Spindeler & Hoyer Gmbh	Goettingen

Spatial Holodynamics (India) Pvt. Ltd.	Worli. Bombay
Spectrogon Ab	Taby
Spectrum Corporation	Ahmedabad
Spindeler & Hoyer Gmbh	Goettingen
Star Shop	Hamburg
Starcke, Ky.	Kokemaki
Steinbichler Optotechnik Gmbh	Neubeuern
Steuer Kg Gmbh & Co.	Leinfelden-Echterdingen
Stichting Voor Holographie En Laseropiek	1017 JT Amsterdam
Stoltz Ag	Baden Dattwil
Studio Fur Holographie	Eichenau
Studio Weil-Alvaron	Malmo
Superbin Co. Ltd	Taipei
Swede Holoprint	Malmo
Swiss Federal Inst Of Technology	Zurich
Tama Art Umversity	Tokyo
Technical University @ Eindhoven	Eindhoven
Technical University Of Budapest	Budapest
Technical University Of Wroclaw.	Wroclaw
Technische Fachhochschule Berlin	Berlin
Technoexan Ltd	St. Petersburg
Technolas Laser Technik Gmbh	Graefelfing
The Foreign Dimension	Hong Kong
Tjing Ling Industrial Research.	Taipei
TNO Institute of Applied Physics	Ad Delft
Tokai University	Hiratsuka City
Tokyo Institute Of Technology	Yokohama
Topac GmbH Holography	Gutersloh
Topcon Inc.	Tokyo
Toppan Printing Co., Ltd.	Tokyo
Toppan Printing Co., Ltd.	
Toyama National College Of Marit	Shinminato
Trend	Zagreb
Universidade Do Porto	Porto
Universita Di Roma	Rome
Universitat Erlangen - Nurnberg	Erlangen
Universite De Neuchatel	Neuchatel
University Of Alicante	Alicante Apdo
University Of Bologna	Belricetto
University Of Munich	Munich
University Of Munster	Munster
University Of Stuttgart	Stuttgart
University Of Tokyo	Toyko
Volkswagen Ag	Wolfsburg
Volvo-Flygmotor	Trollhattan
Waseda University	Tokyo
Wonders of Holography Gallery	Jeddah
X-IAL	Illkirch

Businesses That Sell Holograms

Country	Business	Whole-saler	Retail Shop	Museum /Gallery	Catalogue /Mail Order	Touring Show
Australia	Australian Holographics Pty Ltd.					X
Australia	Holograms Fantastic & optical Illusions	X	X	X	X	X
Australia	Lazart Holographics.	X	X	X		
Australia	Moonbeamers	X				
Australia	New Dimension Holographics.	X	X			
Austria	Holography Center of Austria	X				
Canada	Dec-Art Inc	X				
Canada	Deep Space Holographics	X			X	
Canada	Dimension 3	X			X	
Canada	General Holographics, Inc.	X				
Canada	Leitfoil Inc.		X		X	
Canada	Melissa Crenshaw	X				
Canada	Ontario Science Centre			X		
Canada	Royal Holographic Art Gallery	X	X	X	X	
Canada	Silverbridge Group Inc.		X		X	
Canada	Starlight Holocom	X	X		X	
Canada	The Hologram Store, Ltd	X	X	X	X	
Finland	Starcke, Ky.	X			X	
France	Holomedia France	X				
France	Magic Laser	X				
Germany	3D Vision	X			X	
Germany	Armin Klix Holographie				X	
Germany	Artbridge Light Studies	X				
Germany	Die Dritte Dimension.		X	X		
Germany	ETA-Optik Gmbh	X				
Germany	HOL 3, Galerie fur Holographie			X		
Germany	Holo Gmbh	X				
Germany	Holographie Fachstudio Bad Rothenfeld	X				X
Germany	Holovision	X	X			
Germany	Krystal Holographics Vertriebs GmbH	X			X	
Germany	Lauk Kommunikation			X		
Germany	Phantastica		X			
Germany	Star Shop	X	X		X	
Greece	Cavomit	X				
Hong Kong	Far East Holographics	X				
Hong Kong	Foreign Dimension	X				
Hong Kong	The Foreign Dimension		X			
Hungary	Ap Holografika Studio	X				
India	Ojasmit Holographics	X				
Israel	Holography Israel					X
Italy	Holographic Service	X				
Japan	Brainet Corporation	X				
Japan	Hyogo Prefectual Museum			X		
Japan	Japan Communication Arts Co.	X				
Japan	Light Dimension, Inc.	X	X	X		X
Japan	SAM Museum			X		
Mexico	Hologramas, S.A. de C.V.	X			X	
Netherlands	Dutch Holographic Laboratory BV				X	
Norway	Interferens Holografi D.A.			X		
Pakistan	Dimensions	X				
Poland	Hololand S.C.	X				
Russia	Technoexan Ltd	X				
Saudi Arabia	Wonders of Holography Gallery	X	X	X		
Spain	Karas Studios S.L.			X		
Spain	Museu D' Holografia	X	X	X	X	X
Sweden	HoloMedia Ab/Hologram Museum.	X	X	X	X	X
Switzerland	Galerie Illusoria			X		
Taiwan	Holo Images Tech Co., Ltd.	X			X	
Taiwan	Holo Impressions Inc	X	X			
Taiwan	Infox Corporation	X				
UAE	Hololaser Gallery	X	X	X		
UK	3D Images	X			X	
UK	A.H. Prismatic, Ltd.	X				

Country	Business	Whole-saler	Retail Shop	Museum /Gallery	Catalogue /Mail Order	Touring Show
UK	Amazing World Of Holograms.		X		X	
UK	Applied Holographics, Plc.	X				
UK	Astor Universal - AHL Holographics Division	X				
UK	Embossing Technology Ltd	X				
UK	Holograms 3D			X		X
UK	Holography Studio			X		X
UK	Laza Holograms Ltd.	X		X	X	
UK	Light Impressions Europe Ltd.	X				
UK	New Clear Imports Ltd.		X			
UK	Op-Graphics (Holography) Ltd.	X				
UK	Oxford Holographics	X	X		X	
UK	Spatial Imaging Ltd.	X		X		X
UK	The Holographic Image Studio	X			X	X
USA	3 Deep Hologram Company	X				
USA	3-D Worldwide Holograms, Inc	X			X	
USA	A.H. Prismatic, Inc	X				
USA	American Propylaea Corporation	X				
USA	Another Dimension	X	X		X	
USA	Art, Science & Technology Inst			X		
USA	Artigliography Co.			X		
USA	Astor Universal - AHL Holographics	X				
USA	Bridgestone Graphic Technologies, Inc.	X				
USA	Cherry Optical Holography	X				
USA	Dimensional Arts	X			X	
USA	Direct Holographics	X				
USA	Edmund Scientific Company		X		X	
USA	Elusive Image		X			
USA	Envision Enterprises	X				X
USA	Fantastic Holograms		X			
USA	Fisher Scientific				X	
USA	Flatiron Studio	X		X		
USA	FLEXcon Company, Inc.	X				
USA	FoilMark Holographics	X				
USA	Forth Dimension		X	X	X	
USA	Frank DeFreitas Holography Studio	X		X	X	
USA	Helios Holography		X			
USA	Holo/Source Corporation	X				
USA	Holografica		X			
USA	Hologram Land		X		X	
USA	Hologram World, Inc.	X			X	
USA	Holograms International		X		X	
USA	Holographic and Photonic World Center			X		
USA	Holographic Design Systems	X				
USA	Holographic Dimension	X	X	X	X	X
USA	Holographic Industries, Inc	X	X		X	
USA	Holographic Marketing, Inc.	X				
USA	Holographic Technologies	X			X	
USA	Holographics North Inc.	X		X		
USA	Images Company	X			X	
USA	Integraf	X				
USA	Interactive Industries	X			X	
USA	Krystal Holographics International	X				
USA	Lasart Ltd.	X				
USA	Lasersmith, Inc.	X				
USA	Lazer Wizardry	X	X		X	
USA	Light Impressions Europe	X			X	
USA	Light Wave Gallery		X			
USA	Lightrix	X				
USA	Lone Star Illusions		X			
USA	Lopez's Gallery International			X		X
USA	M.I.T. Museum		X	X		
USA	Metrologic Instruments, Inc.				X	

Country	Business	Whole-saler	Retail Shop	Museum /Gallery	Catalogue /Mail Order	Touring Show
USA	Multiplex Moving Holograms	X			X	
USA	Museum Of Holography/Chicago			X		
USA	MWK Industries	X			X	
USA	New York Hall Of Science			X		
USA	Pennsylvania Pulp & Paper Co.	X				
USA	Polaroid Corporation	X			X	
USA	Portson, Inc. (Laser Images)	X			X	
USA	Rainbow Symphony Inc.	X				
USA	Reva'S Holographic Illusions		X			
USA	Richard Bruck Holography	X			X	
USA	Scharr Industries	X				
USA	Science Kit & Boreal Labs				X	
USA	Star Magic #1		X			
USA	Star Magic #2		X			
USA	Star Magic #3		X			
USA	Star Magic #4		X			
USA	Star Magic #5		X			
USA	The Hologram Company #1		X			
USA	The Hologram Company #2		X			
USA	The Hologram Company #3		X			
USA	The Hologram Company #4		X			
USA	The Hologram Company #5		X			
USA	The Hologram Company #6		X			
USA	The Hologram Company #7		X			
USA	Third Dimension Arts Inc.	X	X		X	
USA	Three-D Light Gallery			X		
USA	U.K. Gold Purchasers, Inc.	X			X	
USA	U.S. Holographics	X				
USA	Van Leer Metallized Products	X				
USA	White Light Works, Inc.	X			X	
USA	Whole Picture			X		
USA	Zero Gravity		X			
USA	Zero Gravity		X			
USA	Zero Gravity		X			
USA	Zero Gravity		X			

Hologram Mastering Businesses

Country	Business Name	Model: Sculpture	Model: Computer Animation	Master: Silver Halide	Master: Dichromate	Master: Photopolymer	Master: Photoresist	Master: Stereograms	Master: Pulsed Portrait
Australia	Australian Holographics Pty Ltd.			X					X
Australia	Holograms Fantastic & optical Illusions	X	X	X			X	X	
Australia	Moonbeamers	X		X		X	X		
Austria	Holography Center of Austria			X					
Canada	Cossette, Marie Andree			X					
Canada	Deep Space Holographics	X	X		X				
Canada	Duston Holographic Services			X					
Canada	Fringe Research Holographics			X					
Canada	General Holographics, Inc.				X				
Canada	Holocrafts: Canadian				X				
Canada	Lasiris Inc.			X					
Canada	Les Productions Hololab!			X					
Canada	Melissa Crenshaw			X					
Canada	Mgd Productions	X		X					
Canada	Mu'S Laser Works			X					
Canada	Photon League Of Holographers.			X					

Country	Business Name	Model: Sculpture	Model: Computer Animation	Master: Silver Halide	Master: Dichromate	Master: Photopolymer	Master: Photoresist	Master: Stereograms	Master: Pulsed Portrait
Canada	Royal Holographic Art Gallery			X					
Canada	Silverbridge Group Inc.				X				
France	Atelier Holographique De Paris			X					
France	Holo-Laser			X					
France	X-IAL			X					
Germany	3D Vision			X					
Germany	AKS Holographie-Galerie GmbH						X		
Germany	Arbeitskreis Holografie B.V.			X					
Germany	Armin Klix Holographie			X					
Germany	Artbridge Light Studies			X					
Germany	ETA-Optik Gmbh	X			X	X	X		
Germany	Holo Gmbh			X					
Germany	Holographie Labor			X					
Germany	Holovision	X	X	X			X	X	
Germany	HRT (Holographie Recording Technologies)			X					
Germany	Krystal Holographics Vertriebs GmbH	X	X			X			
Germany	Laserfilm Eckhard Knuth							X	
Germany	Spindeler & Hoyer Gmbh			X					
Germany	Star Shop	X		X	X	X	X	X	
Germany	Studio Fur Holographie			X					
Germany	Topac GmbH Holography						X		
Greece	Hellenic Institute Of Holography		X	X					
Hong Kong	Foreign Dimension	X		X	X	X	X		
Hungary	Ap Holografika Studio	X		X			X		
India	Holostik India Pvt. Ltd.		X						
India	Ojasmit Holographics	X		X	X	X			
India	Spatial Holodynamics (India) Pvt. Ltd.	X	X	X			X		
Italy	Diaures S.A. Holography Division			X					
Italy	Holofar Lab (Srl)			X					
Japan	Asahi Glass Co.			X					
Japan	Ishii, Ms. Setsuko			X					
Mexico	Hologramas, S.A. de C.V.	X	X				X	X	
Netherlands	Dutch Holographic Laboratory BV		X	X					
Norway	Interferens Holografi D.A.	X		X					
Poland	Hololand S.C.		X				X		
Russia	Technoexan Ltd	X		X					
Spain	Holosco, Ernest Barnes	X		X			X	X	
Spain	Museu D' Holografia			X					
Sweden	Dialectica Ab			X					
Sweden	High Tech Network			X					
Sweden	Holovision AB			X					X
Sweden	Lund Institute Of Tech.			X					
Sweden	Martinsson Elektronik Ab.			X					
Sweden	Swede Holoprint			X					
Switzerland	Holo-Service			X					
Switzerland	Holo-Service.Fries			X					
Switzerland	Hologramm Werkstatt & Galerie			X					

The Foreign

Specialist in manufacturing a whole line of holographic products:

- Watches
- Key Chains
- Calculators
- Glasses

As a custom designer, we provide a full range of holographic services for:

- Advertising
- Promotions
- Displays
- Packaging
- Premium and incentive
- -Gifts

We custom-make any holographic product according to your specifications. We are a French-managed company specializing in holograms with high quality standards, and offer Hong Kong very competitive prices.

Please contact us for details.

Dimension

HOLOGRAPHIC PRODUCTS

Head Office:
1901 Manley Commercial Bldg.,
367-375 Queen's Road, Central
Hong Kong
TEL: (852) 2542-0282
FAX: (852) 2541-6011

Show Room:
The Peak Galleria
Level 2, Shops 29 & 42
The Peak, Hong Kong
TEL: 2849-6361

EXPLORE
THE POSSIBILITIES

Take along an experienced guide

Bobst diecutters and hot-foil stampers have been producing the world's most innovative packaging for over 50 years. When holograms became a commercial reality, Bobst was there with the high-speed application technology to give holograms a viable role in packaging.

Today, there are three Bobst hot foil stamper models available for three sheet sizes: the SP 126-BMA for 32" x 50", the SP 102-BMA for 28" x 40", and the all-new SP 76-BM for 22" x 30". All three models are built on the years of technical expertise that only Bobst can offer.

Explore the possibilities with Bobst

Featured in Bobst's Booth #3707 will be the newest in foil-stamping, diecutting, and folder-gluer technology for the folding carton converter, trade finisher, and commercial printer. Explore your future with an experienced guide.

See the SP 76-BM at Graph Expo'95
Booth #3707 – October 8-11, 1995 – Chicago, IL

BOBST SP 76-BM
HOT-FOIL STAMPER
22" X 30" FORMAT
7,000 SPH OUTPUT

BOBST

Bobst Group Inc., 146 Harrison Ave., Roseland, NJ 07068 USA; phone 201-226-8000
Bobst Canada Inc., 274 Labrosse, Pointe-Claire, Quebec H9R 5L8; phone (514) 426-3030

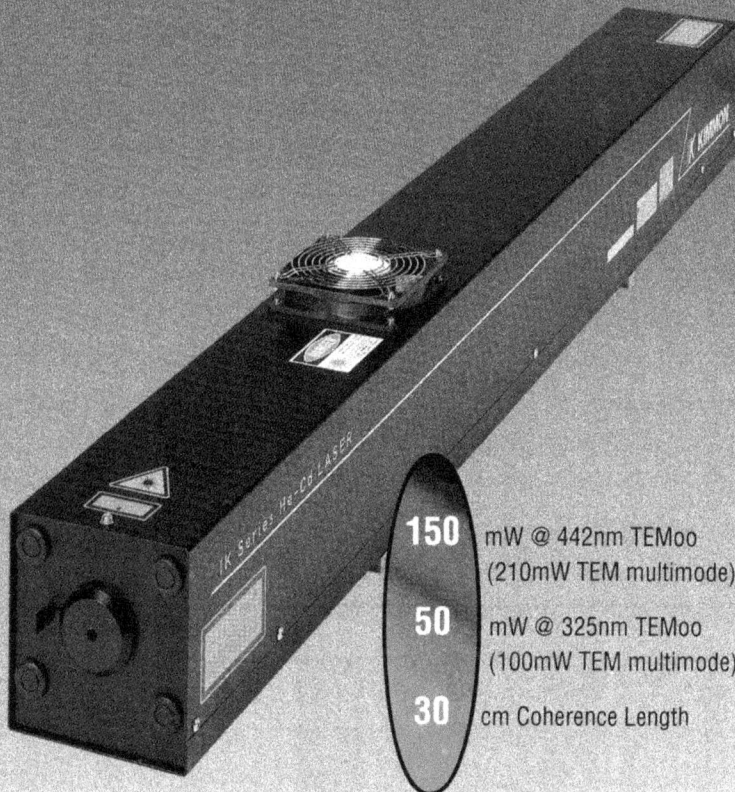

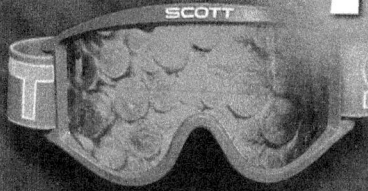

Country	Business Name	Model: Sculpture	Model: Computer Animation	Master: Silver Halide	Master: Dichromate	Master: Photopolymer	Master: Photoresist	Master: Stereograms	Master: Pulsed Portrait
Taiwan	Fong Teng Technology						X		
Taiwan	Holo Images Tech Co., Ltd.	X		X			X		
Taiwan	Holo Impressions Inc	X						X	
Taiwan	Infox Corporation						X		
UK	Applied Holographics, Plc.		X	X			X		
UK	Astor Universal - AHL Holographics Division						X		
UK	Barr & Stroud. Ltd.				X				
UK	Boyd, Patrick			X					
UK	Brighton Imagecraft			X					
UK	De La Rue Holographics Ltd.						X		
UK	Embossing Technology Ltd						X		
UK	Holocrafts Europe Limited.				X				
UK	Holograms 3D			X					
UK	Holographics (Uk) Ltd.			X					
UK	Holography Studio			X					
UK	iC Holographics	X	X	X		X	X	X	X
UK	Jayco Holographics			X			X		
UK	K.C. Brown Holographics			X					X
UK	Laminex/High Tech Uk Ltd.			X					
UK	Laza Holograms Ltd.	X	X	X		X	X		X
UK	Levine, Chris			X					
UK	Light Fantastic Plc			X					
UK	Light Impressions Europe Ltd.	X	X				X		
UK	Op-Graphics (Holography) Ltd.	X		X					
UK	Oxford Holographics			X					
UK	Richmond Holographic Studios			X					
UK	Royal College Of Art.			X					
UK	Severum Design						X		
UK	Spatial Imaging Ltd.	X	X	X	X	X	X	X	X
UK	The Holographic Image Studio	X		X				X	X
UK	Touchwood Holographics			X					
United States of America	General Design		X						
United States of America	James River Products							X	
USA	2Xst Century Finishing Inc.	X					X		
USA	3 Deep Hologram Company			X					
USA	3-D Worldwide Holograms, Inc	X	X	X	X	X	X	X	
USA	A.H. Prismatic, Inc			X					
USA	Acme Holography	X		X		X			
USA	AD 2 , Inc.	X	X			X	X		
USA	American Bank Note Holographics	X	X				X	X	
USA	Another Dimension			X					
USA	Applied Physics Research						X		

Country	Business Name	Model: Sculpture	Model: Computer Animation	Master: Silver Halide	Master: Dichromate	Master: Photopolymer	Master: Photoresist	Master: Stereograms	Master: Pulsed Portrait
USA	Archeozoic Incorporated	X		X					
USA	Art Institute Of Chicago	X	X	X					
USA	Artigliography Co.			X					
USA	Astor Universal - AHL Holographics Division						X		
USA	Astor Universal - AHL Holographics Division						X		
USA	Avant-Garde Studio	X							
USA	Berkhout, Rudie			X					
USA	Bob Mader Photography			X					
USA	Bridgestone Graphic Technologies, Inc.	X	X				X		
USA	Burns Holographics Ltd			X					
USA	Casdin-Silver Holography			X					
USA	Cherry Optical Holography	X		X		X			
USA	Chromagem Inc.	X	X	X		X	X	X	
USA	Classic City Holography Studio	X		X					
USA	Coburn Corporation			X					
USA	Control Module Inc						X		
USA	Deem, Rebecca			X					
USA	Diffraction Ltd.			X					
USA	Dimensional Arts	X	X				X	X	
USA	Ed Wesly Holography			X					
USA	FLEXcon Company, Inc.			X			X		
USA	FoilMark Holographics						X		
USA	Fornari, Arthur David			X					
USA	Forth Dimension	X		X					
USA	Frank DeFreitas Holography Studio			X				X	X
USA	Gerald Marks Studio			X					
USA	Gray Scale Studios Ltd.	X	X						
USA	Harris, Nick			X					
USA	Holage			X					
USA	Holicon Corporation.			X					X
USA	Holo Sciences	X	X	X	X	X	X		
USA	Holo-Spectra	X	X	X			X	X	
USA	Holo/Source Corporation			X			X	X	
USA	Holocraft International			X					
USA	Holographic Applications	X	X						
USA	Holographic Design Systems	X	X	X		X	X	X	X
USA	Holographic Dimension	X	X	X		X	X	X	
USA	Holographic Images Inc.			X					
USA	Holographic Industries, Inc			X					
USA	Holographic Studios			X					
USA	Holographic Technologies		X		X	X	X	X	
USA	Holographics Inc.			X					X
USA	Holographics North Inc.	X		X				X	
USA	Holography Institute			X					
USA	Holomat			X					
USA	Hughes Power Products			X					
USA	Imagen Holography, Inc			X					
USA	Images Company	X				X			
USA	ImEdge Technology			X		X			
USA	Infinity Laser Laboratories	X				X			X
USA	Intrepid World Communications			X		X		X	

Country	Business Name	Model: Sculpture	Model: Computer Animation	Master: Silver Halide	Master: Dichromate	Master: Photopolymer	Master: Photoresist	Master: Stereograms	Master: Pulsed Portrait
USA	Kauffman, John			X					
USA	L.A.S.E.R. Co.			X					
USA	Label Systems/OSI	X					X		
USA	Lasart Ltd.	X			X				
USA	Laser Holography Workshop	X							
USA	Laser Light Designs	X		X					
USA	Laser Reflections			X					X
USA	Lasersmith, Inc.	X	X	X			X	X	
USA	Laserworks			X				X	
USA	Lightrix					X	X		
USA	Lopez's Gallery International	X	X					X	
USA	MacShane Holography	X		X					
USA	Man Environment, Inc.			X					
USA	MasterPrint Holography, Inc.						X	X	
USA	Mgm Converters Inc						X		
USA	Naimark, Michael			X					
USA	Northern Lightworks			X					
USA	O/E Research			X					
USA	Odhner Holographics			X					
USA	OpSec						X		
USA	Optimation Holographics						X	X	
USA	Oriel Corporation			X					
USA	Pennsylvania Pulp & Paper Co.		X				X		
USA	Photon Cantina Ltd.			X					
USA	Pink, Patty			X					
USA	Point Of View Dimensions, Ltd.			X					
USA	Point Source Productions	X	X	X		X		X	
USA	Polaroid Corporation	X	X			X			
USA	Polymer Image			X		X			
USA	Portson, Inc. (Laser Images)	X	X	X		X	X	X	X
USA	Ralcon				X				
USA	Randy James Holography			X					
USA	Red Beam, Inc.			X					
USA	Reynolds Metal Co						X		
USA	Rice Systems			X					
USA	Richard Bruck Holography			X				X	X
USA	Scharr Industries						X		
USA	Sharon McCormack Holography			X					
USA	Simian		X				X	X	
USA	Southern Indiana Holographics			X					
USA	Spectratek Inc.			X					
USA	Stephens, Anait			X					
USA	Three-D Light Gallery			X					
USA	U.S. Holographics			X		X	X		
USA	Unterseher & Associates			X					
USA	Wave Mechanics			X					
USA	Wavefront Technology			X					
USA	White Light Works, Inc.			X					
USA	Zone Holografix			X	X	X			X

Hologram Mass Production

Country	Business	Mass: Silver Halide	Mass: Dichromate	Mass: Photopolymer	Mass: Embossed
Australia	Australian Holographics Pty Ltd.	X			
Australia	Holograms Fantastic & optical Illusions		X		X
Australia	Moonbeamers			X	X
Canada	Deep Space Holographics		X		
Canada	Holocrafts: Canadian		X		
Canada	Melissa Crenshaw	X			
Canada	Royal Holographic Art Gallery	X			
Canada	Silverbridge Group Inc.		X		
France	X-IAL				X
Germany	AKS Holographie-Galerie GmbH				X
Germany	ETA-Optik Gmbh		X		X
Germany	Hologram Company RAQD				X
Germany	Holovision	X			X
Germany	Krystal Holographics Vertriebs			X	
Germany	Laserfilm Eckhard Knuth				X
Germany	Star Shop	X	X		X
Germany	Studio Fur Holographie				X
Germany	Topac GmbH Holography				X
Hong Kong	Foreign Dimension		X	X	X
Hungary	Ap Holografika Studio	X			X
India	Holostik India Pvt. Ltd.				X
India	Ojasmit Holographics		X	X	X
India	Print-M-Boss				X
India	Spatial Holodynamics (India) Pvt.				X
Japan	Toppan Printing Co., Ltd.				X
Mexico	Hologramas, S.A. de C.V.				X
Netherlands	Dutch Holographic Laboratory	X		X	X
Norway	Interferens Holografi D.A.	X			
Russia	Technoexan Ltd	X			
Spain	Holosco, Ernest Barnes	X			X
Spain	Museu D' Holografia	X			
Sweden	Holovision AB	X			
Taiwan	Blue Bond Sales Promoting Inc.				X
Taiwan	Fong Teng Technology				X
Taiwan	Holo Impressions Inc				X
Taiwan	Infox Corporation				X
UK	Applied Holographics, Plc.				X
UK	Astor Universal - AHL Holographics Division				X
UK	Beddis Kenley (Machinery) Ltd.				X
UK	British Technology Group (BTG)				X
UK	De La Rue Holographics Ltd.				X
UK	Embossing Technology Ltd				X
UK	Holocrafts Europe Limited.		X		
UK	Holographics (Uk) Ltd.				X
UK	Jayco Holographics				X
UK	Laza Holograms Ltd.	X		X	X
UK	Light Fantastic Plc	X			X
UK	Light Impressions Europe Ltd.				X
UK	Markem Systems Ltd.		X		
UK	Op-Graphics (Holography) Ltd.	X			
UK	Richmond Holographic Studios	X			
UK	Spatial Imaging Ltd.	X			X
UK	The Holographic Image Studio	X			
USA	2Xst Century Finishing Inc.				X
USA	3-D Worldwide Holograms, Inc	X	X	X	X
USA	AD 2 000 , Inc.			X	X

Country	Business	Mass: Silver Halide	Mass: Dichromate	Mass: Photopolymer	Mass: Embossed
USA	American Bank Note Holographics			X	X
USA	Applied Physics Research				X
USA	Astor Universal - AHL Holographics Dividion				X
USA	Astor Universal - AHL Holographics Dividion				X
USA	Bridgestone Graphic Technologies, Inc.				X
USA	Burns Holographics Ltd				X
USA	Cfc/Applied Holographics				X
USA	Cherry Optical Holography	X		X	
USA	Chromagem Inc.	X		X	X
USA	Coburn Corporation				X
USA	Control Module Inc				X
USA	Creative Label				X
USA	Diffraction Company				X
USA	FLEXcon Company, Inc.				X
USA	FoilMark Holographics				X
USA	Frank DeFreitas Holography Studio	X			
USA	Gardener Promotion Marketing				X
USA	Holicon Corporation.	X			
USA	Holo-Spectra				X
USA	Holo/Source Corporation			X	X
USA	Holographic Design Systems	X		X	X
USA	Holographic Dimension	X		X	X
USA	Holographic Label Converting (HLC)				X
USA	Holographic Studios	X			X
USA	Holographic Technologies		X	X	X
USA	Holographics Inc.	X			
USA	Holographics North Inc.	X			
USA	Hughes Holography Unit		X		
USA	Images Company		X	X	
USA	Infinity Laser Laboratories			X	
USA	Intrepid World Communications	X		X	X
USA	James River Products				X
USA	Label Systems/OSI				X
USA	Lasart Ltd.		X		
USA	Lasersmith, Inc.				X
USA	Lightrix			X	X
USA	Mgm Converters Inc				X
USA	Odhner Holographics	X			
USA	OpSec				X
USA	Optimation Holographics				X
USA	Pennsylvania Pulp & Paper Co.				X
USA	Photon Cantina Ltd.	X			
USA	Polaroid Corporation			X	
USA	Polymer Image			X	X
USA	Portson, Inc. (Laser Images)	X		X	X
USA	Reynolds Metal Co				X
USA	Richard Bruck Holography	X			
USA	Scharr Industries				X
USA	Sillcocks Plastics International				X
USA	Simian				X
USA	U.S. Holographics	X	X	X	X
USA	Van Leer Metallized Products				X

Abeudroth, Detler, AKS Holographie-Galerie GmbH, Germany

Abouchar, Natalalie, Foreign Dimension, Hong Kong

Agehall, Christe r, High Tech Network, Sweden

Aites, Edward, Northern Lightworks, USA

Albrecht, Gerd M., Phantastica, Germany

Allen, Jeff, Envision Enterprises, USA

Anders, Ulrich, Holographie Konzept GmbH, Germany

Andersen, Chad, Meredith In struments., USA

Anderson, Margie, Zero Gravity, USA

Anderson, Mike, Holomex Ltd., UK

Ando, Hiroshi , Nippondenso Co. , Ltd. , Japan

Ankin, Robert, Holo-Spectra, USA

Anoff, Mark, Another Dimension, USA

Aprile, Silvio, New Horizons (Thailand), Ltd., Thailand

Arkin, , Holo Spectra, USA

Attardi, Luigi, Societa Olografica Italia (Soi), Italy

Back, Jonathan, Three-D Light Gallery, USA

Baghdadi, A.M. , Wonders of Holography Gallery, Saudi Arabia

Baghdadi, Abdul Wahab, Hololaser Gallery, United Arab Emerates

Bagley, Shiela, A.H. Prismatic, Inc, USA

Baier, Burckhardt, Melles Griot Gmbh, Germany

Balogh, Tibor, Ap Holografika Studio, Hungary

Bar, Edgar, Holo-Service, Switzerland

Barefoot, Paul D. , Holophile, Inc ., USA

Barnhardt, Donald, Holoflex Company, USA

Barre, Pascal, Holos Art Galerie, Switzerland

Bassin, Barry, Infrared Optical Products, Inc., USA

Bazargan, Kaveh, Focal Image Ltd. , UK

Bear, Sol, Hologram World, Inc., USA

Beauregard, Alain, Lasiris Inc. , Canada

Beeching, Dave , Cfc/Applied Holographics, USA

Beeck, Dr. M.A. , Volkswagen Ag, Germany

Begleiter, Eric, Dimensional Foods Co. , USA

Benedict, Dea, American Propylaea Corporation, USA

Benito, Ramon, Karas Studios S.L. , Spain

Bennett, Sally-Anne, Severum Design, UK

Benton, Stephen, M.l.T. , USA

Benyon, Dr. Margaret, Holography Studio, UK

Berkhout, Rudie, Berkhout, Rudie, USA

Bevans, Heath er, Diffraction Ltd. , USA

Bianchi , Herman-Josef, Arbeits kreis Holografie B.V., Germany

Billeri, Ralph, Control Module Inc, USA

Billings, Loren, Museum Of Holography/Chicago, USA

Billings, Loren, School Of Holography, USA

Billings, Robert, Holographic Design Systems, USA

Birenheide, Richard , HRT (Holographie Recording Technologies Gmbh), Germany

Bjelkhagen, Dr. Hans, Holicon Corporation., USA

Bjelkhagen, Dr. Hans, Northwestern University, USA

Bjelkhagen, Hans , American Propylaea Corporation, USA

Blosvern, Moss, Kinetic Systems, Inc. , USA

Blyth, Jeff, Brighton Imagecraft, UK

Bobeck, Paula, E.I. Dupont De Nemours & Co. USA

Boissonnet, Philippe, Centre D'Art, Canada

Boone, Pierre, Laboratory Vin ckiner, Belgium

Booth, Roberta, Booth, Roberta, USA

Borklund, Willliam, Harland Check Printers, USA

Botos, Steve A., Aerotech Inc. , USA

Bowman, Jim, L.A.S.E.R. Co. , USA

Boyd, Patrick, Boyd, Patrick, UK

Bradshaw, Roy, Reel Image, USA

Braginton, Mary, Coulter Optical Company, USA

Brandtjen, Hank A. , Brandtjen & Kluge, Inc., USA

Brill , Louis, Illuminations, USA

Broadbent, Donald C., Holographic Technologies, USA

Broeders, Jan M., Optische Fenomenen, Netherlands

Brown , Clare , Sil ve r Dragon Holography, USA

Brown, George, Barr & Stroud, Ltd. , UK

Brown, Gordon, Ford Motor Company, USA

Brown, John, Light Impressions Europe Ltd. , UK

Brown, Kerry J., Arti gl iography Co., USA

Brown, Kevin, Holographic Dimension, USA

Brown, Kevin c. , K.C. Brown Holographics, UK

Bruck, Richard, Richard Bruck Holography, USA

Buell, Richard, James River Products, USA

Bunts, Frank, Flatiron Studio, USA

Burn s, Joseph, Burns Holographics Ltd, USA

Bussard, Jan, Holography Presses On (HPO), USA

Bussard, Jan, Texti le Graphi cs, Inc. , USA

Butteriss, Tony, New Dimension Holographics., Australia

Cambell, Jeff, Hughes Holography Unit, USA

Cantos, Brad, Holage, USA

Capucci , Pier Lui gi, University Of Bologna, Italy

Casdin-S ilver, Harriet, Casdi n- Silver Holography, USA

Cason, Thad, Infinity Laser Laboratories, USA

Castagna, Luigi, Holomedia France, France

Chambard, , Holo 3, France

Cheimets, Alex, 3 Deep Hologram Company, USA

Chen, Alex c.T. , Infox Corporation, Taiwan

Chen, Dr. Hsuan, Saginaw Valley State University, USA

Cherry, Greg, Cherry Optical Holography, USA

Chiang, Mark, Fong Teng Techno logy, Taiwan

Chiarot, Roy, Photon Cantina Ltd., USA

Chiou, Billy, Holo Impressions, Taiwan

Chiou, Craig, Holo Images Tech Co. , Ltd., Taiwan

Christakis, Anne-Marie, Magic Laser, France

Christakis, Anne-Marie , Musee De L'Holographie, France

Clare, George W. , Touchwood Holographics, UK

Clark, David, Ealing Electro-Optics Inc. , USA

Clarke, Walter, Global Images, UK

Claytor, Linda H. , Fresnel Technologies Inc, USA

Cohnen, Claus, Star Shop, Germany

Colgate, Gilb e rt, American Bank Note Holographics, USA

Conklin, Don, Glass Mountain Optics, USA

Connors, Betsy, Acme Holography, USA

Cooke. D. J., Oxford University, UK

Cooper, Nick , Oxford Holographies, UK

Cope, Jonathan, Laza Holograms Ltd., UK

Cossa, Rich, Planet 3-D, USA

Course n, Dan, G.M. Vacuum Coating Lab, Inc., USA

Cox, Dr. J. Allen, Honeywell Technology Center, USA

Crenshaw, Melissa, Melissa Crenshaw, Canada

Crist, William, Holocraft International, USA

Cullen , Karoline , Holocrafts: Canad ian , Canada

Cullen, Ralph, Ralph Cullen Holographics, UK

Cullen, Ralph, Uk Optical Supplies, UK

Curiel, Yoram, OpSec, USA

Cvetkovich, Thomas J ., Chromagem Inc. , USA

D' Entremont, Joseph P. , Lenox Laser, USA

Da-Hsiung, Hsu, Beijing In stitute Of Posts, China

Dainty, J., Imperial College Of Science, UK

Damer, Cynthi a, AD 2000, Inc ., USA

Dandliker, Rene, Universite De Neuchatel, Switzerland

Dausmann, Gunther, Holographic Systems Munchen, Germany

Davis, Ernie, MWK Industri es, USA

Davis, Gene, Helios Holography, USA

Dayus, Ian, A.H. Prismatic, Ltd., UK

de Roos, Mark, Deep Space Holographics, Canada

Deem, Rebecca, Deem, Rebecca, USA

Deem, Rebecca, Zone Holografix, USA

DeFreitas, Frank, Frank DeFreitas Hol ography Studio, USA

del-Prete, Sandro, Galerie Illusoria, Switzerland

Denisyuk, Yuri ., Deni syuk, Yuri N. , Russia

Desai , Yogesh, Spatial Holodynamics (India) Pvt. Ltd. , India

Deutschmann, Gunter, Association for Hologram Techniques (AHT), Germany

Deutschmann, Gunter, Holographie Fachstudio Bad Rothenfeld, Germany

Deutschmann, Sven, Topac GmbH Holography, Germany

Dietrich, Edward, OpSec, USA

Dion, David, FoilMark Holographics, USA

Dondi, Alesandro, Diavy sri, Italy

Drake, Francis P. , University Of Wisconsin, USA

Drinkwater-Welch, Deni , Wave Mechanics, USA

Driscoll, John, New York Hal l Of Science, USA

Duesterberg, Richard, Vincennes University, USA

Duffey, Will iam, Regal Press Inc, USA

Duston , Deborah A. , Duston Holographic Services, Canada

Dutton, Keith , Laser International, UK

Duvic, Gene, Zero Gravity, USA

Dyens, George, Mgd Productions, Canada

Ed House, Simon, Aust ralian Holographics Pty Ltd., Australia

Eichler, Dr. Jurgen , Technische Fachhochschule Berlin , Germany

Faklis, Dean, Rochester Photonics Co., USA

Farina, Joseph A. , Laser Hol ography Workshop. USA

Fattel, Issac, Crystal Holographics In c. , USA

Fee, Renee, Holografica, USA

Feinberg, Jack, University Of Southern Calif., USA

Feingold, M. , P.S.A Peugeot Citroen, France

Feitisch, Alfred, Spectra-Physics Lasers, Inc., USA

Fimia, A .. University Of Alicante, Spain

Fischer, Julian , Holovision, Germany

Fisher, Gary, Man Environment, Inc. , USA

Fisk , Mi chael , Liconix. USA

Fitzpatrick, Dr. Collee n, Rice Systems. USA

Florence, Mike, James River Products, USA
Fornari, Arthur Dav id, Fornari, Arthur David, USA
Forsberg, Mona, HoloMedia Ab/Hologram Museum. , Sweden
Fournier, Jean-Marc, Rowland Institute For Science, USA
Fox, Mandy, Los Angeles School Of Holography, USA
Fox. , Mandy, White Light Works, Inc., USA
Frankmark., Robert, Volvo-F1ygmotor, Sweden
French, Dr. Willima, 3M Optics Technology Center, USA
Frieb, M.T. , Holographie Anubis, Germany
Fries, Urs, Holo-Service.Frie s, Switze rland
Frisk, E.O., Optical Works Ltd. , UK
Fukuda, Akinobu, SAM Museum, Japan
Fukuma, Mineko, Japan Communication Arts Co. , Japan
Gabri elson, Dan, Pennsylvania Pulp & Paper Co., USA
Gallagher, Dan, Total Register Inc., USA
Gallagher, John, Total Register Inc. , USA
Galon , Derek, Royal Holographic Art Gallery, Canada
Gardener, John, Gardener Promotion Marketing, USA
Garrett, Steve, Midwest Laser Products, USA
Gauchet, Pascal, Atelier Holographique De Paris, France
Gellert, Andrew, Dimension 3, Canada
Gibson, Dr. J.A., Ag Electro-Optics Ltd. , UK
Gillespie, John, Jodon In c., USA
Gleason, Tom, Laser Technology, Inc., USA
Gorglione, Nancy, Cherry Optical Holography, USA
Gorglione, Nancy, Laser Affiliates, USA
Green, Keith, LaminexlHigh Tech Uk Ltd ., UK
Greguss, Technical University Of Budapest, Hungary
Gunning, Stephan, Studio Creaius, USA
Gunther, John E., Hughes Holography Unit, USA
Gustafson, Glenn, Dimensional Cinematography Co., USA
Gustafsson., Jonny, Holovision AB, Sweden
Gutekunst, Horst, Hologranun Werkstatt & Galerie, Switzerland
Guy, Desmet, Multifacet, Belgium
Halkes, Adrian J., Far East Holographics, Hong Kong
Hall, Patricia M. , Hallmark Capital Corp, USA
Halotek, John, Label Systems/OSI, USA
Hannigan, Francine, Silhouette Technology Inc., USA
Hansen, Matt, Holomat, USA
Hard en, Jim, Light Wave Gallery, USA
Harris, Nick, Harris, Nick, USA
Harrison, Ann Mari e, Intrepid World Communications, USA
Hashimoto, Chikara, Central Glass Co., Ltd., Japan
Hashimoto, Reiji , Topcon Inc. , Japan
Haskell, Dara, Third Dimension Arts Inc., USA
Hassen, Chuck, Holo Sciences, USA
Hatton, Keith, Checkpoint, UK
Haynes, Ken, Simian, USA
Heck, David, Spectra-Physics Lasers, Inc., USA
Heide, Dr. Biren, HRT Holographic Recording Technologies GMBH, Germany
Heidemann, Scott, Oriel Corporation, USA
Hein, Elke, Die Dritte Dimension., Germany
Hell, Dean, Diffraction Company, USA
Hell, Mikael, Martinsson Elektronik Ab., Sweden

Henrion , Mitch, Optitek, USA
Hepburn, James, Leitfoil Inc. , Canada
Hepburn, James, Silverbridge Group Inc., Canada
Herman Weil, Lektor H., Studio Weil-Alvaron, Sweden
Hess, Bob, Point Source Productions, USA
Hillard, Bill, Neovision Productions, USA
Hoefer, Dan, American Laser Corporation, USA
Holden, Dr. Laurence, Astor Universal - AHL Holographics Division, UK
Hollinsworth, T.R., Expanded Optics Limited, UK
Holtzb latt, Mark, Lopez's Gallery International, USA
Hopkins, Anthony, New Clear Imports Ltd., UK
Hopkin s, Jim, Coherent, Inc., USA
Horn, Rolf, HoloMedia Ab/Hologram Museum., Sweden
Horne, Peg, Scharr Industries, USA
Horvath, Josef, Czechos lovak Academy Of Science, Czech
Hoskins, Gregory, Robert Sherwood Holo Design, USA
Hsu, Jonathan, Holo Impress ions In c, Taiwan
Hubel, Paul, Uni versity Of Oxford, UK
Hudson, P.M.G., De La Rue Holographics Ltd., UK
Huff, Lloyd, Uni versity Of Dayton, USA
Hurst, Andrew, Pi lkington Optronics, UK
Hwang, Edward, Superbin Co. Ltd, Taiwan
ide, Makoto, Nippon Polaroid K.K., Japan
Inagaki, Takehumi, Fujitsu Laboratories Ltd., Japan
Ineichen, Beat, Stoltz Ag, Switzerland
Infantes, Mrs. , Karas Studios S.L., Spain
Inoue, Yutaka, Brainet Corporation, Japan
Ishihara, Dr. Satoshi, Ministry Of International Trade, Japan
Ishii, Setsuko, Ishii, Ms. Setsuko, Japan
Ishikawa, Kazue, Sophia University, Japan
Itch, B J, Whiley Foils Limited, UK
Iverson, Mark, Datacard Corporation, USA
Iwata, Fujio, Toppan Printing Co., Ltd. , Japan
Jacobs, Randy, Van Leer Metallized Products, USA
Jain, Anil, APA Optics Inc., USA
Jamison, Pamela, Light Impressions Europe, USA
Jeong, T.H. , Integraf, USA
Jeong., Dr. Tung H. , Lake Forest College, USA
Johann , Larry, Southern Indiana Holographics, USA
Jorgensen, Dean, Optimation, USA
Joyce, Shannon, Wavefront Technology, USA
Jung, Ni ko lau s, Krystal Holographics Vertriebs GmbH, Germany
Jung, Prof. Dieter, Academy of Media Arts Cologne, Germany
Junger, Mr., CI-IIRON Technolas GmbH, Germany
Juptner, Prof. Werner, BIAS, Germany
Jurewicz, Arlene, Synchronicity Holograms, USA
Kane, Brian, General Design, USA
Kasprzak, Henryk, Technical University Of Wroclaw., Poland
Katsuma, Hidetoshi, Tan1a Art Umversity, Japan
Kauffman, John, Kauffman, John, USA
Kauffmann, J., Imac International, Inc., USA
Keilholz, Mr., Spindeler & Hoyer Gmbh, Germany
Kelley, Orick, American Holographic Inc., USA
Kelly, Ed, Keystone Scientific Co., USA
Kelly, Jim, Whole Picture, USA

Kendall, M., Kendall Hyde Ltd. , UK
Kenny, Mike, MWK Industries, USA
Kilpatrick, Jack, Laser Resale Inc., USA
Kinoshita, Dr. Kenji , Toyama National College Of Marit, Japan
Kirk, Roland L., Holo vision Systems Inc., USA
Klix, Armin, Armin Klix Holographie, Germany
Kluepfel, Brian, Baldar, USA
Knuth, Eckhardt, Laserfilm Eckhard Knuth, Germany
Koch, Mr. , Optical Coating Laboratory GmbH, Germany
Koizumi , Fumihiko, Asahi Glass Co. , Japan
Kollin, Joel S. , Holonix, USA
Kooi , A. , Optilas B.V, Netherlands
Koril, Jerry, Creative Label, USA
Kortner, Frau, HOL 3, Galerie fur Holographie, Germany
Kottova, Alena, Ontario Science Centre, Canada
Kreischer, Cody, Kreischer Optics, Ltd., USA
Krick, Reva, Reva'S Holographic Illusions, USA
Krick, Robert, Reva's Holographic Illusions, USA
Krueger, Dave, Holograms Intemational, USA
Krueger, Jean, Holograms International, USA
Kubitzek, Rudiger, ETA-Optik Gmbh, Germany
Kuwayama, Tetsuro, Canon Inc. R&D Headquarters, Japan
Labelle, Scott, Holographic Label Converting (HLC), USA
Lacey, Lee, Holo/Source Corporation, USA
Laczynski, Andrew, Holography Development Group, Canada
Ladiges, Rofi n-Sinar Laser Gmbh, Germany
Lancaster, Tan, Reconnaissance, Ltd. , UK
Lancaster, Ian M., Holography News, UK
Lancaster, Ian M. , International Hologram, UK
Lansing, Joseph, Electro Optical Industries, Inc., USA
Larim, Jim, Laser Optics, Inc., USA
Larson, Ann, Portson , Inc. (Laser Images), USA
Larson, Steve, Ponson, Inc. (Laser Images), USA
Lauren, Wendy, Photonics Di rectory, USA
Leafloor, Stephen, Starlight Holocom, Canada
Lefloc H. C., Aerospatiale, France
Leith, Emmet, Uni versity Of Michigan , USA
Lembessis, Alkis, Hellenic Institute Of Holography, Greece
Lembessus, Alkis, Cavomit, Greece
Leseberg, Dr. Detlef, Leseberg, Dr. Detlef, Germany
Lessard, Roger, Laser Inspeck, Canada
Lessard, Roger A., Universite Lava l, Canada
Leuchs, Gerd, Univers itat Erlangen - Nurnberg, Germany
Lev, Steven , Chromagem Inc. , USA
Levine, Chris, Levine, Chris, UK
Levine, Jeffery, Mesmerized Hologrphic Marketing, USA
Levy, Rob, Holo/Source Corporation, USA
Levy, Uri, Holo-Or Ltd, Israel
Liberatori, Pablo, Catalyst Strategic Design, Inc., USA
Lieberman, Dan, Hologramas, S.A. de C.V, Mexico
Lieberman, Larry, Holographic Images Inc., USA
Liegeois, Dr. Christian D., X-1A L, France
Lifshen, Alan, Lone Star Illusions, USA
Lind, Michael, Batelle Pacific Northwest, USA
Lindquist-Dihn, Jinna, Zero Gravity, USA
Lissack, Selwyn, Laserworks, USA
Lill, Michael, MasterPrint Holography, Inc., USA

Lkegami, Dr. Koji, Numazu College Of Technology, Japan

Lopez, Argelia, Elusive Image, USA

LoSardo, Sal , Towne Technologies, USA

Loucks, Bruce, Leitfoil Inc. , Canada

Loucks, Bruce, Sil verbridge Group Inc ., Canada

Loushin, Sharilyn, Advanced Optics , Inc., USA

Love, Val e rie , Op-Graphics (Holography) Ltd. , UK

Lovygin, Igor, Technoexan Ltd , Russia

Lubetsky, Nea l, Point Of View Dimens ions, Ltd , USA

Lucy, Thomas, Holo Gmbh, Germany

Luton, Chris, Holocrafts Europe Limited. , UK

MacArthur, Anna, Ana MacArthur, USA

Macshane, Jim, MacShane Holography, USA

Magarinos, Dr. Jose R. , Holographic Optics Inc, USA

Malmqvist, Sven, Saab-Scania, Sweden

Malott, Michael, Laser Light Designs, USA

Margolis, Mark, Rainbow Symphony Inc., USA

Marks, Doug, Polaroid Corporation , USA

Marks, Gerald, Gerald Marks Studio, USA

Martinez, Guillermo, Catalyst Strategic Design, Inc. , USA

Masamori, Ichiro, Mazda Motor Corp. , Japan

Maslenkov, Michael , Technoexan Ltd, Russ ia

Mathers, John, Optical Surfaces Ltd. , UK

Mathias, Laug, Lauk KO"l11munikation, Germany

Mathieu , Marie-Chri stiane, Les Productions Hololab l , Canada

Mccain, Richard , McCain Marketing & Graphic Design, USA

McCormack, Sharon , Sh a ron Mccormack Holography, USA

McGahan, Norman , Astor Universal - AHL Holographics Division. USA

McGaw, Trevor, Holograms Fantastic & optical Illusions, Australia

McGrew, Steve, New Light Industries, USA

McKay, Dave, Smith & McKay Printing Co. Inc., USA

McLeod, Don, Corion Corp. , USA

McMahan, Robert, Mcmahan Electro-Optic, USA

McNaughton, James, Kaiser Optical Systems, Inc. , USA

Metz, Barbra, Images Company, USA

Metz, Michael , ImEdge Technology, USA

Meulien, Odile, Artbridge Light Studies, Germany

Meulien-Ohlmann, Odile, Art, Science & Technology Inst, USA

Meulien-Ohlmann , Odile, Holographic and Photonic World Center, USA

Meuse, Ron, Mu'S Laser Works, Canada

Meyer, Steve, Mgm Converters In c, USA

Meyrueis, P., Photonics Systems Laboratory, France

Miller, Peter, Crown Roll Leaf, Inc." USA

Mistry, Rohit, Jayco Holographics, UK

Mitamura, Shunsuke, University Of Tsukuba, Japan

Mitchell, Astrid, Applied Holographics, Pic., UK

Mizuno, Toru, Nippondenso Co., Ltd., Japan

Molin , Nils-Erik , Lulea University Of Technology, Sweden

Monaghan, Brian, Pennsylvania Pulp & Paper Co. , USA

Moore, Dr. Duncan, University Of Rochester, USA

Moore, Lon, Lightri x, USA

Moore, Lon, Red Beam, Inc., USA

Monison, Dan, Laser Technical Services, USA

Morterud, Alan P., Imagen Holography. Inc, USA

Morthier, Mr. Frank, Agfa N.V, Belgium

Mouroulis, P., Rochester Inst. Of Technology, USA

Muller, Joa chin, Gresser, E. , Kg, Gennany

Mulvaney, Mark, Letterhead Press, Inc., USA

Munday, Rob, Spatial Imaging Ltd., UK

Munzer, Hubert, OWIS Gmbh/ FOSTEC, Germany

Murata, M. , Mitsubishi Heavy Industries Ltd., Japan

Murra, Maria, Imad, In c, USA

Murray, Jeffrey, Holography Institute, USA

Murray, Jeffrey, L.A.S.E.R. News, USA

Naeve, Ambjorn, Dialectica Ab, Sweden

Naimark, Mi chael , Naimark, Michael, USA

Nakajima, Dr. Masato, Keio University, Japan

Nelson, Drew, Laser Ionics Inc ., USA

Nicholson, Anna-Marie, Holographics Inc., USA

Nishioka, Teiichi, Toppan Printing Co., Ltd., Japan

Norton , Dan, Polymer Image, USA

O' Brien, Roger, Holotek, USA

O' Connor, Tom, Lasersmith, [nc. , USA

Oahlmann, Dietmar, Artbridge Light Studies, Germany

Odhner, Jeff, Odhner Holograph ics, USA

Oishi, Mariko. Light Dimension, Inc. , Japan

Oliver, Jane, Markem Systems Ltd. , UK

Olmo, Anthony, 21st Century Finishing Inc., USA

Olmo, George, 21st Century Finishing Inc., USA

Olson, Bernadette, Laser Reflections, USA

Olson, Ron, Laser Reflection s, USA

Orazem , Vito , Deutsche Gesellschaft fur Holografie, Germany

On, Edwina, Richmond Holographic Studios, UK

Orszag, Alain, Quantel, France

Osada, RB, Fantastic Holograms, USA

Osada, Richard M. , Lazer Wizardry, USA

Oteri , Lance , Holographic Label Converting (HLC), USA

Owen, Harry. Kaiser Optical Systems, Inc., USA

Page, Michael. Ontario College Of Art, Canada

Pahnue, Roland , Kolbe-Dru ck mit Tochtergesellschaften , Germany

Paletz, Jim, Hologram World , Inc. , USA

Panico, John , Images Company, USA

Parker, Bill, Diffract ion Ltd , USA

Parker, Dr. Steve, British Aerospace PIc., UK

Parker, Ric, Rolls-Royce PIc, UK

Patterson, Rich, Reynolds Metal Co, USA

Pattersson, Sven-Goran, Lund Institute Of Tech., Sweden

Payne, Patty, Burleigh Instruments, [nc., USA

Pecheux , Patrice, Laser Movement, France

Pepper, Andrew, Creative Holography Index, The, Germany

Perry, John, Holographics North Inc., USA

Petersen, Joel, Wavefront Tec hnology, USA

Pfiel, Larry, Dimensional Arts, USA

Phillips, Jacque. Direct Holographics, USA

Phillips. Prof. Nick, Loughborough Univ. Of Tech. , UK

Phillips. Ronald, Interac ti ve Industri es, USA

Pink, Patty, Pink, Patty. USA

Pitt , Royal Photographic Society, UK

Pook, Jim , 3M Optics Technology Center, USA

Price, Stu, Shipley Chemical Co. , USA

Pricone, Robert, Holographic Industries, Inc, USA

Pryputniewic z, Rysza rd, Worcester Polytec hnic Institute, USA

Qualls, Steve, Laser Photoni cs, In c., USA

Racey, Carl. Amazing World Of Holograms., UK

Rallison, Richard, Ralcon, USA

Ralston, Mary, Bobst Group, USA

Rankin , Bill, Holo-Spectra, USA

Rankin, Kevin, Omnichrome, USA

Rankin, Kevin. Physical Optics Corporation., USA

Rapke, Mark, Holographic Marketing, Inc., USA

Ratcliffe, David, Australian Holographics Pty Ltd. , Australia

Rayfield , Dave, U.S. Holographics, USA

Redzikowski, Mark, Agfa Div of Bayer Corp. , USA

Reichert. Uwe, 3D Vision, Germany

Reingruber, Adlas G.M.B.H. & Co Kg., Germany

Reinhart, Werner, Leonhard Kurz GmbH, Germany

Rezny, Abe, Laser Light Ltd. , USA

Rhody, Alan, Ross Books, USA

Richardson, Dr. Martin. The Holographic Image Studio, UK

Ridol. Graham, Embossing Technology Ltd, UK

Rincon, Angelika, Hologramas, S.A. de c.v., Mexico

Rizzi , M. Luciana, Cise Spa Technologie Innovative, Italy

Roberts, Judy, Jr Hol ographics, USA

Robinson, D., National Physical Laboratory, UK

Rob inson , Deborah, Lightri x, USA

Robinson , George, Hologram Land, USA

Rodia, Carl M., Carl M. Rodia And Associates, USA

Ross, Franz, Ross Books, USA

Ross, Jonathan, Holograms 3D. UK

Ross, Michael , IBM Almaden Research Center, USA

Rossing, Thomas, Northern Illinois University, USA

Roth, Stephen, Mind'S Eye, USA

Rothau s, Valli, Space Age Designs Inc., USA

Ruey-Tung, Hung, Ruey-Tung, Miss. Hung, Japan

Ryden, Hans , Karolinska Institutet, Sweden

Sachs, Scott, Zero Gravity, USA

Sakai, Miss Tomoko, HODIC Holographic Display Artists & Engineers Club, Japan

Santis, Paolo De, Universita Di Roma, Italy

Sapan, Jason, Holographic Studios, USA

Sato, Shunichi, Sharp Corp. , Japan

Saxby, Graham, Wolverhampton Polytechnic, UK

Scheir, Peter, AD 2000, In c., USA

Schipper, Wilfried, Hologram Company RAQD GmbH, Germany

Schmelzer, Dr. Carlo, Studio Fur Holographie, Germany

Schumann, Walter, Swiss Federal Inst Of Technology, Switzerland

Schwartz, Jeffery, 3-D Worldwide Holograms, Inc, USA

Schwartzman, Frederic, Foreign Dimension, Hong Kong

Sciammarella, Cesar, Illinois Institute Of Technology, USA

Scriff, Joe, Mitutoyo Measuring Instruments, USA

Seamans, Warren, M.I.T. Museum, USA

Seitz, Herr, Steuer Kg Gmbh & Co., Germany

Sekulin, Robert, Rutherford & Appleton Labs, UK

Seymour, Bill , Bobst Group, USA

Shah, Kail esh, Oja smit Holographics, India

Shahjahan, Mr. , Dimensions. Pakistan

Sharma, Govind, Holostik India Pvl. Ltd. , India
Sharpe, Frank, Datasights Ltd. , UK
Sherwood, Robert, Robert Sherwood Holo Design, USA
Shimon, Hameiri, Holography Israel , Israel
Shun, Prof. Zhu De, Shandong Academy of Sciences, China
Shvarts, Dr. Kurt, Physics In stitute. Latvian, Latvia
Siberine, Rod, Holopak Technologies, USA
Siegel, Steven, Lasart Ltd. , USA
Sikorsky, Zbigniew, Institute Of Plasma Physics, Poland
Simson, Bernd, General Holographics, Inc., Canada
Simson, Paula, General Holographics, Inc. , Canada
Singh , Ravinder, Print-M-Boss, India
Sivy, George, Gray Scale Studios Ltd ., USA
Skipnes, Olav, Interferens Holografi D.A. , Norway
Smelzer, Geert T. A., Technical University @ Eindhoven, Netherlands
Smith, Jeff, Ion Laser Technology Inc., USA
Smith, S.D., Beddis Kenley (Machinery) Ltd. , UK
Smith, Steven L. , Lasersmith, Inc. , USA
Soales, Bob, CVI Laser Corporation, USA
Soares, Oliverio, Universidade Do Porto, Portugal
Song, Chung, Dan Han Optics, South Korea
Solt, Gudrun, AKS Holographie-Galerie GmbH, Germany
Souparis, Hughes, Hologram Industries, France
Sowdon, Michael , Fringe Research Holographics, Canada
Spalding, Jean, Norland Products, Inc. , USA
Spanner, Dr. Karl , Phys ik In strumente (Pi) Gmbh & C, Germany
Spierings, Walter, Dutch Holographic Laboratory BV, Netherlands
Springer, Greg, Excitek Inc., USA
St. Cyr, Susan, Holographic Applications , USA
Starcke, Ari-Veli , Starcke, Ky., Finland
Stehle, Robert, Sopra, France
Steinbichler, Dr. H., Labor Dr. Steinbichler, Germany
Stein bichler, Dr. H. , Steinbichler Optotechnik Gmbh, Germany
Steinfeld, Belle, Dell Optics Company, Inc, USA
Stephens, Anait, Stephens, Anait, USA
Stepien, Pawel, Hololand S.C. , Poland
Stetson., Dr. Karl A., United Technologies, USA
Stich, Boguslaw, Holografia Pol ska, Poland
Stockier, Len, OlE Research, USA
Stockton, Jobn, Tamarack Storage Devices, USA
Stooss, Richard, Krystal Holographics Vertriebs GmbH, Germany
Stotesbery, William, Microelectronics and Computer Technology Corp., USA
Styns, Erik, Free University Of Brussels., Belgium
Su, Dr. J.J., Industrial Technology Research In st., Taiwan
Sugarman, Stephen, Holographic Products, USA
Surana, Rajendra, Ojasmit Holographics, India
Svennson , Lennart, Royal In stitute Of Technology, Sweden
Sweeney, Dr. Eugene, British Technology Group (BTG), UK
Synowiec, George, Lumonics Ltd, UK
Talbott, M., Laser Arts, USA

Taylor, Rob, Forth Dimension , USA
Thiemon, Ms., Daimler Benz Aerospace, Germany
Tholen, Morien , 3M, USA
Thoma, John , Astor Universal - AHL Holographics Di vidion, USA
Thomas, Jackie, Laser In stitute Of America, USA
Tidmarsh, David, Applied Holographics, Pic., UK
Tiziani. , Hans, University Of Stuttgart, Germany
Tobin, John, Moonbeamers, Australia
Todd, Sandra, Coherent, Inc. , USA
Tolia, Dr. Arun, Spatial Holodynamics (India) Pvt. Ltd., India
Tong, Chen Guo, Morning Light Holograms, China
Townsend, Patrick, ACI Systems, USA
Trackeberry, John, Optimation Holographics, USA
Trayner, David, Richmond Holographic Studios, UK
Tribillon, Dr.Jean Louis, Holo-Laser, France
Trolinger, James D. , MetroLaser, USA
Trolinger, James D. , Metrolaser, USA
Tsujiuchi. , Jumpei, Chiba University, Japan
Tzong, Tang Yaw, Institute Of Optica l Science, Taiwan
Unbehaun, Klaus , Holopublic Unbehaun, Germany
Unterseher, Fred, Zone Holografix, USA
Unterseher. , Fred, Unterseher & Associates, USA
Upatnieks, Juri s, Environmental Research (Erim), USA
Uram, Marvin, U.K. Gold Purchasers, Inc., USA
Uwe, Saurda, Holographie Labor, Germany
Uyemura, T. , University Of Tokyo, Japan
Valdivia, Allison, Optics Plus Inc, USA
Van Renesse, Ruud L. , TNO Institute of Applied Physics, Netherlands
Varney, Chris, Electro Optics Developments Ltd. , UK
Vikram , Chandra, University Of Alabama, USA
Vince, Bill, Spectrolab Intemational Ltd., UK
Vogel , Jon, Holographics (Uk) Ltd., UK
Von Bally, Gert, University Of Munster, Germany
Vukicevic, Dalibor, Trend, Yugoslavia
Wada, Takashi, Dai Nippon Printing Co., Ltd., Japan
Waddell, P. , University Of Strathclyde, UK
Wahlberg, Bjorn, Swede Holoprint, Sweden
Waitts, George, Crown Roll Leaf, Inc." USA
Wale, R. D., Galvoptics Ltd. , UK
Wanlass, Mike, Spectratek Inc. , USA
Wanyun, Huang, Beijing Normal University., China
Ward, John, Optical Corporation Of America, USA
Wells, Karan, Deep Space Holographics, Canada
Wesley, Ed, Art Inst itute Of Chicago, USA
Wesly, Ed, Ed Wesly Holography, USA
Wesly, Ed, Lopez's Gallery International, USA
Westphal , Carlo, Die Dritte Dimension. , Germany
Westphal, Paul , Amity Photonics Co. , USA
White, John, Coburn Corporation, USA
White, Steve, Electro Optical Industries, Inc., USA
Wichert, Yvonne, Holo Gmbh, Germany
Wilbur, Fred, Elusive Image, USA

Williams, Betty, Metrologic Instruments, Inc., USA
Wilson, Brett, Laza rt Holographics., Australia
Windeln, Dr. Wilbert, ETA-Optik Gmbh, Germany
Wober, Irmfried, Holography Center of Austria, Austria
Wollenwerber, Andreas, Magick Signs Holographie, Germany
Woodd, Peter H.L. , Light Fantastic Pic, UK
Woodd, Peter H.L. , Optical Security Industry, UK
Woodman, Bety, Applied Physics Research, USA
Woods, Nicola, Photon League Of Holographers., Canada
Wyant, James, Wyko Corp., USA
Wyman, Ken, Aerotech Inc ., USA
Yamaguchi , Masahiro, Tokyo Instit ute Of Technology, Japan
Yamazaki, Hitoshi , Hyogo Prefectual Museum, Japan
Yannuzzi, Paul E. , Archeozoic Incorporated, USA
Yokota, Hideshi, Tokai University, Japan
Yoon, Menssa, M.LT. , USA
Yoshikawa, Dr. Hiro shi , Nihon University, Japan
Zellerbach , Gary, Holart Consultants, USA
Zinth, Prof. , University Of Munich, Gemlany
Zucker, Richard, Bridgestone Graphic Technologies, Inc., USA
Zurek, Mr. , University Of Munich, Germany

Equipment Providers - Recording Materials

Country	Business Name	Silver Halide	Dichromate	Photopolymer	Photoresist
Canada	Royal Holographic Art Gallery	X			
Germany	HRT (Holographie Recording Technologies Gmbh)	X			
Germany	HRT Holographic Recording Technologies GMBH	X			
Germany	Newport Gmbh	X			
Japan	Nippon Polaroid K.K.			X	
Mexico	Hologramas, S.A. de C.V.				X
Russia	Slavich Joint Stock Company	X			
Russia	Technoexan Ltd	X			
Taiwan	Infox Corporation				X
UK	Agfa Ltd	X			
UK	Brighton Imagecraft		X	X	
UK	Laza Holograms Ltd.	X		X	X
UK	Light Impressions Europe Ltd.			X	X
UK	Markem Systems Ltd.				X
UK	Pilkington Optronics		X	X	
USA	3 Deep Hologram Company	X			
USA	Agfa Div of Bayer Corp.	X			
USA	American Bank Note Holographics				X
USA	E.I. Dupont De Nemours & Co			X	
USA	Eastman Kodak Company	X			
USA	Holo/Source Corporation	X		X	X
USA	Holographic Dimension				X
USA	Holographic Technologies			X	X
USA	Ici Americas	X			
USA	Images Company	X	X	X	
USA	James River Products				X
USA	Keystone Scientific Co.	X			
USA	Laser Reflections	X			
USA	Lightrix			X	X
USA	Pennsylvania Pulp & Paper Co.				X
USA	Polaroid Corporation			X	
USA	Ralcon		X	X	X
USA	Shipley Chemical Co.				X
USA	Towne Technologies				X

Holographic Computer Memory/Storage

USA	Microelectronics and Computer Technology Corp.
USA	Mitutoyo Measuring Instruments
USA	Optitek
USA	Tamarack Storage Devices

Stereogram Printers - Custom Made To Order

Man Environment, Inc.	USA

Stereogram Printers - Complete "ready-to-run" Systems

USA	Dimensional Arts
Netherlands	Dutch Holographic Laboratory BV
USA	Holo Sciences
Germany	Holovision
USA	Man Environment, Inc.
UK	Spatial Imaging Ltd.

Equipment Providers - Embossing Equipment

Country	Business Name	Foil	Shims	Electroplating Equipment	Hot Stamp Machine	Embossing Machine
Australia	Lazart Holographics.				X	
Canada	Leitfoil Inc.				X	
Germany	Hologram Company RAQD GmbH				X	X
Germany	Kolbe-Druck mit Tochtergesellschaften	X				
Germany	Steuer Kg Gmbh & Co.					X
Greece	Cavomit				X	
India	Holostik India Pvt. Ltd.	X	X			
India	Ojasmit Holographics		X	X		
India	Print-M-Boss	X	X	X		
India	Spectrum Corporation		X	X		X
Mexico	Hologramas, S.A. de C.V.	X	X	X	X	X
Taiwan	Fong Teng Technology	X	X			X
Taiwan	Holo Images Tech Co., Ltd.	X	X	X	X	X
Taiwan	Infox Corporation	X	X			
UK	Applied Holographics, Plc.	X	X			
UK	Beddis Kenley (Machinery) Ltd.	X	X		X	X
UK	British Technology Group (BTG)	X				
UK	Dimuken					X
UK	Global Images		X	X	X	X
UK	Light Impressions Europe Ltd.	X	X	X	X	X
UK	Markem Systems Ltd.	X				
UK	Spatial Imaging Ltd.		X			
UK	Whiley Foils Limited	X				
USA	3M	X				
USA	American Bank Note Holographics	X	X	X	X	X
USA	Bobst Group				X	X
USA	Burns Holographics Ltd		X			
USA	Chromagem Inc.		X			
USA	Coburn Corporation		X			
USA	Crown Roll Leaf, Inc.,	X	X			
USA	Dazzle Equipment Company		X	X	X	X
USA	Holo/Source Corporation	X	X	X		X
USA	Holographic Dimension	X	X	X	X	X
USA	Holographic Technologies	X	X	X	X	
USA	Holopak Technologies	X	X			
USA	James River Products		X	X		X
USA	Lightrix	X	X			
USA	Pennsylvania Pulp & Paper Co.		X			
USA	Silver Dragon Holography	X	X	X	X	X
USA	Total Register Inc.				X	X
USA	Tyler Group				X	

Laser Equipment Suppliers

Country	Business Name	Manufacturer of Lasers	Distributor of Lasers	Laser Repair	Reconditioned Lasers	Holography Kits With Laser
Australia	Australian Holographics Pty Ltd.	X				
Canada	Royal Holographic Art Gallery		X		X	X
Germany	Adlas G.M.B.H. & Co Kg.	X				
Germany	Gresser, E., Kg		X			
Germany	Metrologic Gmbh	X	X			
Germany	Newport Gmbh	X	X	X		
Germany	Physik Instrumente (Pi)		X			
Japan	Fuji Electric Co. Ltd	X				
Japan	Kimmon Electric Co., Ltd.	X				
Mexico	Hologramas, S.A. de C.V.					X
Russia	Technoexan Ltd	X				X
Saudi Arabia	Wonders of Holography Gallery		X			
Sweden	Martinsson Elektronik Ab.		X			
UK	Ag Electro-Optics Ltd.	X	X	X		
UK	Laser International		X			
UK	Light Impressions Europe		X			
UK	Lumonics Ltd	X				
UAE	Hololaser Gallery		X			
USA	ACI Systems	X				
USA	Aerotech Inc.	X				
USA	American Laser Corporation	X				
USA	Cambridge Laser Labs			X	X	
USA	Coherent, Inc.	X	X	X	X	
USA	Excitek Inc.		X	X	X	
USA	Fisher Scientific		X			
USA	Frank DeFreitas Holography Stiudio		X			X
USA	Holo Spectra			X	X	
USA	Holo-Spectra		X	X	X	X
USA	Holographic Technologies		X		X	
USA	Images Company		X			X
USA	Ion Laser Technology Inc.	X				
USA	James River Products		X			X
USA	Jodon Inc.	X				
USA	Keystone Scientific Co.					X
USA	Laser Innovations			X		
USA	Laser Ionics Inc.	X				
USA	Laser Photonics, Inc.	X				
USA	Laser Reflections	X				
USA	Laser Resale Inc.		X		X	
USA	Laser Technical Services		X	X	X	
USA	Lasermetrics, Inc.			X		
USA	Lexel Laser, Inc.	X				
USA	Liconix	X				
USA	Lumenx technologies	X		X		
USA	Meredith Instruments.				X	
USA	Metrologic Instruments, Inc.	X		X		X
USA	Midwest Laser Products		X	X	X	
USA	MWK Industries		X			
USA	Omnichrome	X	X		X	
USA	Panatron Inc.			X	X	
USA	Science Kit & Boreal Labs		X			X
USA	Siemens (USA)	X				
USA	Spectra-Physics Lasers, Inc.	X		X		

Lab Equipment Providers

Country	Business Name	Isolation Tables	Optical Mounts	Front Surface Mirrors	Beam-splitters	Lenses/Prisms/Windows	Collimators	Pinhole/Filters/Polarizers	Gratings	Darkroom Chemicals	Dark room (lights/timer)	Safety Equipment	Fully Equipped Studio for Rent
Australia	Australian Holographics.	X	X	X	X	X	X	X					X
Canada	Harvard Apparatus, Canada		X	X		X	X	X					
Germany	CHIRON Technolas					X							
Germany	Fostec Gmbh Feinmechanik	X											
Germany	HRT Holographic Recording Technologies		X	X						X			
Germany	Melles Griot		X	X	X	X	X	X					
Germany	Moeller Wedel Optische Werk			X	X	X							
Germany	Newport	X	X	X	X	X	X	X	X	X	X	X	
Germany	Optical Coating Laboratory			X	X	X	X						
Germany	Physik Instrumente	X	X	X	X	X	X						
Germany	Spindeler & Hoyer Gmbh	X	X	X	X	X	X	X	X				
India	Ojasmit Holographics	X	X										
Russia	Technoexan											X	
Sweden	Martinsson Elektronik Ab.	X	X	X	X	X	X	X					
UK	Ag Electro-Optics Ltd.			X	X	X	X						
UK	Brighton Imagecraft									X			
UK	CSI		X	X	X	X	X						
UK	Datasights .		X										
UK	Ealing Electro-Optics	X	X	X	X	X	X	X					
UK	Galvoptics .		X	X	X	X	X	X					
UK	Holomex Ltd.										X	X	
UK	Kendall Hyde			X		X	X						
UK	Laza Holograms.									X	X		X
UK	Light Impressions Europe Ltd.	X	X	X	X	X	X	X	X	X	X	X	
UK	Optical Surfaces Ltd.		X	X	X	X	X						
UK	Optical Works		X	X	X	X	X	X	X				
UK	Ralph Cullen Holographics		X	X	X	X	X						
UK	Siemens Ltd.		X	X	X	X	X						
UK	Spectrolab International	X	X		X			X					
UK	The Holographic Image Studio												X

Country	Business Name	Isolation Tables	Optical Mounts	Front Surface Mirrors	Beam-splitters	Lenses/ Prisms/ Windows	Colli mators	Pinhole/ Filters/ Polarizers	Grati ngs	Darkro om Chemic als	Dark room (lights /timer)	Safety Equip ment	Fully Equip ped Studio for Rent
UK	Uk Optical Supplies	X	X	X	X	X	X	X	X	X	X	X	
UK	Vinten Electro Optics Ltd		X	X	X	X	X	X					
USA	3-D Worldwide Holograms	X	X	X	X	X	X	X	X	X			X
USA	Advanced Optics, Inc.		X	X	X	X	X						
USA	Aerotech Inc.		X	X	X	X	X						
USA	American Holographic		X	X	X	X	X	X	X				
USA	Amity Photonics Co.		X	X	X	X	X						
USA	Classic City Holography Studio												X
USA	Ealing Electro-Optics		X	X	X	X	X						
USA	Eastman Kodak									X			
USA	Edmund Scientific		X	X	X	X	X	X			X	X	
USA	Electro Optical Industries, Inc.		X	X	X	X	X	X					
USA	Frank DeFreitas Holography Stiudio			X	X	X							X
USA	Fresnel Technologies				X		X						
USA	G.M. Vacuum Coating Lab			X	X								
USA	Glass Mountain Optics			X			X						
USA	Holo-Spectra	X	X				X						X
USA	Holographic Technologies	X	X	X	X	X	X	X	X		X		
USA	Holography Institute												X
USA	Images Company	X	X	X	X	X	X	X		X	X		
USA	James River Products	X	X	X	X	X	X	X		X	X		
USA	Keystone Scientific Co.									X			
USA	Kinetic Systems, Inc.	X											
USA	Kreischer Optics, Ltd.		X	X	X	X	X						
USA	Laser Optics,		X	X	X	X	X	X	X				
USA	Laser Reflections												X
USA	Laser Resale.	X	X										
USA	Lenox Laser							X	X				
USA	Melles Griot	X	X	X	X	X	X	X	X				

Country	Business Name	Isolation Tables	Optical Mounts	Front Surface Mirrors	Beam-splitters	Lenses/ Prisms/ Windows	Colli mators	Pinhole/ Filters/ Polarizers	Grati ngs	Darkro om Chemic als	Dark room (lights /timer)	Safety Equip ment	Fully Equip ped Studio for Rent
USA	Newport Corporation	X	X	X	X	X	X	X	X	X	X	X	
USA	Norland Products, Inc.		X							X			
USA	Odhner Holographics							X					
USA	Optical Corporation Of America			X			X						
USA	Optics Plus Inc		X	X	X	X	X						
USA	Optimation							X					
USA	Ralcon								X				
USA	Rolyn Optics		X	X	X	X	X	X	X				
USA	Science Kit & Boreal Labs			X	X	X		X					
USA	Spectra-Physics Lasers, Inc.			X	X	X							
USA	Towne Technologies									X			

Speciality Businesses - HOE, NDT, CGH

Country	Business Name	HOE: Custom Work	HOE: Research Projects	NDT: Fluid Analysis	NDT: Metal Stress	NDT: Research Projects	CGH: Stereogram Animation Businesses	CGH: Master/Mfg Stereograms
Australia	Holograms Fantastic & optical Illusions						X	X
Australia	Moonbeamers	X	X					
Belgium	Laboratory Vinckiner					X		
Canada	Duston Holo. Services					X		
Canada	Melissa Crenshaw		X					X
China	Beijing Normal University.					X		
Czech	Czech Academy Of Science		X			X		
France	Aerospatiale					X		
France	Holo 3		X			X		
France	P.S.A Peugeot Citroen					X		
France	Photonics Systems Lab					X		
Germany	3D Vision						X	
Germany	BIAS					X		
Germany	Daimler Benz Aerospace		X			X		
Germany	ETA-Optik Gmbh	X						
Germany	Holopublic Unbehaun		X					
Germany	Labor Dr. Steinbichler					X		
Germany	Leseberg, Dr. Detlef		X					
Germany	Steinbichler Optotechnik					X		
Germany	Tech. Fachhochschule Berlin			X		X		
Germany	Univ. Erlangen - Nurnberg		X			X		
Germany	University Of Munster			X		X		
Germany	University Of Stuttgart					X		
Germany	Volkswagen Ag					X		
Greece	Hellenic Institute Of Holo					X		X
Hungary	Ap Holografika Studio		X					
Israel	Holo-Or Ltd		X					
Israel	Holography Israel		X					
Italy	Cise Spa Tech. Innovative		X			X		

Country	Business Name	HOE: Custom Work	HOE: Research Projects	NDT: Fluid Analysis	NDT: Metal Stress	NDT: Research Projects	CGH: Stereogram Animation Businesses	CGH: Master/Mfg Stereograms
Japan	Asahi Glass Co.		X					
Japan	Central Glass Co., Ltd.		X					
Japan	Hyogo Prefectual Museum		X					
Japan	Keio University		X					
Japan	Lab. of Image Information		X					
Japan	Mazda Motor Corp.					X		
Japan	Mitsubishi Heavy Ind. Ltd.					X		
Japan	Nihon University					X		
Japan	Nippondenso Co., Ltd.		X					
Japan	Nissan Motor.		X					
Japan	Tama Art Umversity		X					
Japan	University Of Tokyo					X		
Mexico	Hologramas, S.A. de C.V.						X	X
Netherland	Dutch Holographic Lab BV	X	X			X	X	X
Poland	Hololand S.C.	X	X					X
Poland	Institute Of Plasma Physics		X					
Portugal	Universidade Do Porto					X		
Russia	Technoexan Ltd		X					
Sweden	Karolinska Institutet		X					
Sweden	Lulea University Of Tech					X		
Sweden	Royal Institute Of Tech					X		
Sweden	Saab-Scania					X		
Sweden	Spectrogon Ab		X					
Sweden	Volvo-Flygmotor					X		
Switzerland	Stoltz Ag					X		
Switzerland	Swiss Federal Inst Of Tech					X		
Switzerland	Universite De Neuchatel					X		
Taiwan	Institute Of Optical Science		X					
UK	Ag Electro-Optics Ltd.		X			X		
UK	British Aerospace Plc.		X			X		
UK	CSI		X					
UK	Datasights Ltd.		X					
UK	Ealing Electro-Optics (Uk)		X					
UK	Electro Optics Dev.		X					
UK	Expanded Optics Limited		X					
UK	Galvoptics Ltd.		X					
UK	Holography Group		X			X		
UK	Imperial College Of Science		X			X		
UK	Kendall Hyde Ltd.		X					
UK	Laza Holograms Ltd.							X
UK	National Physical Lab					X		
UK	Optical Surfaces Ltd.		X					
UK	Optical Works Ltd.		X					
UK	Rolls-Royce Plc				X	X		
UK	Rutherford & Appleton Labs			X		X		
UK	Spatial Imaging Ltd.						X	X
UK	Uk Optical Supplies		X					
UK	Ultrafine.					X		
UK	University Of Aberdeen					X		
UK	University Of Strathclyde					X		
UK	Vinten Electro Optics Ltd		X					
USA	3-D Worldwide Holograms,						X	
USA	3M Optics Technology Ctr		X					
USA	Acme Holography							X
USA	Advanced Tech Program		X			X		
USA	American Bank Note Holo	X						
USA	American Holographic Inc.		X					

Country	Business Name	HOE: Custom Work	HOE: Research Projects	NDT: Fluid Analysis	NDT: Metal Stress	NDT: Research Projects	CGH: Stereogram Animation Businesses	CGH: Master/Mfg Stereograms
USA	American Propylaea Corp		X				X	
USA	Amity Photonics Co.		X					
USA	APA Optics Inc.	X	X					
USA	Art Institute Of Chicago						X	X
USA	Chromagem Inc.	X	X					
USA	Continental Optical		X					
USA	Control Optics		X					
USA	Corion Corp.		X					
USA	Coulter Optical Company		X					
USA	Dell Optics Company, Inc		X					
USA	E.I. Dupont De Nemours		X					
USA	Ealing Electro-Optics Inc.		X					
USA	Electro Optical Industries,.		X					
USA	Environmental Research					X		
USA	Ford Motor Company					X		
USA	Frank DeFreitas Holo Studio		X					X
USA	Fresnel Technologies Inc		X					
USA	G.M. Vacuum Coating Lab		X					
USA	General Design						X	
USA	Holo Sciences						X	
USA	Holo-Spectra							X
USA	Holo/Source Corporation	X						
USA	Holoflex Company					X		
USA	Holographic Design Systems						X	X
USA	Holographic Dimension	X	X			X		X
USA	Holographic Optics Inc		X					
USA	Holographic Studios		X					
USA	Holographic Technologies	X	X			X	X	X
USA	Holographics North Inc.							X
USA	Holography Institute	X	X					
USA	Holonix	X	X					
USA	Holotek		X					
USA	Hughes Holography Unit		X			X		
USA	IBM Almaden Research Ctr		X					
USA	Ici Americas		X					
USA	Illinois Institute Of Tech		X					
USA	Imagination Plantation						X	
USA	ImEdge Technology	X	X					
USA	Infrared Optical Products		X					
USA	Inrad, Inc		X					
USA	Intrepid World Com	X	X					
USA	Kaiser Optical Systems, Inc.		X					
USA	Kreischer Optics, Ltd.		X					
USA	Laser Optics, Inc.		X					
USA	Laser Technology, Inc.					X		
USA	Lasersmith, Inc.						X	X
USA	Lawrence Berkeley Lab		X			X		
USA	Mcmahan Electro-Optic					X		
USA	Metrolaser					X		
USA	MetroLaser			X		X		
USA	Northern Illinois University		X			X		
USA	Northwestern University					X		
USA	Odhner Holographics		X					
USA	Optical Corp Of America		X					
USA	Optical Research Services	X	X					
USA	Optics Plus Inc		X					
USA	Pennsylvania Pulp & Paper		X					

Country	Business Name	HOE: Custom Work	HOE: Research Projects	NDT: Fluid Analysis	NDT: Metal Stress	NDT: Research Projects	CGH: Stereogram Animation Businesses	CGH: Master/Mfg Stereograms
USA	Physical Optics Corporation.		X					
USA	Point Source Productions	X					X	X
USA	Ralcon	X	X					
USA	Rice Systems			X		X		
USA	Rochester Inst. Of Tech		X					
USA	Rochester Photonics Co.	X	X					
USA	Rowland Inst. For Science					X		
USA	Sandia National Laboratories					X		
USA	Silhouette Technology Inc	X	X					
USA	Titan Spectron					X		
USA	University Of Alabama					X		
USA	University Of Arizona					X		
USA	University Of California		X			X		
USA	University Of Dayton		X			X		
USA	University Of Michigan					X		
USA	Univ Of North Carolina			X		X		
USA	University Of Rochester					X		
USA	Univ Of Southern Calif.					X		
USA	University Of Wisconsin					X		
USA	Worcester Polytechnic Inst					X		
USA	Wyko Corp.					X		

Appendix

This appendix assumes that you have no prior knowledge of holography. Discussed are the major types of holograms and how they are made; also included are brief introductions to the lasers and recording materials used to make them. More detailed information is contained in the relevant chapters of the book.

Holography: A Day in the Life

You wake up in the morning and go down to the kitchen. Cracking open the cereal, a bright image jumps out at you from the front of the box. You walk to your mailbox, expecting that letter from your cousin Claude in Canada. Affixed in the place where a regular stamp would be is yet another of those odd three-dimensional images you just saw on your cereal box. The magazine which you subscribe to also has one of these bright, reflective pieces of foil stamped on it, with an image seemingly moving across the cover.

Still bleary-eyed, you dress and get in your car. Not pay- ing close attention to the road at this early hour, you run a stop sign. The friendly California Highway Patrolman who pulls you over takes your license, which also has a hologram on it, and runs it over a scanning device to make sure that it's the authentic item.

This scenario is an example of the frequency with which we see holograms in our everyday life. Rarely a month or two passes without some publication using a hologram to attract attention. Examples of low volume to high volume uses are: tickets to Presidential Inaugurals, Super Bowl tickets, and Microsoft's Windows 95, which sells in the tens of millions. They all have holograms on their surfaces to make it tough on counterfeiters and also look attractive to the consumer.

Holography and Photography

Although often compared with photography, holography is really a completely different medium with different applications. They are the same only because they both are ways of capturing an image, and, at times, similar chemical processing is used in making both items. There are, however, major differences between a photograph and a hologram.

3D, Or Not 3D?

A photograph can be made under any ordinary lighting condition but the resulting image is only two-dimensional. That is, if you move a photograph from side to side you do not see around the image; you will always see nothing but a flat image, even though many gimmicks (e.g. 3-D glasses) have been marketed over the years to trick the eye into thinking it is viewing a three-dimensional image.

A hologram, however, actually is a 3-D image. With some holograms, the image actually forms in air in front of the holographic plate and you can look around the object just as you would in real life. The distance the image forms in front of or behind the holographic plate and the degree to which you can look around the object depends on how the hologram was made; there are a large variety of holograms, each with good and bad points.

Although some holograms may be viewed in ordinary daylight, they all require a very narrow, almost single, wavelength of light to be made. Because of this lasers, which can put out a single wavelength of light, are almost universally used to make holograms.

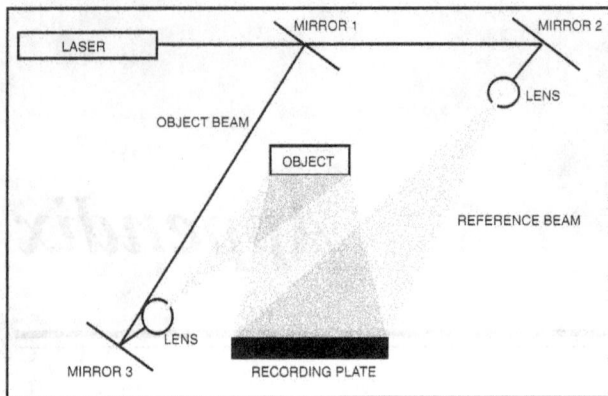

Figure 1.1. Transmission Hologram

To borrow a phrase, a picture (or in this case, a hologram) is worth a thousand words. Therefore, to fully grasp what holograms look like, one should go to a local shop or gallery which displays them or order some from one of the many mail-order houses found in this publication.

Perhaps the best way to proceed from here is to explain, step-by-step, the making of a simple hologram. After discussing some of the fundamentals of holography, the more exotic varieties of holograms will be addressed.

Making a Transmission Hologram

Suppose one enters a studio where a simple hologram is about to be made. There is a special vibration-free table in the room. On the table is a laser, some mirrors and a photo-sensitive plate in a plate holder. Everything on the table is arranged in a carefully measured manner. In the center of this holographic set-up is the object to be holographed.

To make the hologram, we turn on the laser. The laser beam strikes Mirror 1. Due to the fact that Mirror 1 is only partially reflective, part of the beam is reflected toward Mirror 3, and the other part passes through Mirror 1 to Mirror 2. For this reason, Mirror 1 is referred to as a beamsplitter.

The beam which passes through Mirror I is called the reference beam because it never actually strikes the object being holographed. After the reference beam strikes Mirror 2, it is reflected through a lens toward the photographic recording plate. The lens' function is to spread the beam so that it will cover the entire plate (in some cases, the lens is placed in front of Mirror 2; in either case its function is the same--to spread the beam).

At the same time, the other beam (which we call the object beam) reflects off Mirror 3 and passes through a lens. This lens spreads the beam out so that it illuminates the entire object. The beam then reflects off the object (hence the name object beam) and strikes the photographic recording plate.

The two beams must travel exactly the same distance so the light waves in the beams will be in sync with each other. After exposure (exposure time depends on the laser and type of film or emulsion used), the photographic plate is developed, and the resulting developed plate is a hologram.

Holding the developed plate up to light, we see that the plate is semi-transparent. On closer inspection we see that the darkness of the plate is caused by developed emulsion. The plate seems to have countless swirls of thread-like developed emulsion which are called fringes. The fringe patterns look like the swirls that make up your fingerprints or the boundaries of topographic maps. There appears to be no order to the swirls.

Viewing the Finished Hologram

To see the completed hologram, the photographic recording plate is developed and put back in the plate holder on the table in exactly the same place it was for the exposure. The object and Mirrors 1 and 3 are then removed from the table.

The laser is turned on again and the reflection beam illuminates the plate. When you look through the plate, an image is seen of the original object, in its original place and at its original depth.

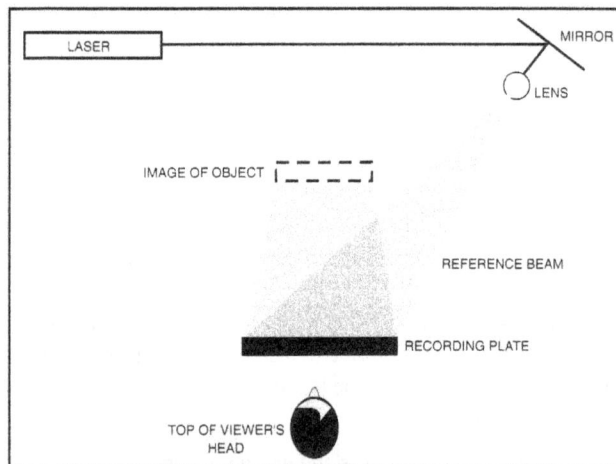

Figure 1.2. Viewing a transmission hologram

A detailed explanation of why this happens would occupy many pages. A simple explanation might go like this: the object and reference beam strike the photosensitive recording material at the same time and create a pattern as they expose the chemicals in the emulsion. Since both beams strike the plate at the same time, the light waves in the beams naturally interfere and combine with each other. Therefore the pattern of exposed emulsion we get when we develop the plate is called an interference pattern.

After development, we aim only the reference beam at the plate, at exactly the same angle that originally exposed the plate. As the reference beam goes through the plate, the interference pattern naturally causes part of the reference beam to change direction. This phenomenon is called diffraction.

The interference pattern, because it was created by light from the original object, diffracts the reference beam pass- ing through it back in the direction of the original object. This deflected beam has exactly the same form as the beam that was originally reflected from the object, because the object beam made the patterns in the emulsion.

So, the diffracted beam recreates an image of the original object in the same place it was at exposure time and it looks just like the original object. In other words, a hologram mimics the way light reflects from an object, without the object being there.

In order to record a clean and clear interference pattern of our object and reference beam, we use a laser, which has a single beam of light at one wavelength. We cannot use just any light as our source, because the light from common light bulbs contains many constantly-changing wave-lengths. If we make the exposure using regular light, the interference patterns would be completely blurred and useless because of the changing wave-lengths and multitude of interference patterns that would be created.

Reflection Holograms

The hologram just discussed is called a **transmission hologram** because the light passes through the plate to us.

It is also possible to make a hologram where the light reflects off the surface and back to our eyes for us to see the image. This is called a **reflection hologram**.

How are reflection holograms made? Look at Figure 1.1. If we transfer the reference beam around with mirrors so it illuminates the recording plate from the back instead of from the same side as the object beam, we create a reflection hologram, as shown below. It is that simple.

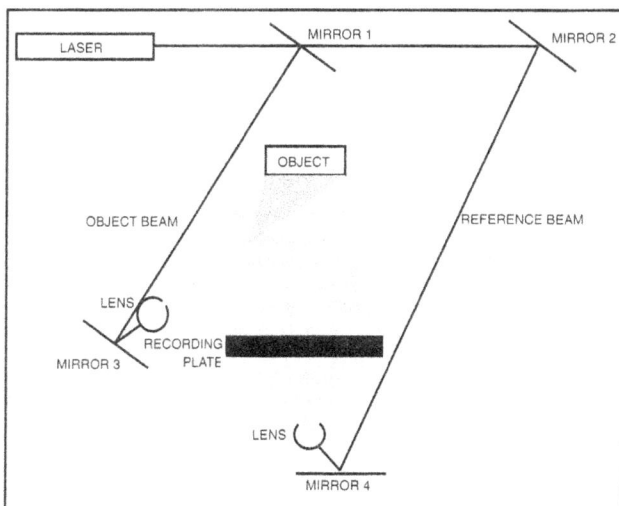

Figure 1.3. Reflection Hologram

As their name implies, all reflection holograms are illu-

minated from the same side as the viewer. In other words, when viewing a reflection hologram, the reflection light comes from your side of the plate, strikes the plate and reflects back to your eyes.

It should be pointed out that all reflection holograms can be viewed in sunlight (frequently called daylight-viewable or white-light viewable holograms) whereas transmission holograms, depending on how they are made, can be either daylight-viewable or laser-viewable (only viewable with a laser light). We will discuss these categories shortly.

Within the two major divisions of holograms (reflection and transmission), there are many variations. Like any other specialized field, holography has its own lingo, and in some cases the same hologram can be described using more than one name.

We will try to clarify the sub-divisions of holograms next -- but remember, each of these subsets can always be classified into the larger families of transmission or reflection holograms.

Thick and Thin Holograms

Another broad classification of holograms is made when differentiating between "thick" (sometimes called "volume") holograms or "thin" holograms. One of the reasons the words "thick" or "volume" and "thin" are used in conversation is that they allow one to instantly get an idea of some of the properties of the hologram. Very thin holograms provide little depth to their object upon reconstruction. Embossed holograms, such as the images on bank charge cards, are examples of thin holograms. Thick holograms have the ability to replay or reconstruct the image with considerable depth or projection.

How do you decide if a hologram is a thick or a thin hologram? If you look closely at the surface of the hologram you see the fringes caused by the interference pattern. A hologram is considered to be "thick" if the thickness of the recording medium is considerably greater than the spacing between the interference fringes. Otherwise the hologram is considered a "thin" hologram.

The distance between interference fringes will depend on a number of things, such as the wavelength of light being used, and the density of particles in the emulsion.

These interference fringes are called Bragg planes after Sir William Henry Bragg and Sir William Lawrence Bragg, who did much of the early work on the subject. As you would expect, Bragg planes actually go all the way through the medium, but are visible to our eye only where they meet the surface.

It should also be pointed out that in a reflection holo- gram, where the reference beam and object beam strike the plate from opposite sides, the Bragg planes slice through the medium at very shallow angles.

Conversely, in a transmission hologram, where the reference beam and object beam strike the plate from the same side, the Bragg planes cut the emulsion at much sharper angles and thus are further apart.

The H-1 and Master Hologram

It is important that we cover the topic of the H-1 and master in this introduction because it is a fundamental procedure in the making of almost every commercial holo- gram.

H-1 stands for "hologram one", which simply means it is the first hologram you make on the path to your desired final holo- gram. Sometimes the H-1 is the master hologram from which you make multiple copies. Frequently, though, there is more than one hologram that needs to be made before you get the desired master hologram from which you will make copies. If this is the case, the next hologram in the sequence is called the H-2, and then H-3, and so forth.

The laser-viewable transmission hologram is most often used as the master hologram (H-1). Transfer copies (making another hologram using the image on the master as the subject) can be made in quantity from the master. These transfer holograms can either be other laser-viewable transmission holograms, white-light viewable transmission holograms or reflection holograms (which are always white- light viewable).

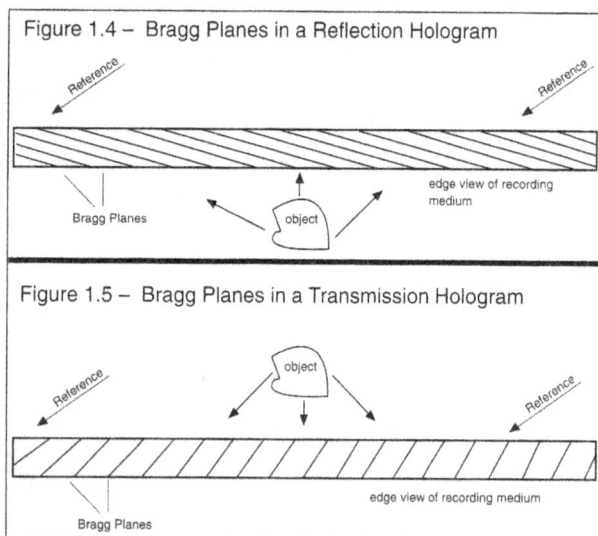

Figure 1.4 – Bragg Planes in a Reflection Hologram

Figure 1.5 – Bragg Planes in a Transmission Hologram

A question that immediately comes to mind is, "Why would anyone want to make an H-2?" Well, historically one of the big problems that holographers used to have was placing the object to be holographed exactly where they wanted it. Suppose, for example, you want the object in the final hologram to appear half in front and half behind the recording plate. How would you do it? You obviously can't do it on your first shot because the object would have to be going right through your photographic plate.

This problem was solved by the following procedure:

* Make a transmission hologram. We call this H-1 because it is our first hologram.

* Since the H-1 hologram creates an image of the object, why not use the image made by our H-1 as our subject and

make a hologram (H-2) of the image made by the H-1? In other words, make a hologram of our hologram. This H-2 hologram can be a transmission or reflection hologram, depending on your need.

* It sounds strange, because you are making a hologram of an image and not an object. But it works.

* Now, since you can make a hologram of the H-1 's image, take time to move the image around to wherever you want it positioned. In this case, adjust the H-2 recording plate so that the image of the object is half in front and half behind the plate and then make your H-2 . The problem of getting half of the object in front of the plate, and half behind, is solved.

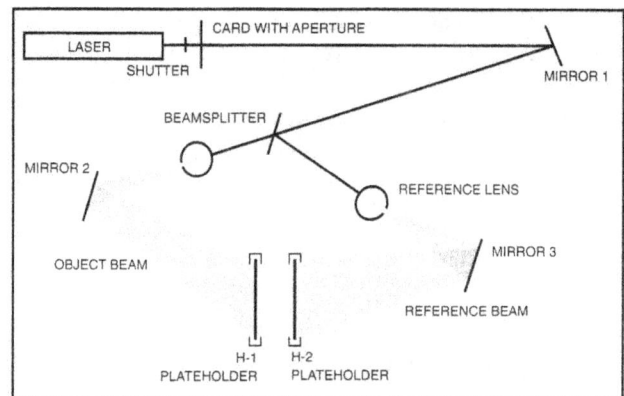

Figure 1.6 – Reflection H-2 being made from H-1

In short, there are at least three good reasons why an H-2 should be made:

1) The H-2 allows you to reposition the image of your subject. When you reposition your image from the H- 1, you may make your subject focus out in front of the recording plate, behind the plate, or anywhere within the limits of your equipment (you are usually limited by the laser's ability and the quality of the optics). The creative potential here is enormous because you are able to move solid objects around like they are ghosts. You can have two objects occupying the same space, etc. The process of moving the image around to make the H-2 is called image planing.

2) It gives the holographer a chance to brighten up the image. Since you may move your image anywhere, you can focus the image right at the recording plate. This concentrates the light directly on the recording material and brightens up the image considerably. This is commonly done in silver halide reflection holo- grams .

3) It saves time on remakes. If you develop the H-2 and decide you don't like the position of your subject astride the recording plate, you don't have to find the original subject and set it up again. This can be important if there are large costs in arranging the H-1 shot.

Going through the pains of making H-1, H-2, etc. to get a good master is necessary for creating most holograms. It is technically possible to get some desired holograms by skipping this process, but the results are generally very inferior. A master is almost always used for commercial jobs of value.

Sunlight-Viewable and Laser-Viewable Holograms

We mentioned earlier that although it is necessary to use a laser to make a transmission hologram, it is not always necessary to use a laser to see a transmission hologram. In fact, most transmission holograms can be seen in sunlight.

This may seem confusing, because we have said that in order to see a holographic image you have to shine the reference beam that made the hologram on the plate. This is true, but sunlight contains a multitude of wavelengths, including the one we used to make our exposure. The sun is such a great distance from earth that it appears to be a single beam of light shining on our plate. It would seem that we have only to position the plate at the proper angle, and we should see our image.

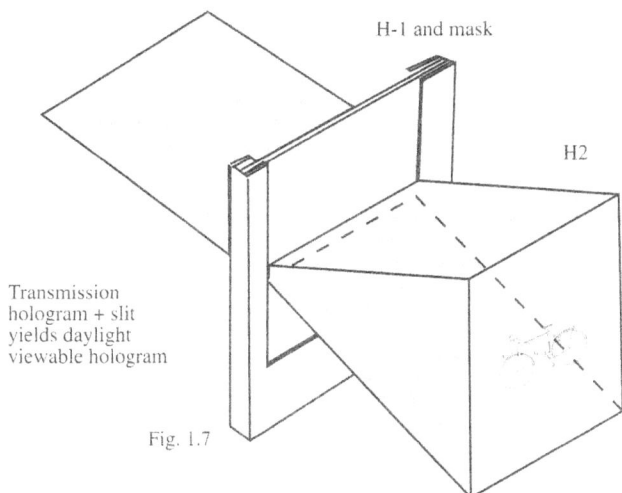

H-1 and mask

H2

Transmission
hologram + slit
yields daylight
viewable hologram

Fig. 1.7

This is logical, but it also stands to reason that if sunlight passed through a transmission hologram we would also get images being formed by all of the other wavelengths that are somewhat close to the wavelength of your reference beam. These other frequencies of light would diffract at a somewhat different angle than the original reference beam. The result would be a multitude of images forming right next to each other, creating a blur instead of a clear, crisp image.

That's exactly what does happen and it took a while for a solution to be developed. Around 1969 Dr. Steven Benton came up with a solution. The resulting hologram is sometimes referred to as a Benton hologram or, more frequently, a rainbow hologram.

Rainbow Holograms

Benton reasoned that since our problem is too much mag-

ery at the point of reconstruction for our object, why not block off some of it? In other words, suppose we put up an opaque mask against the transmission hologram, with a long, narrow horizontal slit through which we view our transmission hologram. This would certainly clean out a lot of the annoying secondary images that are blurring the primary image's reconstruction.

This "cleaning" comes at a price, however, because the mask causes loss of vertical parallax (the ability to be able to see over and under our object). We would, however, still have our horizontal parallax (ability to see side-to-side around the object). Humans, with feet fixed on the ground and eyes on a horizontal plane, are actually more accustomed to horizontal parallax than vertical.

The procedure to produce this masked hologram is as follows:

1) First a normal transmission hologram is made.

2) Next, a transfer copy of the transmission hologram is made, but an opaque card with a horizontal slit is placed between H-1 and H-2.

If the copy hologram is viewed in the frequency of our laser light, the eyes must be positioned at the real image of the slit to see the holographic image.

Now, imagine viewing this H-2 hologram in two different colors of light. A hologram of the image made through the slit will be played back, but each of the two wavelengths of light will diffract through the hologram fringes at a slightly different angle. There will be two different images of the object, each a different color and each at a slightly different vertical position.

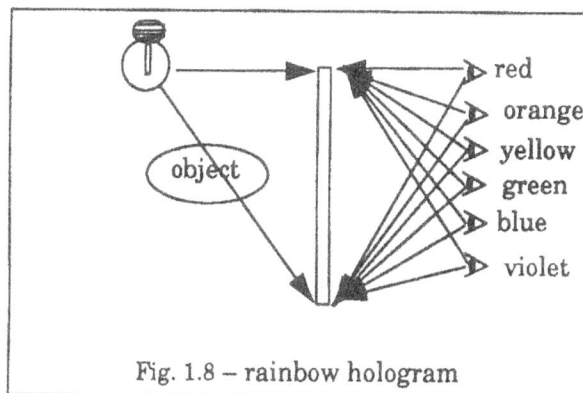

red
orange
yellow
green
blue
violet

object

Fig. 1.8 – rainbow hologram

Next, think of the image in white light or sunlight. All of the wavelengths will reconstruct their own image, all slightly displaced vertically with respect to one another. We are faced with the same problem we had with the original transmission hologram, except the images being recreated have no vertical parallax.

As you move up and down in front of the plate the color of light will change but the image will be the same (hence the name "rainbow hologram"). As you move from side to side you will have horizontal parallax because nothing has been done to destroy it. By careful planning, the image may be made any desired color, or even a combination of colors (a multi-color rainbow hologram).

In effect, the hologram is filtering the white light, while all that is sacrificed is vertical parallax, which, as we mentioned, our two horizontally-positioned eyes usually don't miss anyway. Rainbow images are often extremely bright, because all of the frequencies in white light are being used to form the image. So the rainbow hologram technique is a way of making a transmission hologram sunlight-viewable. Other names for this are daylight-viewable or white- light viewable. They all mean the same--a hologram you can see without the need of a laser.

Some rainbow transmission holograms are displayed in art galleries on glass plates and on film. However, they are much more popular in two other forms. The two most common forms in which rainbow transmission holograms are seen are as embossed holograms and holographic stereograms.

In an embossed hologram, the light goes through a rain- bow transmission hologram that has been embossed in clear plastic, strikes a mirrored backing and reflects back through the rainbow transmission hologram to the viewer's eyes. We will discuss embossed holograms later in this appendix .

Holographic stereograms are made from motion picture film. There is a little more detail on these holograms later in this appendix and a whole chapter is devoted to the subject in this book.

Lighting

Sunlight is not the only source of light that works with white-light viewable holograms. There are a whole range of light sources with which to view a white light hologram; some are better then others.

Just remember that white-light holograms require a light source which contains the original exposure wavelength and enough intensity to replay the hologram. Ideal sources cast sharp shadows, like a spotlight or an average clear light bulb with a single filament. Some very bad light sources, such as fluorescent lights, are extremely diffuse and in some circumstances render white light holograms unviewable .

Whether or not a white light hologram is viewable in bad lighting depends on how the hologram was made. Especially vulnerable are holograms which project an image far out in front of the plate or have great depth. If a hologram like this is illuminated with spotlights coming from multiple sources at different angles, the hologram forms projected images at all the different angles dictated by the spotlights. The mixture of projected images "blurs out" the image the holographer is attempting to recreate.

Hence, holograms made to be viewed in a wide range of lighting, including light from multiple sources, use subjects that have very shallow depth. This is because if there is little or no depth to the object, all the images being created by the different sources appear to be focusing in the same place.

Thus, if you go into a shop that specializes in holograms, one usually finds that the shop has subdued, overhead lighting with spotlights focused on the holograms. This serves the dual purpose of creating a pleasant lighting environment, as well as providing a single spotlight for holograms with considerable projection.

People who display holograms in their homes find an inexpensive way to illuminate them is with a clear light bulb having a single filament. These bulbs are available at any shop with a large selection of light bulbs.

Image Projection of Holograms

Although transmission holograms seem to be naturally designed to create a hologram with considerable projection, one can also make reflection holograms that have a great deal of projection. In fact, reflection holograms with considerable projection are a favorite among artistic holographers and the buying public. They are favored because they can be hung on the wall and illuminated just like a painting, whereas transmission holograms need to be lit from behind, often requiring a much larger viewing area.

Laser-viewable transmission holograms demonstrate amazing depth and projection when the correct equipment is used. It should be noted that the depth of the holographic image is not so much a function of the power of the laser as it is the coherence length of laser light (you can read more about coherence length in the laser chapter).

Theoretically, the maximum image projection in front of the hologram plate can be as great as the projection in back of the plate (depth of the image). Unfortunately, it is difficult for our brains to make sense of greatly projected images. Because of this and the fact that there are some optical distortions in the image planing process, projected distances in transmission holography are usually kept under four feet.

Laser transmission holograms have the widest parallax (the ability to see around an object from side to side) and resolve the greatest depth of objects. There are laser transmission holograms, for example, of people and objects in a 4000 cubic foot room, made by pulsed lasers.

Not surprisingly, projected hologram images like this generate one of the highest shock and thrill responses from viewers. A good percentage of first-time viewers respond by waving their hand back and forth through projected images in disbelief.

Pseudoscopic and Orthoscopic

Both laser-viewable and white light transmission holograms share a fascinating property. As you recall, after the developing stage, the hologram is put it back in the plateholder. The original object has been removed from the table, yet the viewer can look through the holographic plate and see the object appearing deep within the plate at its original position on the table.

Now comes the interesting feature. Take the transmission hologram out of its plateholder, flip it over, and put it back in the plateholder. Step back and look at the plate. You see the image forming out in front of the plateholder (between you and the plateholder). It focuses in air the same distance in front of the plateholder as it originally sat behind the plateholder. You also see that the image is a pseudoscopic image.

What is pseudoscopic? An image as normally seen in everyday life is an orthoscopic image. A pseudoscopic image is the opposite of this. For example, if one's viewpoint moves to the right, you do not see more of the right view of the image but the left view, and when the viewer raises his viewpoint, the lower part of the subject comes into view instead of the upper part.

A pseudoscopic image yields an exciting effect, but it can be confusing to the viewer. Some artists, however, have produced exciting pseudoscopic work using geometric shapes like wide spirals, pyramids and cones.

Recording Materials

Although we have discussed how holograms are made, we have not discussed the material that they are recorded on. In photography, the most common item used to capture images is a silver halide emulsion on a film base. In holography, there are a number of items used to record your image. The most common recording media are:

1) silver halide

2) dichromated gelatin

3) photopolymer

4) photoresist

Note that we use the phrase "recording materials" instead of emulsions. That is because not all the items used to capture holographic images are emulsions. In an embossed hologram, for example, the holographic image is literally stamped into clear plastic. A discussion of the different recording materials requires a chapter of its own, which you will find in Chapter 5.

Lasers

We will discuss lasers in depth in Chapter 4 of this publication, but it is important that we touch on the subject in this appendix. The type of laser you use affects what subjects you can holograph and we want to discuss that next.

There are two major kinds of lasers; the Continuous Wave (CW) laser and the pulsed laser. The CW laser emits a steady wave of laser light, whereas the pulsed laser emits laser light in bursts.

Continuous Wave Laser: The power of a CW laser is measured in watts (w). The CW laser is by far the most common laser used in holography. In holography labs, most of these lasers fall in the 5 to 50-mw (milliwatt) range. One of the great problems with CW lasers is that they cannot make the extremely short exposures necessary to capture a live subject.

Consequently, there must be absolutely no motion at all during the exposure with a CW laser. An exposure with a CW laser can take a fraction of a second to many seconds. Because there cannot be any motion at all during the exposure, we need eliminate any vibration coming from the ground. To do this we make or buy a vibration isolation table on which to put our laser, optics, and objects. Since it is absolutely critical that we have no motion at all, the subjects that we holograph with CW lasers have to be "dead" objects. Feathers might move in the breeze and living things simply move too much.

Remember that what we are recording on the plate is the reference beam and the object beam converging (or interfering) with each other at the plate. If the object moves even a microscopic amount (on the order of wavelengths) from one moment to the next, we will record two different interference patterns and the hologram fuzzes out or doesn't even show. The effect is like some photographic daguerreo- types of the 1800's - only much less forgiving.

Pulsed Laser: Pulsed lasers, quite the opposite of CW, emit extremely quick bursts of very powerful laser light. Consequently, the exposure time is much shorter than a CW laser. Exposures can be made in nanoseconds (one nanosecond is one billionth of a second).

You don't need a vibration isolation table for the pulsed laser. What can you shoot? Anything you want. You can shoot an entire room of people belly dancing in costumes of paper with feathers in their hair, and birds flying around the room. Why such freedom? Because your subject can't move significantly in a nanosecond.

What are the drawbacks of pulsed lasers? Why doesn't everyone buy one? The answer is money. They cost about $60,000 and require a lot of extra overhead and care. Lasers don't last forever and when a pulsed laser bums out it is expensive to fix. Holographers are anxiously awaiting, with cash in hand, a low-cost, easily-maintained pulsed laser.

The reason this discussion on lasers is important is that who you go to for making your hologram depends on what you want holographed. If you have a corporate logo or some other object that does not move, a CW laser can do just as good a job as a pulsed laser. There is no need to pay extra to have your hologram made by pulsed laser if you do not need it. On the other hand, if you have a live subject, you must use a pulsed laser.

Choosing a Subject

One last topic we should cover in this section is what kind of images you can use to make a holograms.

As with any creative art form there are many choices avail- able, and much depends on what you want to accomplish. To simplify things, we will list some of the most common items that are used for holographic subjects and then comment on each item.

1) 3-D model (like a sculpture).

2) 2-D model - just like a graphic arts layout for a printing job, complete with overlays.

3) Movies - specially-shot motion pictures for a holo-gram that moves.

Although some of the above topics are covered in more detail in the book, we will give an overview of each now.

3-D model: This is the most common way to create a subject for a hologram. What your model is, of course, depends on the type of laser you are going to use for your exposure. With a pulsed laser, as we have already mentioned, you can shoot live subjects and just about any- thing you wish. Most holograms, however, are made with a CW laser and the models used have a dramatically different look depending on the type of paint used to cover the item.

In order to get some idea of what a hologram will look like, many artists paint the subject, then look at it under the light of the laser they are going to use for exposure. Several repaints may be necessary to get exactly what you want. It is possible, with some of the latest techniques, to shoot a hologram of a subject and then reduce the subject in the copying process. Consequently you can have portraits of people reduced and still preserve much of the parallax that holograms have to offer.

2-D model: You will see this method being used with great abundance in embossed holograms. You have the ability to create camera-ready art for an embossed hologram much the same way you would create artwork for a printer, with the different overlays designated to be different colors and at different depths in the final hologram.

You can also have a combination of photos, line art and 3-D objects in your final embossed hologram, although the depth of the 3-D object is limited because we are dealing with a thin hologram. Coloration for this process has come a long way in the last few years. Enclosed in this book is an ad featuring a sample of the type of 2D holo- gram we are discussing. This topic will be discussed in much more depth in the embossed holography chapter.

Movies: Motion picture footage is used in the making of holographic stereograms, which is covered extensively in the chapter on stereograms. Suffice it to say that you make holograms of motion picture footage, if it is shot correctly (with the view in mind of using it in a holographic stereo- gram). Popular usage of this format includes plastic cylinders which rotate; looking into the plastic, one sees the image that was filmed. It is suspended in mid-air in the center of the cylinder, performing whatever was done in the movie.

It is also possible to make flat stereograms and emboss them, so that as you move the hologram with your hands, the image moves as well.

We have now had a look at some of the basic principals behind making holograms. Because of its wide use, we should give an overview here of how embossed holograms are made.

Embossed Holography

Embossed holograms are created in several steps. We will present an overview of the steps involved . The most common steps, once the model (not a simple step) is made, are as follows:

1) The hologram master is made.

2) A rainbow transmission hologram is made on photoresist (photosensitive emulsion) from the master.

3) The photoresist is etched to relieve the hologram patterns.

4) The photoresist is plated with silver and nickel. The photoresist now behaves like a metal mold (information on coating photoresist plates for holography is included in our chapter on recording materials).

5) The metal mold, or shim, is removed from photoresist and now has holographic patterns on it.

6) This shim is used as a stamping die and stamps the holographic patterns into plastic.

7) The plastic with the holographic pattern stamped in it has a mirror-like backing so light comes through the plastic, strikes the mirror-like backing and, reflecting back out, displays the white light rainbow transmission hologram.

Let's look at these steps in a little more detail:

Steps 1-2: There is a possible shortcut here. Sometimes flat art can be directly exposed onto the photoresist mate- rial. This is recommended for simple projects only.

Photoresist is a very tricky medium to record on holographically. It is nowhere near silver halide in responsiveness and requires long exposures even with lasers of several watts of power. A small holography studio would have to be equipped with an expensive laser and heat-resistant optics. New alternatives to this medium of photoresist are being researched.

The typical turn-around time is four to six weeks to receive excellent photoresist plates that are fit for metallizing. The holographer should make several photoresist plates of good quality to cover any problems that might occur in the metallizing phase.

Steps 3-4: After the photoresist hologram is checked for clarity, brightness and overall quality it goes to the metallizing stage of production. A thin layer of silver is deposited on the photoresist. Silver by itself cannot withstand the stamping pressure, so additional coats of a nickel-based material are deposited to reinforce the back of the silver. When it achieves the desired thickness, the nickel-silver shim is pulled from the photoresist plate and this becomes a metal mother die.

Steps 5-7: The first shim must be perfect. This first shim can have several shims made from it and in turn several shims made from those. The heat and pressure of embossing thousands of holograms wears out the shims so extras should be made. Any deterioration becomes very obvious if the shims are not changed regularly.

Producing the final hologram is the job of the embosser. Embossed holograms are made by stamping the shim onto aluminum. Heated polyester material which has a metallized backing is used less often. Aluminum is softer and less destructive of the nickel master. Although not used widely, colored metallized backing is an option. Most often the silver color is selected.

Index

A

absorption 56
AD 2000 Inc. 44, 47, 48
Adobe Photoshop 24
Adobe Premiere 22
Agfa-Gevaert 73, 75
Alias !Wavefront's Power Animator 22, 24
American Bank Note Holographics 22, 31,47
American holographers 12
amplification 56
amplitude 52
Anderson, Steven 36
angular drift 62
animation 19, 22, 25, 26, 34, 38, 39
aperture plane 27
Applied Holographics 47
artistic holography 11 , 62, 63
Asia 12
Astor Universal 47

B

bandwidth 53, 60, 63, 66
Bartolini 76
Benton, Steven 151
Beta SP 31
birefringence 54
bonds 51
Bragg, Henry 149
Bragg, Lawrence, 149
Bragg planes 75
Brewster windows 59, 60, 63, 66
Brewster's angle 55
Bridgestone Technologies 47
Bums 76
Byte by Byte 38

C

California 147
Canada 147
categories of merchandise 16
catephoresis 66
chain of distribution 12
chemical potential energy 51
China 78
cinematographer 19
cinematography 30, 34
Clay 76
coefficient of expansion 59

coherence 53, 62
coherence length
53, 63 , 66, 76, 153
collision 56
color centering 60
color-tuning film 71
computer aided holographic design 37
computer animation 3, 30
computer generated hologram 37
computer graphics 13, 24, 25
computer imagery 39
computer technology 19
computer workstation 37
conservation of energy 49
constructive interference 53, 62
continuous wave 52,58, 153
continuous wave transmission 58
copyright 13, 14, 16
Cross hologram 19
Cross, Lloyd 19, 20, 39
CybersculptureTM 22, 26
cycle 52

D

depth perception 21
destructive interference 53, 62
dichromate 12, 17, 40,73, 153
diffraction 55, 149
digital camera 39
digital origination 4
Dimensional Cinematography Corporation 30, 31
Disney 68
display holography II
distributor 14
DOS 28
DuPont 4, 17, 68, 71

E

Einstein, Albert 53, 56
electric fields 50
electric potential energy 50
electromagnetic radiation 49, 51 , 55, 59
electromagnetic waves 51 ,53, 54
electronic imaging 4
embossed holograms 3, 17, 40, 43, 152, 154, 155
embossed holography 25, 44, 63, 66
embossed stereogram 19
energy 49, 55
English holographers 12
etalon 62, 63 , 68, 76
Europe 12

F

film recorder 39
Fischler, Ben 22
Fisk, Michael 49
Foire transforms 37
Ford, Henry 31
Frankenstein 36
Freehand 37
frequency 52
full parallax 3, 28
full parallax grid 22

G

gas lasers 57
Gaussian energy distribution 61
General Design 22, 25
giftware industry II
Graphic Workshop 28
gravitational potential energy 50
Gunther, John 29
Gustafson, Glen 30

H

Haines, D. 39
Hassen, Chuck 4, 25
heat exchanger 63
Henderson, Rickey 30, 31, 35
Hertz 52
Hess, Bob 4,27, 25, 29
high resolution CRT 39
high-reflector 57
Hijack 28
Holo Sciences 22, 25
Holobank 47
Holographic Design 48
holographic optical printer (HOP) 20, 38, 39
holographic stereogram 3,19, 21 , 30, 37, 38, 39,152, 154
HOP transfer 20
horizontal parallax 20, 25, 27, 151
horizontal slits 22
Hughes Power Products 4, 29
Hurwitz, Noah 22

I

Ilford 73
Illustrator 37
Image Capture 39
image planing 150
Imagination Plantation 22
integral holograms 19
interference 53, 55, 60
interference pattern 148, 149, 154
inverse square law 36

J

Japan 12
Joule, James 49

K

Kane, Brian 4,22, 27
Karioff, Boris 35, 36
kinetic energy 51
kiosk 14
Klug 39
Kodak 38, 73, 75

L

laser beam 54, 57
Laser Reflections 67
Lasersmith Inc. 43,37,38
latent images 74
LCD 5, 21, 38, 39
lenticular 38, 67
lenticular screen 31
licensed products 12, 14
Liconix 49
Light Amplification from Stimulated
Emission of Radiation 56
Light Impressions 47
linewidth 60, 68
Lino 300 image setter 37
liquid lasers 57
Live Entertainment 22
longitudinal mode spacing 60
longitudinal modes 60, 62
Los Angeles 36

M

Macintosh 22, 28
mail order 14, 148
mastering 13
matter 49
McCOlmack, Sharon 31
McGrew, Steve 3
mechanical potential energy 50
metastable states 67
Microsoft's Windows 95 147
Miller, Doug 48
MIT 38, 39
mode hopping 61, 63
model maker 19
motion blur 21, 22
Mullen, Charlie 31
Multiplex Company 21 , 39
Multiplex hologram 19

N

New York 37

Nobel Prize 56
nodal lines 61
North America 12
nsLooksTM 68
nuclear potential energy 51
Nykvis, Sven 35

O

Oakland A's 35
Oakland Museum 34
Olson, Ron 49, 58, 67
OmniDex 71
Ondrejik, Charlie 78
Open Render tools 24
optical feedback 57
optical pump 57
orthoscopic 153
oscillation 51
output coupler 57

P

Pantone 25
parallax 22, 39, 40, 153
PC 28
PCX format 28
Pentium 38
Pepsi 68
phase 53
phase relationship 53
photon 27, 51 , 55,
56, 57, 58, 59, 60, 62
photopolymer 12, 17, 28, 29, 68,
71, 153
photoresist 46, 62, 63, 66, 71 ,
76, 77, 78, 153 , 155
Photoshop 31, 37, 38
pictorial holography II
Pizza Hut 30, 31
Planck's constant 55
Point Source Productions 22, 25,
27, 29
polarization 54, 62
Polaroid 17, 39, 71
population inversion 56, 67
Positive Light 67
potential energy 50
PowerMac 38
primary wavelength 58
Prince 30, 31 , 35
profit margins 14
pseudoscopic 153
pulse 52
pulsed holography 58, 67
pulse length 58
pulsed laser 49, 58,153 , 154

Q

Q-switch 58, 67, 68
quanta 55
quantum mechanics 55, 56
Quark Express 37

R

Rearden, Craig 36
reference beam 148, 151 , 154
reflection hologram 25 , 39,75, 149, 153
resonator 58, 59
retailers 14

S

S.I.R. studios 35
San Francisco 20, 36
Saxby 76
Scattering 54
Scheir, Peter 44, 48
Sculpt 4D 38
Sculpty 40
secondary wavelength 58, 66
security II, 37
shim 43 , 62, 76, 77, 155
Shipley 76
Silicon Graphics Inc. 22, 24, 38
silver halide 29, 73,75, 150, 153
silver halide film 16, 74
silver halide glass plates 16, 40
Simeon Hologrphics 38, 39
Sivy, George 39, 40
smear 43
Smith and McKay Printing 47
Smith, Steve 19, 37
solarization 59, 60
sole rights to sell 14
Solid LightTM 22, 26, 27, 28
solid state lasers 57, 68
speed of light 53
spontaneous emission 56
Star Trek, the Next Generation 35
stati-up costs 18
stereo vision 19, 21
stereo grams 19
stimulated emission 56
Super Bowl 39, 147
Sweden 35

T

telemarketing 14
TEM mode 61
thermal lensing 59
tilt plates 59
time smear 21, 22, 26, 34

Towne Labs 77
transmission hologram 21 , 38, 75,
149, 150, 151 , 153, 155
transmission master 5
triangulation 21

U

unit cost 13
Universal Studios 30, 31, 36
UNIX 24
Unterseher, Fred 3
Upper Deck Corporation 22

V

vertical parallax 26, 151
vertical slits 22
viewing zone 39
voltage 50, 57

W

wall decor 11, 16, 17
Warner Brothers 35
Watt, James 58
watts 58, 153
wavelength
52, 55, 56, 62, 67, 75, 147, 151, 152
wholesaler 14
Windows 28
work 49
work for hire 13

DID YOU BORROW THIS COPY?

Get Your Own!

Now is the time to order your personal copy of the Holography MarketPlace. This international directory is the first and only resource of its kind. You will refer to the HMP day after day, so don't you want one of your own?

Standing Order Plan

☐ If you would like to receive the *Holography Marketplace* each year, check this box and you will be put on our Standing Order Plan. We will ship each new edition to you upon publication.

To Order by Check:

Mail check or money order to:

ROSS BOOKS
HOLOGRAPHY MARKETPLACE
P.O. BOX 4340
BERKELEY, CA 94704 USA

To Order by Credit Card:

Credit card customers can order by phone, FAX, or e-mail using Visa, MasterCard or American Express.
Telephone: (800) 367 0930 or (510) 841 2474
FAX: (510) 841 2695
e-mail: rossbooks@aol.com

Cost

For USA continental delivery:
$25.00 (Includes UPS shipping costs).

For Canada/Mexico/Alaska/Hawaii delivery:
$30.00 (Includes Air Delivery).

For all other world delivery:
$40.00 (Includes Air Delivery).

FAX your order to: (1) (510) 841 2695

For orders by mail or fax, fill out the information below. Include a check or credit card information.

NAME:_____

COMPANY:_____

ADDRESS:_____

CITY/STATE/PROVINCE:_____

COUNTRY/POSTAL CODE:_____

PHONE/FAX:_____

Credit Card Information

☐ Visa   ☐ MasterCard   ☐ American Express

Card #_____ Name on card:_____

Expiration Date:_____ Signature:_____

☐ Please mail me a form to list my company in the next edition of Holography MarketPlace.